Geophysics: Theory and Applications

Geophysics: Theory and Applications

Editor: Zachary Russell

RCallisto Reference

www.callistoreference.com

Callisto Reference,
118-35 Queens Blvd., Suite 400,
Forest Hills, NY 11375, USA

Visit us on the World Wide Web at:
www.callistoreference.com

ISBN: 978-1-64116-259-3 (Hardback)

Cataloging-in-Publication Data

Geophysics : theory and applications / edited by Zachary Russell.
   p. cm.
Includes bibliographical references and index.
ISBN 978-1-64116-259-3
1. Geophysics. 2. Earth sciences. 3. Physics. I. Russell, Zachary.
QE501 .G46 2020
550--dc23

Table of Contents

Preface

This book has been an outcome of determined endeavour from a group of educationists in the field. The primary objective was to involve a broad spectrum of professionals from diverse cultural background involved in the field for developing new researches. The book not only targets students but also scholars pursuing higher research for further enhancement of the theoretical and practical applications of the subject.

The study of physical properties and processes of the Earth and its surrounding space environment is called geophysics. Quantitative methods are used for their analysis. Geophysics is a subject of natural science that is applied to societal needs such as environmental protection, mineral resources and mitigation of natural hazards. It is an interdisciplinary subject that significantly contributes to Earth sciences. Electromagnetic waves, gravity, vibrations, heat flow, electricity, magnetism, mineral physics and radioactivity are the various phenomena that are studied under geophysics. Modern geophysics is an interdisciplinary field which delves into solar-terrestrial relations, fluid dynamics of the oceans, and magnetism in magnetosphere and ionosphere. From theories to research to practical applications, case studies related to all contemporary topics of relevance to geophysics have been included in this book. It presents researches that have transformed this discipline and aided its advancement. Coherent flow of topics, student-friendly language and extensive use of examples make this book an invaluable source of knowledge.

It was an honour to edit such a profound book and also a challenging task to compile and examine all the relevant data for accuracy and originality. I wish to acknowledge the efforts of the contributors for submitting such brilliant and diverse chapters in the field and for endlessly working for the completion of the book. Last, but not the least; I thank my family for being a constant source of support in all my research endeavours.

Editor

Depth and Lineament Maps Derived from North Cameroon Gravity Data Computed by Artificial Neural Network

**Marcelin Mouzong Pemi ⓘ ,[1,2] Joseph Kamguia,[3]
Severin Nguiya,[4] and Eliezer Manguelle-Dicoum[1]**

[1]*Faculty of Science, University of Yaounde 1, P.O. Box 812, Yaounde, Cameroon*
[2]*Department of Renewable Energy, Higher Technical Teachers' Training College (HTTTC), University of Buea,
  P.O. Box 249, Cameroon*
[3]*National Institute of Cartography (NIC), P.O. Box 157, Yaounde, Cameroon*
[4]*Faculty of Industrial Engineering, University of Douala, P.O. Box 2701, Cameroon*

Correspondence should be addressed to Marcelin Mouzong Pemi; mouzong.pemi@ubuea.cm

Academic Editor: Filippos Vallianatos

Accurate interpretation of geological structures inverted from gravity data is highly dependent on the coverage of the recorded gravity data. In this work, Artificial Neural Networks (ANNs) are implemented using Levenberg-Marquardt algorithm (LMA) to construct a background density model for predicting gravity data across Northern Cameroon and its surroundings. This approach yields statistical predictions of gravity values (low values of errors) with 97.48%, 0.10, and 0.89, respectively, for correlation, Mean Bias Error, and Root Mean Square Error for two inputs (latitude, longitude) and 97.08%, 0.13, and 1.14 for three inputs (latitude, longitude, and elevation) for a set of anomalies as output. The model validation is obtained by comparing the results to other classical approaches and to the computed Bouguer, lineaments, and Euler maps obtained from measured gravity data. The depth of most of the deep faults and their orientation are in agreement with those obtained from other studies. The results achieved in this study establish the possibility of enhancing the quality of the analysis, interpretation, and modeling of gravity data collected on sparse grid of recording stations.

1. Introduction

The Northern Cameroon and its surroundings, the subject of this study, have prompted many researchers and prospectors to identify superficial and deep structures and to indicate their geodynamics and tectonic implications [1–5]. The vast majority of these studies used the data from surveys carried out by French Research Institute for Development (IRD) and other private and public institutions. Unfortunately, these data are scattered and unevenly distributed over the Cameroonian territory since they were mostly collected along available roads. Researchers have used interpolation techniques to extract relevant information from uncovered areas, such as Generic Mapping Tool (GMT) [6], minimum curvature [7], kriging [8, 9], the least-squares method [10], and finite element approach through cubic B-spline function

[11]. These methods have yielded interesting results and allowed in specific cases extracting relevant information from uncovered areas. However, the reliability and robustness of these conventional interpolation techniques are hampered by a limited number of input variables and a small sample size. These limit the validity of the reconstructed density models to narrow areas. In order to construct models that cover large areas more robustly, we propose to use ANN technique through Levenberg-Marquardt algorithm.

Haykin [12] defined Neural Network as a massively parallel distributed processor made up of simple processing units that has a natural propensity for storing experiential knowledge and making it available for use like the brain. They have recently gained in popularity in geophysics and have been applied to a variety of problems. Geophysical problems such as seismic waveform recognition [13], first-break picking

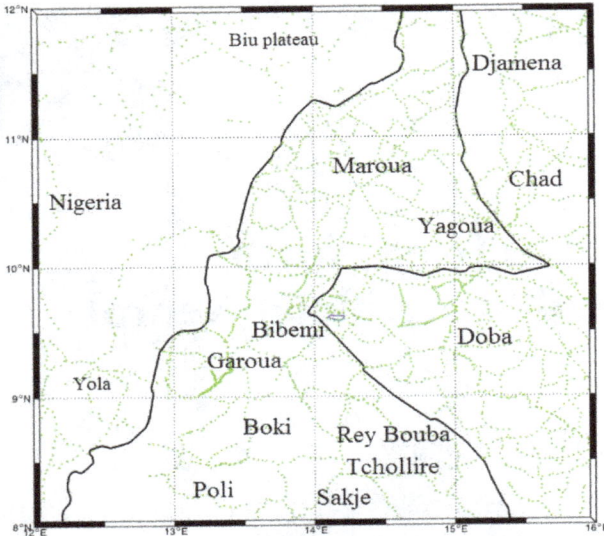

FIGURE 1: Spatial distribution in the study area.

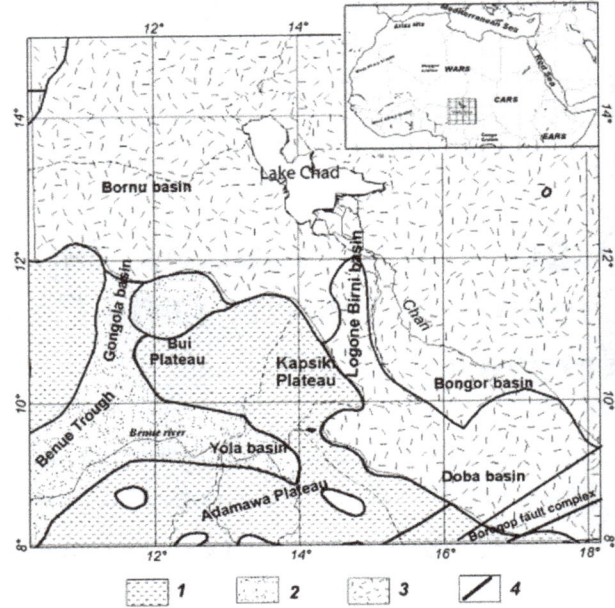

FIGURE 2: Geological map of the study area: (1) Precambrian; (2) Early Cretaceous; (3) Tertiary-Quaternary; and (4) faults or major structural elements (modified from Eyike et al., 2010).

and trace editing [14], earthquake prediction [15, 16], electromagnetic [17], magnetotelluric [18], seismic inversion [19–21], shear-wave splitting [22], well-log analysis [23], seismic deconvolution [24, 25], seismic discrimination [26], and seismic signals detection and classification [27] used ANNs to forecast the unknown. Grêt et al. [28] and Ghalambaz et al. [29] performed gravity interpretations in two dimensions and event classification, respectively, to discriminate bodies of a similar kind of anomaly and approximate shape parameters like depth, vertical extension, and radius. They use a Hybrid Neural Network and Gravitational Search Algorithm (HNGSA) to solve Wessinger's equation [29]. In this paper, we develop a background density model using ANN to extract gravity anomalies with their geographical variables (latitude, longitude, and elevation). The model integrates input and output data of an area covering North Cameroon and its surroundings (Figure 1) measured by IRD. In addition, a model validation is carried out by establishing Bouguer and residual maps of the studied area and then comparing them to those obtained by ANN. With this approach, we expect the estimation of gravity values in uncovered areas during data acquisition. The results show that ANNs can be used to interpolate data of uncovered area and increase the resolution of geological structures in poorly covered areas.

2. Geology of the Area

The data used in this study cover three main domains: northern extension of Benue trough in the East of Nigeria, Northern Cameroon, and West of Chad (Doba basin). Fairhead and Okereke [30] listed four important events in West Central Africa Rift System: extensional, compressional, tectonics, and subsidence. The surface geology (Figure 2) is mainly composed of the Precambrian basement made up of migmatites and anatectites of the Mokolo unit [31] and formations of the Pan African Mobile Belt. Old sediments

(Early Cretaceous in age) of 4 km thickness cover the Garoua sedimentary basin [4] while young sediments (lower cretaceous in age) cover the Precambrian basement [32].

3. Material and Method

3.1. Data Collection and Packaging. The measurement of gravity fields can be used to calculate relative and absolute values regarding variations of the fields across earth surface. It is linked to frameworks like Global Positioning System and Digital Terrain Model. The data are treated with respect to the equipment used (for example, Scintrex CG3/CG5 relative gravity meters or Micro-g Lacoste and Romberg as shown in Figure 3).

The data is processed to remove undesirable influences from the surroundings in order to isolate Bouguer anomaly (BA). It represents the difference between the measured and calculated gravity (formula (1)). BA is then modeled to derive geological features for characterizing, quantifying, and interpreting the mass or density distributions in the soil.

$$BA\left(\varphi, \lambda, h\right) = g_{mes(\varphi,\lambda,h)} - \Big(\gamma_{0(\varphi)} + C_{Free\ air(\varphi,h)}$$
$$+ C_{Bouguer(\varphi,h)_{slab/curvature}} \tag{1}$$
$$+ C_{Terrain(\varphi,\lambda,h)_{topography/bathymetry}}\Big),$$

where φ, λ and h represent the latitude, longitude, and elevation, respectively, $g_{mes(\varphi,\lambda,h)}$ measures the gravity, $\gamma_{0(\varphi)}$ represents the theoretical gravity, $C_{Free\ air(\varphi,h)}$, $C_{Bouguer(\varphi,h)_{slab/curvature}}$, and $C_{Terrain(\varphi,\lambda,h)_{topography/bathymetry}}$ stand,

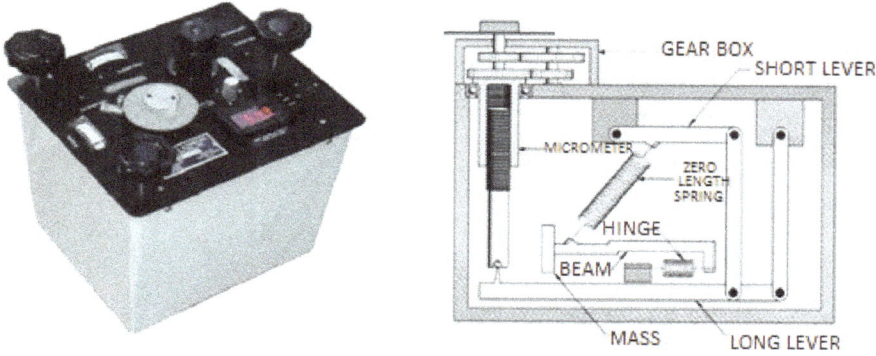

FIGURE 3: Lacoste and Romberg G/D gravity meters (Micro-g Lacoste, USA).

respectively, for free air, slab/curvature, and topography corrections.

We aim to develop a connectionist model (Neural Network) to correlate Bouguer anomalies with the geographical variables. In general, an array of Artificial Neural Networks is a juxtaposition of unitary, functional, and interconnected elements [33, 34]. There are a multitude of possible arrangements [35]. We choose to use the Multilayer Perceptron (MLP). It is mostly used in time series prediction because of its general property of being universal parsimonious approximator [36]. In its architecture, the neurons are organized in layers as shown in Figure 4.

The inputs x_k, $k=1,...,K$ are multiplied by weights w^k_{ji} and summed up together with the constant bias term θ^k_j. The resulting n^1_3 is the input to the activation functions g and f. The activation function is originally chosen to be a relay function, but for mathematical convenience a hyperbolic tangent (tanh) or a sigmoid function is most commonly used.

The output y of the MLP network is given by equation below:

$$y = g\left(\sum_{j=1}^{3} w^2_{ji} g\left(n^1_j\right) + \theta^2_j\right)$$

$$= f\left(\sum_{j=1}^{3} w^2_{ji} g\left(\sum_{k=1}^{K} w^1_{kj} x_k + \theta^1_j\right) + \theta^2_j\right). \quad (2)$$

From (2), the MLP network is a nonlinear parameterized map from input space $x \in R^K$ to output space $y \in R^m$ (here m=3). The parameters are the weights w^k_{ji} and the biases. f and g are activation functions defined in advance. In our study, we have used tansig for the hidden layer and purelin for output layer. Given input-output data, (x_k, y), 1,..., N, finding the best MLP network is formulated as a data fitting problem. The parameters to be determined are (w^k_{ji}, θ^k_j).

From an arbitrary weight (random value) defined at the beginning, the weights are adjusted by backpropagating the error according to the expression:

$$w_{ji}(n) = w_{ji}(n-1) + \Delta w_{ji}(n), \quad (3)$$

where

$$\Delta w_{ji}(n) = -\eta \frac{\partial E(n)}{\partial w_{ji}(n)} = \eta \delta_j(n) y_i(n), \quad (4)$$

where the local gradient δ_j is defined in

$$\delta_j(n)$$
$$= \begin{cases} e_j(n) y_j(n)\left[1 - y_j(n)\right], & \text{if } j \in \text{output layer,} \\ y_j(n)\left[1 - y_j(n)\right]\sum_k \delta_k(n) w_{kj}(n), & \text{if } j \in \text{hidden layer.} \end{cases} \quad (5)$$

In (5), e_j is the difference between the output y and target d values, as shown in

$$e_j(n) = d_j(n) - y_j(n). \quad (6)$$

E(n) is the sum of the quadratic errors observed on the set of output neurons written as

$$E(n) = \frac{1}{2}\sum_{j \in C} e^2_j(n). \quad (7)$$

3.2. Shape of MLP. The data set is composed of 2922 samples which comprise latitude, longitude, elevation, and the corresponding Bouguer anomalies. These data are extracted from a database computed for the whole Cameroon by Collignon [37] and Poudjom-Djomani [38]. They cover the study area, the Northern Cameroon, and its surroundings located between longitude of 12° and 16°E and latitude of 8° and 12°N (Figure 5).

60% of these data (1754 samples) are used for training, 20% of the data (584 samples) are used for the validation, and 20% (584 samples) for testing the models. Data conditioning processes are conducted to speed up the training ANNs, which includes interpolating missing data, normalizing the data, and then randomizing them. Usually, the missing data are calculated imprecisely by averaging the neighboring values. In this study, the missing values are forecasted by ANN.

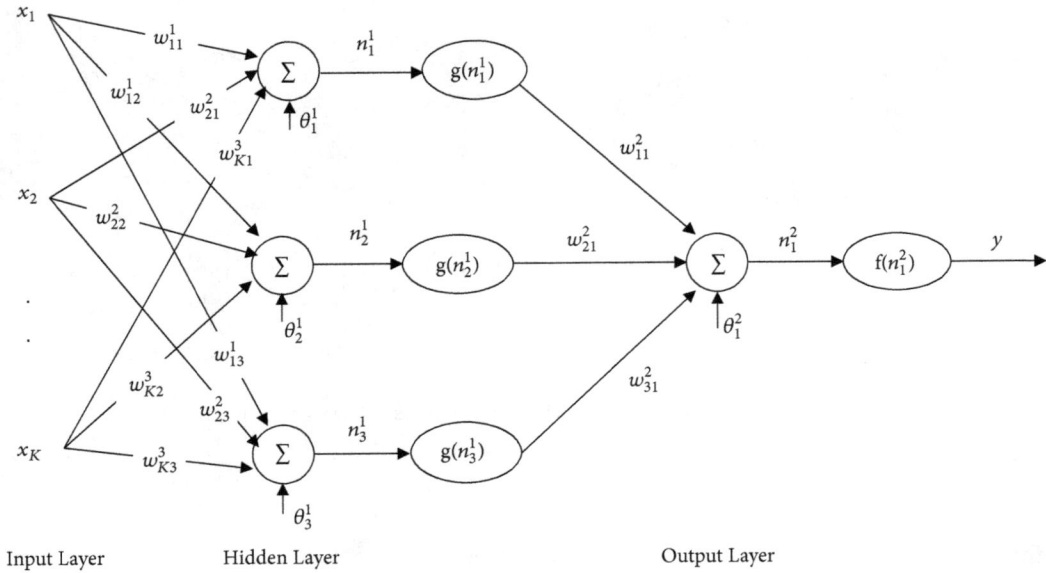

FIGURE 4: Details of a neuron (with $x_1, x_2, \ldots, x_k$ inputs and one output g_j).

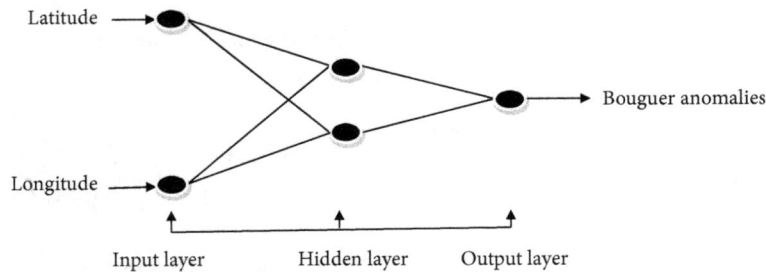

(a)

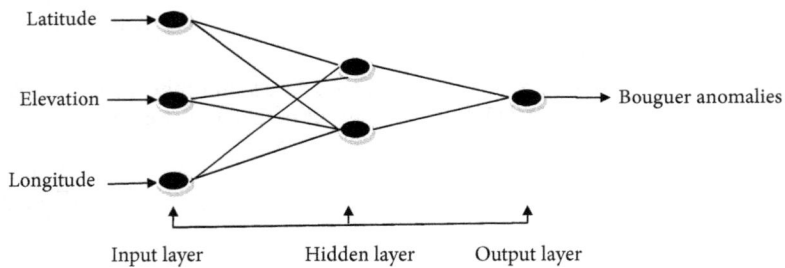

(b)

FIGURE 5: Multilayered Perceptron (MLP) network for (a) 2 inputs and (b) 3 inputs.

3.3. Training Algorithm. The diagram (Figure 6) stresses the steps to follow when implementing this scheme of ANN using Levenberg-Marquardt algorithm scheme:

$$w_{k+1} = w_k - \left(J_K^T J_k + \mu I\right)^{-1} J_k e_k, \qquad (8)$$

The algorithm adjusts the weights according to (8) where J is the Jacobian, μ is positive and called combination coefficient, I is the identity matrix, and e the error vector. This algorithm takes more memory but less time. Training automatically

stops when generalization stop improving, as indicated by an increase in the mean square error of the validation sample.

3.4. Results and Discussion. To quantitatively evaluate the ANN and verify its trend, we conduct statistical analysis involving the coefficient of determination (R^2), the Root Mean Square Error (RMSE), and the Mean Bias Error (MBE). The network structure identification is 2-190-1 and 3-370-1, respectively, for 2 and 3 inputs where the first number indicates number of neurons in the input layer, the last

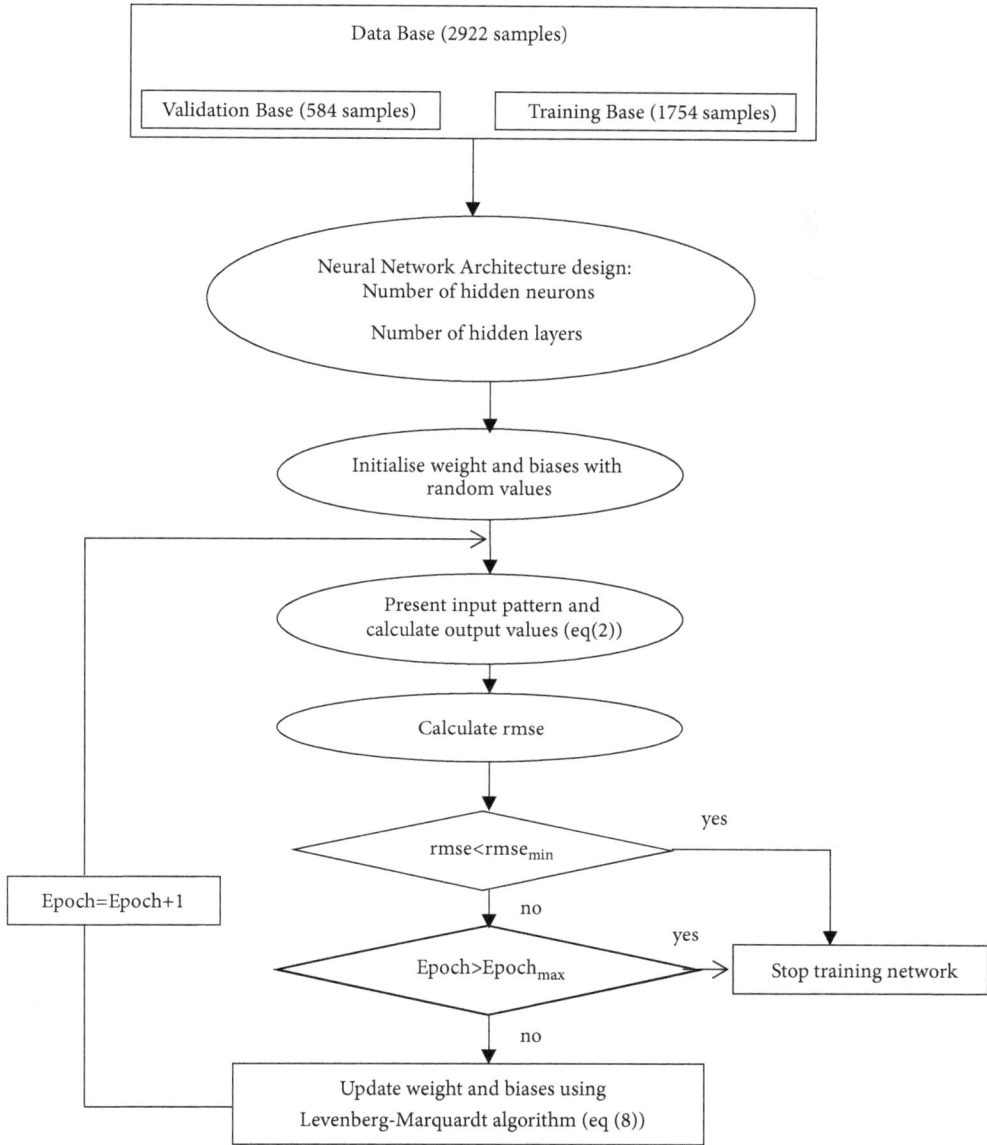

FIGURE 6: General layout of the neural model.

number represents neurons in the output layer, and the numbers in between represent neurons in the hidden layers. We present the best achieved results for the MLP ANN models (Figures 7 and 8).

As shown in Figures 8 and 9, we observe a good match on the plot for regression for all data in both networks, where R^2 has values of 0.95027 and 0.94254, respectively, for two and three inputs; only a few data are not too close to the fitting line. There is a slight difference between the two models; that with two inputs yields a suitable correlation of 97.48% with less neuron in the hidden layer whereas the model with three inputs yields a 97.08%, coefficient of determination.

MBE and RMSE yield very low values as shown below:

(i) For model 1 (model inputs L and l) the network structure is 2-190-1 for 0.95027, 0.10, and 0.89 representing, respectively, R^2, RMSE and MBE.

(ii) For model 2 (model input L-l-h), the network structure is 3-370-1 and the values of statistical errors R^2, RMSE, and MBE are, respectively, 0.94254, 0.13, and 1.14.

The results indicate that, for the test base, there is a very good correlation (Figures 9 and 10). This signifies the

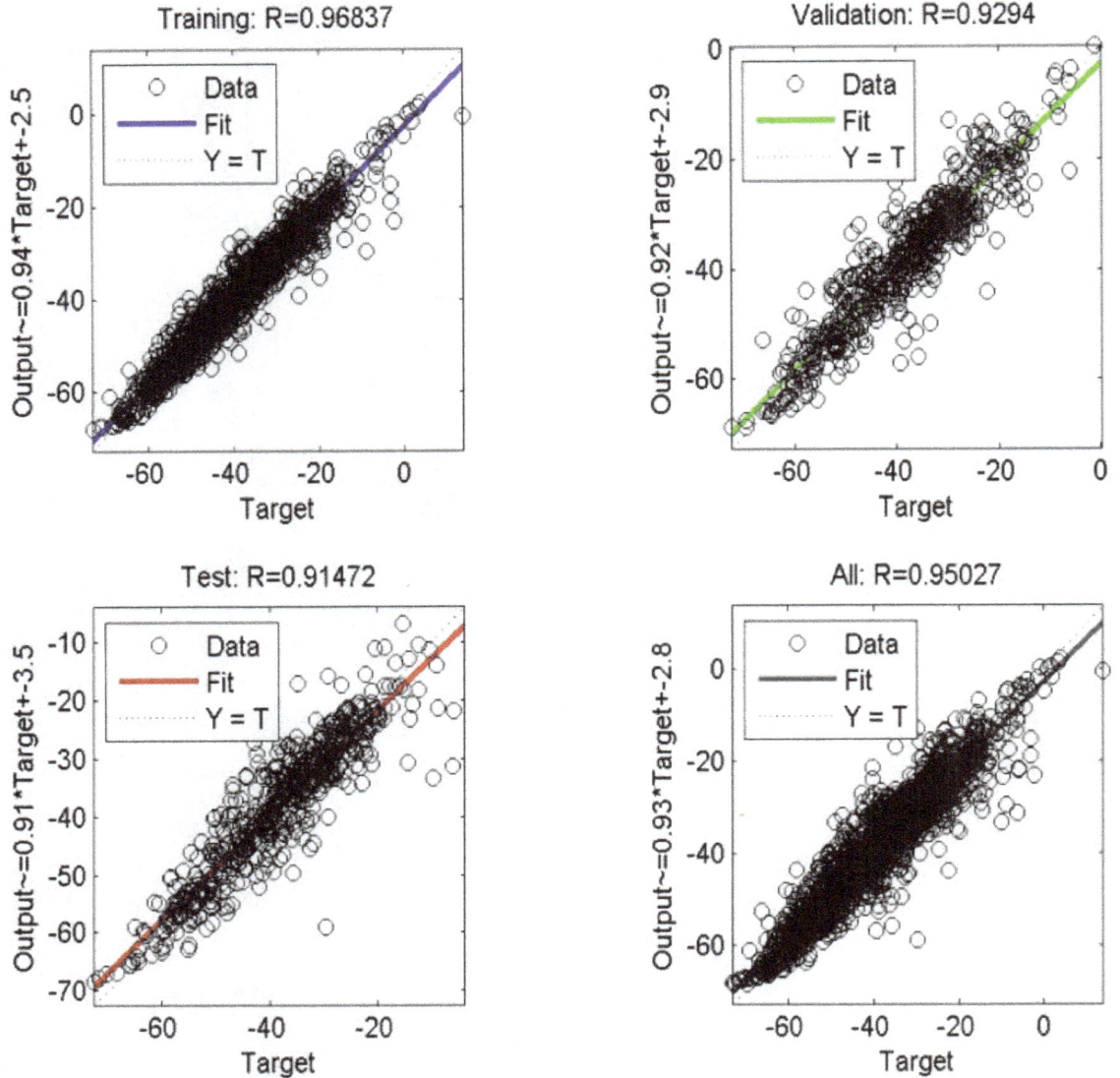

FIGURE 7: Performance (regression) of ANN model for prediction of Bouguer anomalies around Benue trough with 2 inputs.

possibility of having values of anomalies for areas where they do not exist by giving the geographical location.

Comparing observed and simulated data, we can see as shown in Figures 9 and 10 a good match for two and three entries.

Model Validation. We now compare results obtained using ANNs to classical approaches and, in addition, Bouguer, Euler, and lineaments maps for two and three entries.

Comparing ANNs Results to Other Methods. For two entries, we compare Neural Networks with classical methods based on multiple linear regression with specific approaches developed in most software used in geophysics such as Surfer and Oasis Montaj (Table 1). Z, the anomaly, is given by

$$Z = Ax + By + C, \qquad (9)$$

where A, B, and C are constants to be determined; x and y are geographical coordinates of a given point.

For three inputs, we use a classical multiple linear regression implemented through a matrix approach programmed [39] in Matlab or Excel to solve

$$Z = Ax + By + Ch + D, \qquad (10)$$

where $A, B, C,$ and D are constants to be determined; $x, y,$ and h, respectively, are geographical coordinates and elevation of a given point.

For Multiple Linear Regression Analysis, the following constants were obtained:

(i) Multiple Linear Regression Analysis with 2 inputs (MLRA2): $A = -1,6229$; $B = 2,7449$; $C = -42,2721$

(ii) Multiple Linear Regression Analysis with 3 inputs (MLRA3): $A = -0,7526$; $B = 1,4844$; $C = -0,0376$; $D = -29,1101$.

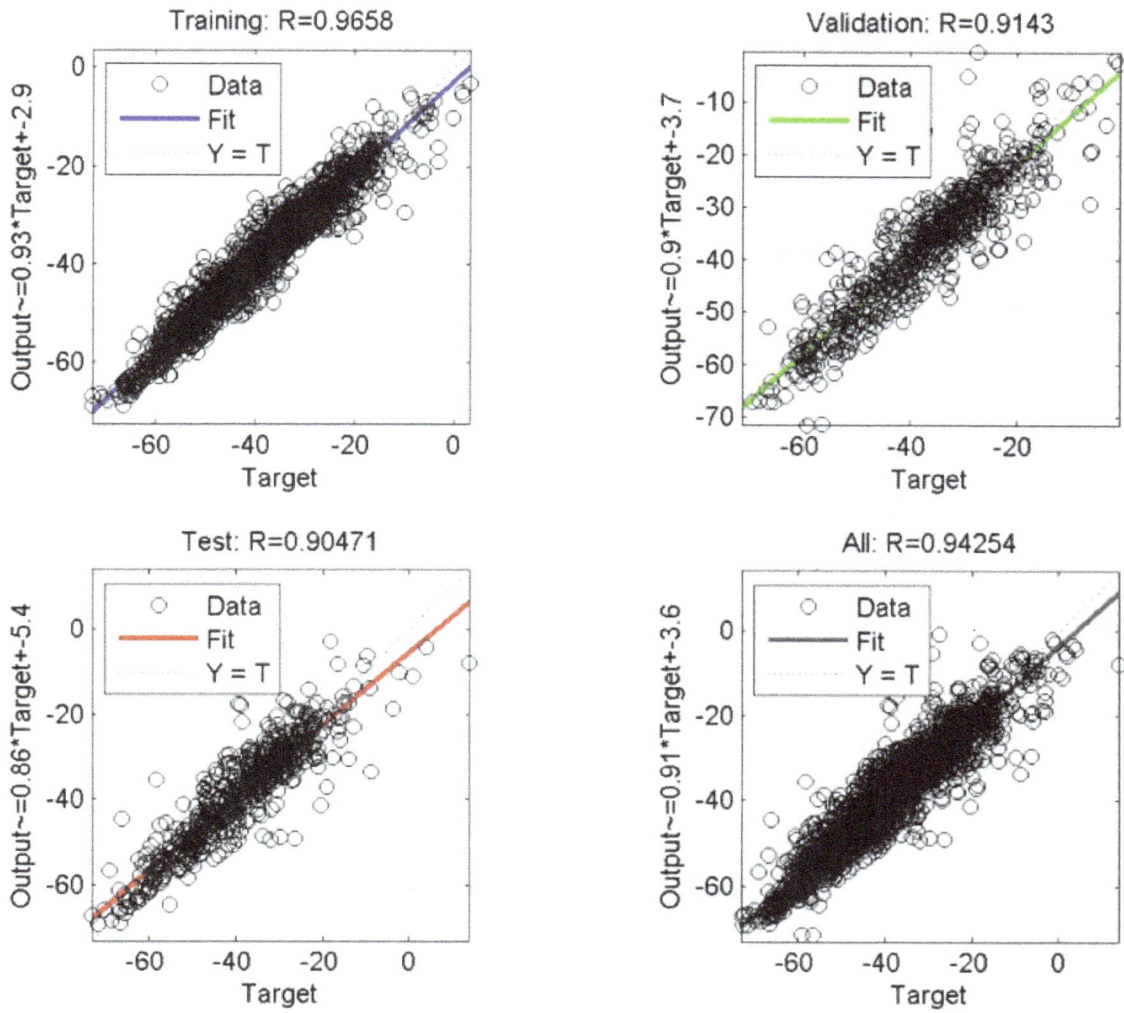

FIGURE 8: Performance of ANN model for prediction of Bouguer anomalies around Benue trough with 3 inputs.

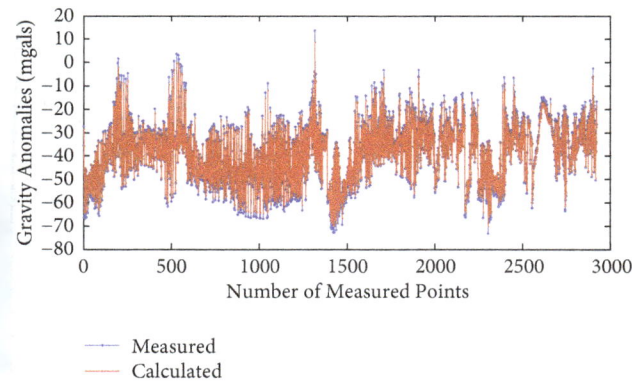

— Measured
— Calculated

FIGURE 9: Comparison of present ANN model (blue color) and measured anomalies (red color) for two inputs (*y*-axis for Bouguer anomalies and *x*-axis for gravity stations).

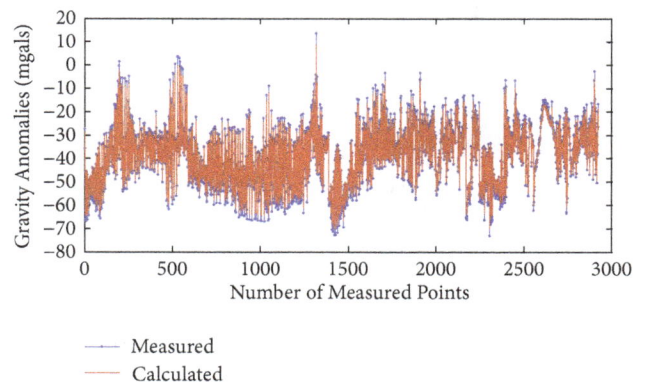

— Measured
— Calculated

FIGURE 10: Comparison of present ANN model (blue color) and measured anomalies (red color) for three inputs.

In Figure 11, we represent Bouguer anomalies versus the outputs inferred from MLRA3. In addition, in Table 2 and

Figure 12, we analyze the evolution of errors and Root Mean Square Error (RMSE) against the number of iterations and neurons in the hidden layer (NI and NNHL, respectively).

TABLE 1: Comparing interpolation methods.

Methods	Correlation factor	Mean Bias Error	Root Mean Square Error
	2 inputs		
ANN2	0,95	0,89	0,10
Kriging	0,06114	-0,00142648	12,2701515
Minimum Curvature	0,06114	-0,00142648	12,2701515
Radial Basis Function	0,06114	-0,00142648	12,2701515
Polynomial Regression	0,06099	0,00476981	12,2701501
Multiple linear regression	0,0610	5,2595e-11	12,2680
Inverse distance to a Power	0,06099	0,00476981	12,2701501
	3 inputs		
Multiple linear regression analysis	0,2273	-9,5263e-11	11,1289
ANN3	0,942	1,14	0,13

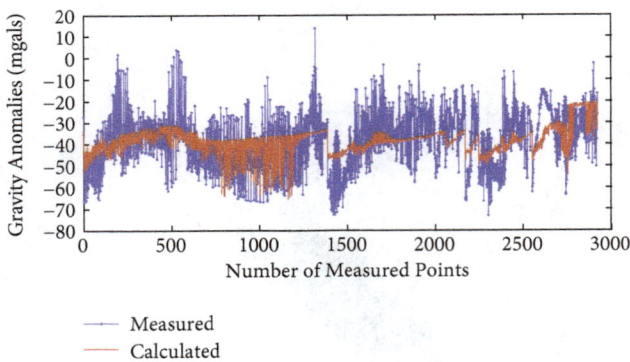

FIGURE 11: MLRA3 (red color) versus Bouguer anomalies measured (blue color).

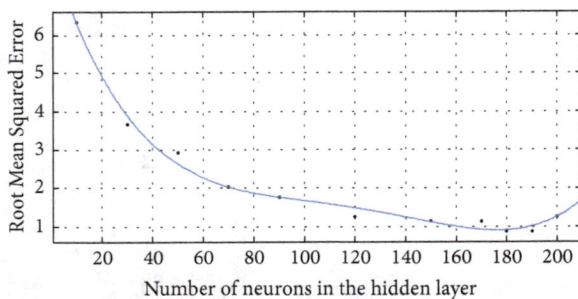

FIGURE 12: Root Mean Square Error (RMSE) versus number of neurons in the hidden layer (NNHL) for two inputs.

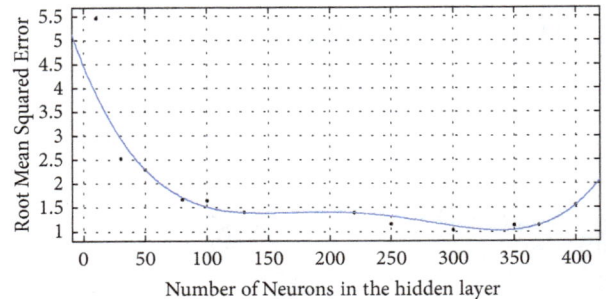

FIGURE 13: Root Mean Square Error (RMSE) versus number of neurons in the hidden layer (NNHL) for three inputs.

TABLE 2: Errors against the number of iterations and neurons in the hidden layer for 2 inputs.

Iterations	NN	RMSE	MBE
34	10	6,33	-0,25
97	30	3,68	-0,17
37	50	2,92	0,24
92	70	2,03	0,039
47	90	1,77	0,049
28	120	1,27	0,05
63	150	1,15	-0,165
21	170	1,14	0,13
64	180	0,88	0,0456
36	190	0,88	0,0456
15	200	1,27	-0,15

Plotting errors against iterations to us did not have any mathematical explanation; instead we plot errors against number of neurons in the hidden layer where it is obvious that one has low value of RMSE with increasing number of neurons in the hidden layer, unless the situation (above 190) where there is an overfitting (increasing RMSE) exists. The same conclusion arises for three inputs (Figure 13).

The results obtained show very good precision for Neural Network compared to classical approaches. Though ANNs can approximate any function, regardless of its linearity, they have some limitations such as their "black box" nature, greater computational burden, increasing accuracy by a few percent which can bump up the scale by several magnitudes (proneness to overfitting), and the empirical nature of model development (needs a lot of data for the training and cases for validation and test).

Comparing Measured and Calculated Bouguer, Euler, and Lineaments Maps. We generate and discuss the data obtained through ANN by establishing Bouguer, Euler (Reid et al.

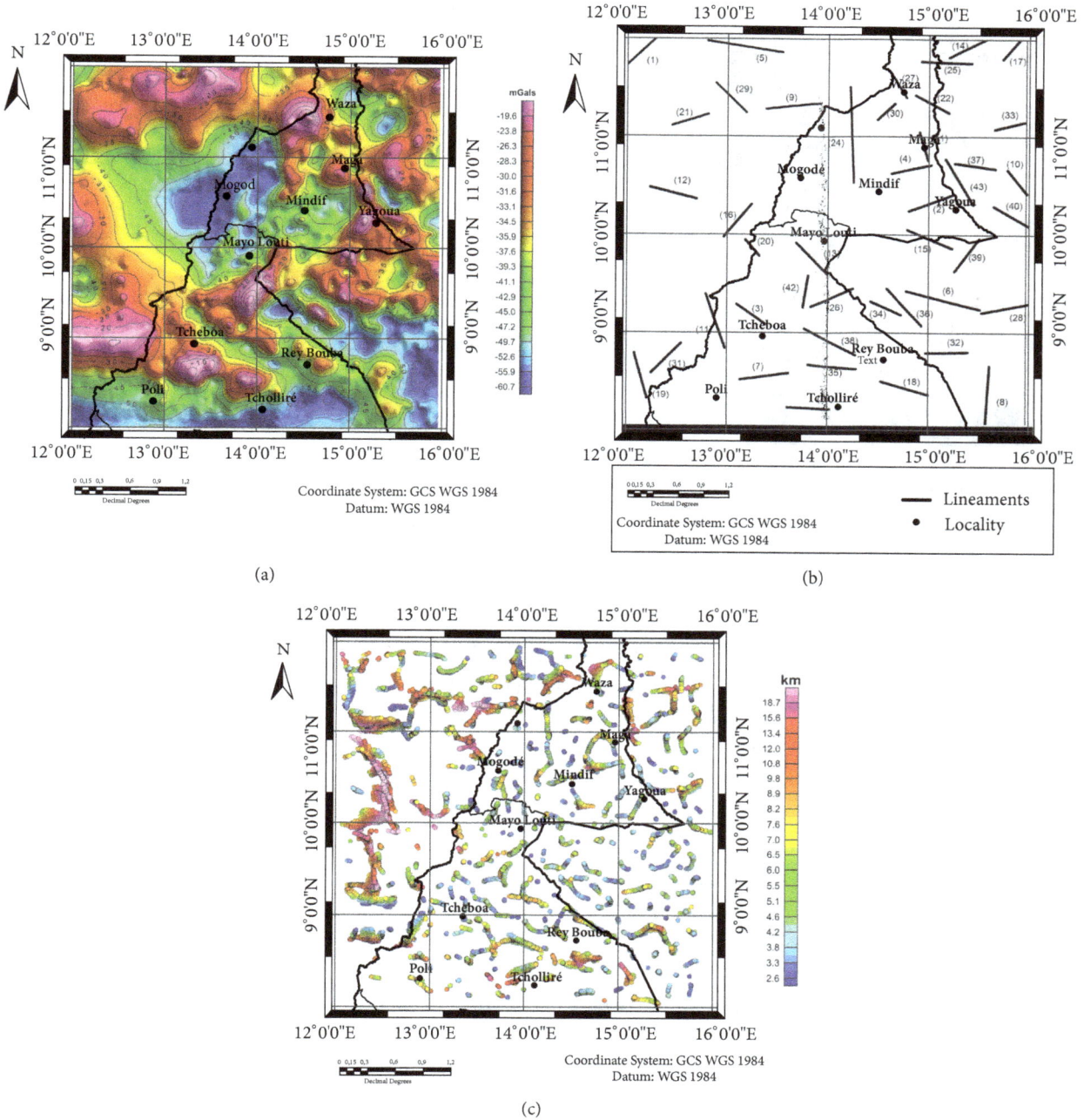

FIGURE 14: Bouguer (a), lineaments (b), and Euler maps (c).

[40]; (formula (11)), and lineament maps from Bouguer data obtained after prediction through inversion using Oasis Montaj 6.3 software.

$$(x - x_0) \frac{\partial T}{\partial x} + (y - y_0) \frac{\partial T}{\partial y} + (z - z_0) \frac{\partial T}{\partial z}$$

$$= N(B - T),$$

(11)

where T is the total field of magnetic or gravity source detected at (x, y, z), B is the regional gravity or magnetic field,

and N is the structural index value that needs to be chosen according to a prior knowledge of the source geometry.

Bouguer, Bouguer 2 and 3 entries maps (Figures 14(a), 15(a), and 16(a)) present structurally the same geological entities. A strict analysis makes it possible to distinguish between these maps: the positive anomaly structures (Waza, Maroua, South of Tcheboa, and Yagoua); the negative anomaly structures (Mogobé, south boundary of the area); and the gradient zones ensuring the transition between these anomalies of different signatures. These different features are the

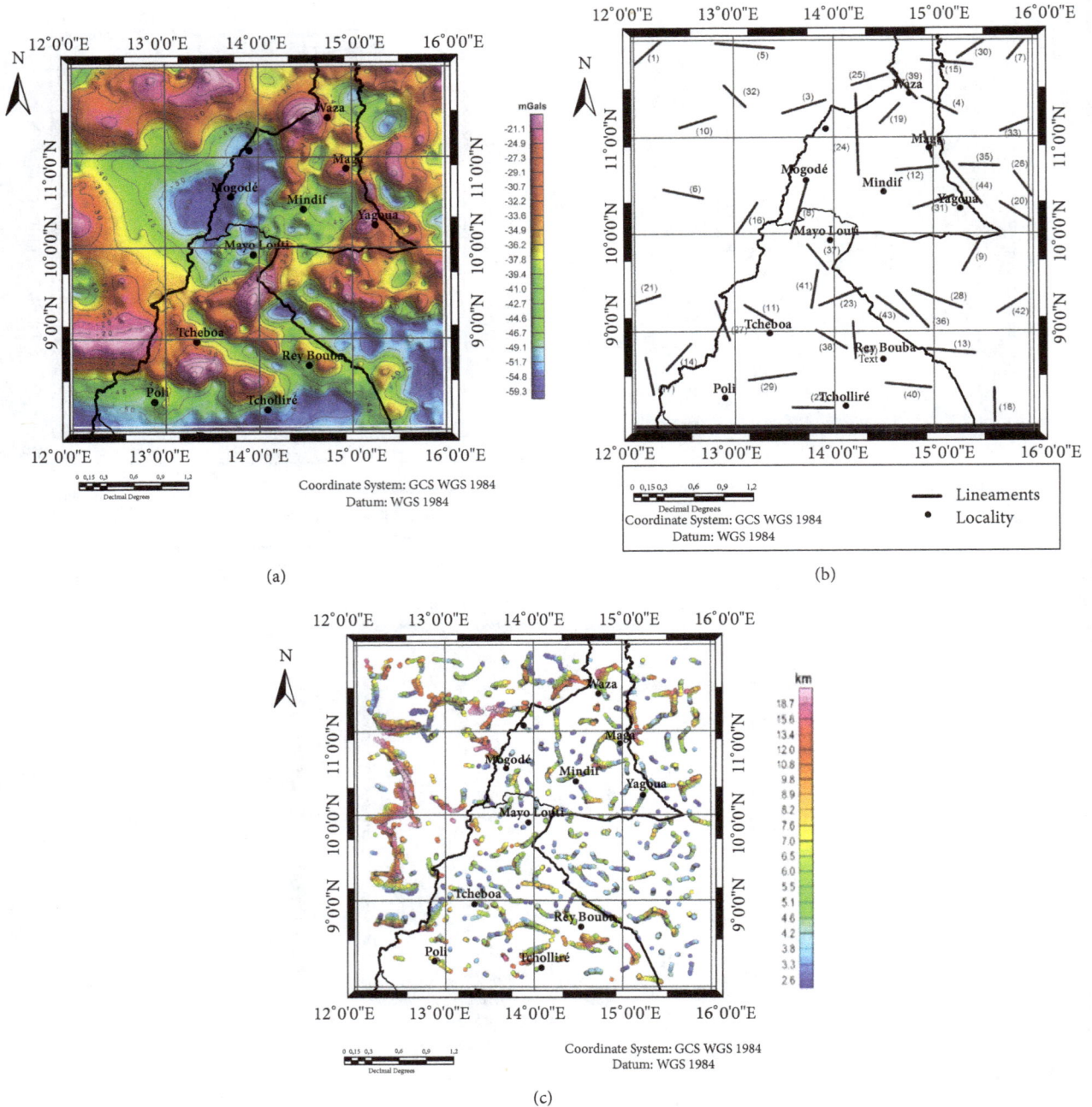

(a)

(b)

(c)

FIGURE 15: Bouguer (a), lineaments (b), and Euler maps (c) for two entries.

signature of the Precambrian, Early Cretaceous, and Tertiary-Quaternary rocks in the studied area. Transitions between structures require better materialization.

By using and comparing also Bouguer, Bouguer 2 entries, and Bouguer 3 entries, there is also a similarity between the lineament maps (Figures 14(b), 15(b), and 16(b)), thus highlighting tectonics in the area. Between Euler maps (Figures 14(c), 15(c), and 16(c)), the depths of the source structures of anomalies are described.

The lineaments obtained are later compared with existing results from other works. From Figures 14–16, we have a network of faults (above forty) with similar strikes. The depths range from about 2.6 km (at Tcheboa, Garoua sedimentary basin) to about 18.7 km (north of Waza). The results obtained by inversion of gravity data match and complete those obtained by Mouzong et al. [41], Eyike et al. (2010), and Kamguia et al. [4].

(a)

(b)

(c)

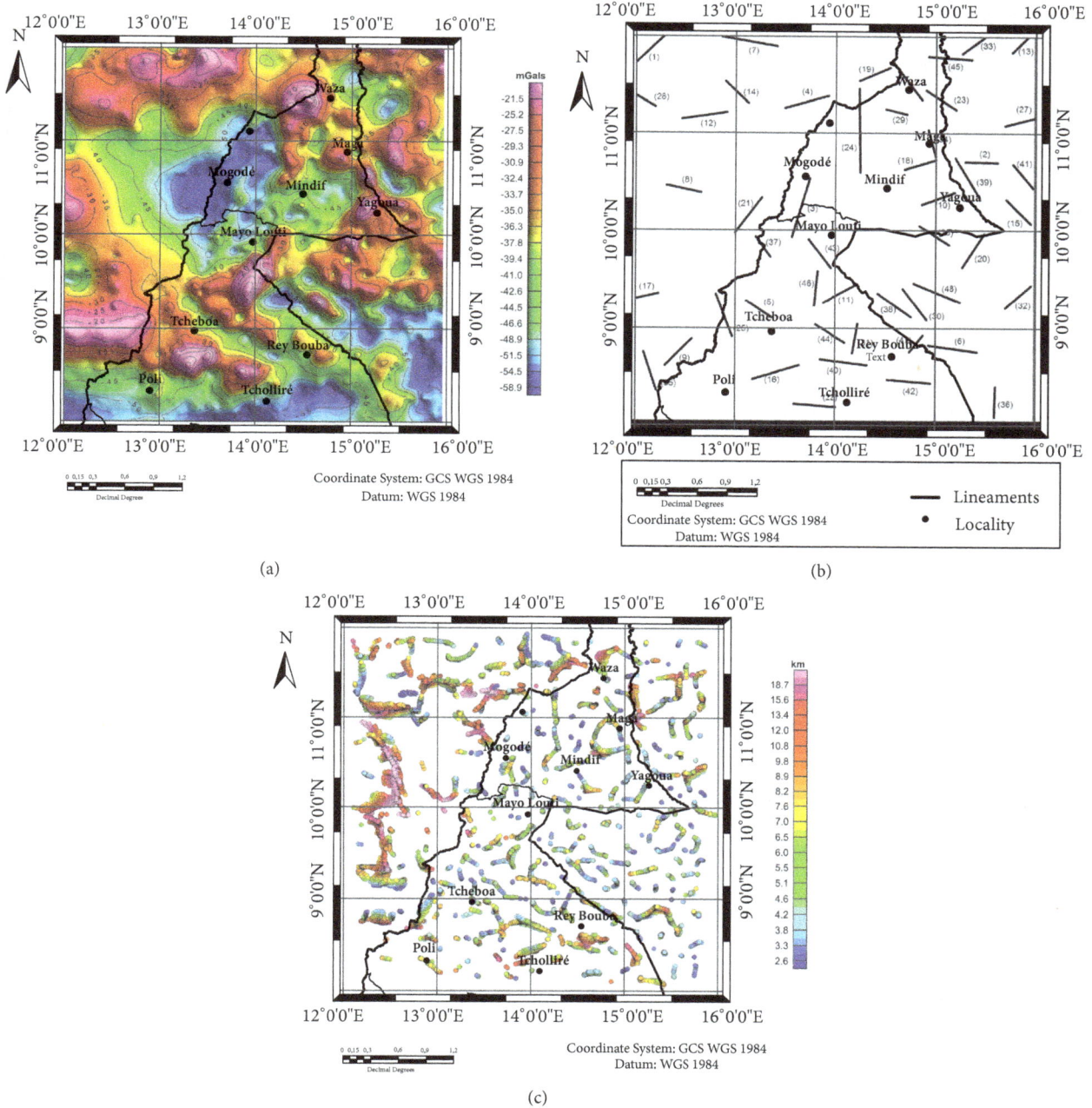

FIGURE 16: Bouguer (a), lineaments (b), and Euler (c) maps obtained for three inputs.

The ANN based model for gravity anomalies is accurate for the prediction of these anomalies in Northern Cameroon and its surroundings.

4. Conclusion

In this paper, an Artificial Neural Network (ANN) model was estimated for the prediction of gravity anomalies using, respectively, two (longitude and latitude) and three (longitude, latitude, and elevation) inputs in Northern Cameroon and its surroundings along with the corresponding anomaly.

Existing gravity data were used for training, validation, and testing of the Neural Network. With each of these inputs, we obtained a good correlation on the plot for regression for all data in both networks, where R^2 has values of 0.95027 and 0.94254, respectively. In order to validate the model, results were compared to those from classical interpolation approaches; in addition Bouguer, Euler, and lineaments maps were compared to our prediction. Low values of MBE and RMSE indicate the effectiveness of the approach. We provide in this work new deep faults for the studied area. The model is promising for evaluating the gravity anomaly at a specific point where there is no measured value. This method can therefore be recommended in geophysics to improve the resolution of geological features for uneven coverage of recorded gravity data and also to reduce the cost of geophysical surveys.

Acknowledgments

The authors are indebted to IRD (Institut de Recherche pour le Developpement) for providing them with the data used in this work.

References

[1] A. Eyike, F. E. Nyam, and C. A. Basseka, "Topography of the Moho Undulation in Cameroon from Gravity Data: Preliminary Insights into the Origin, the Age and the Structure of the Crust and the Upper Mantle across Cameroon and Adjacent Areas," *Open Journal of Geology*, vol. 08, no. 01, pp. 65–85, 2018.

[2] H. E. Ngatchou, G. Liu, C. T. Tabod et al., "Crustal structure beneath Cameroon from EGM2008," *Geodesy and Geodynamics*, vol. 5, no. 1, pp. 1–10, 2014.

[3] A.-P. K. Tokam, C. T. Tabod, A. A. Nyblade, J. Julià, D. A. Wiens, and M. E. Pasyanos, "Structure of the crust beneath Cameroon, West Africa, from the joint inversion of Rayleigh wave group velocities and receiver functions," *Geophysical Journal International*, vol. 183, no. 2, pp. 1061–1076, 2010.

[4] J. Kamguia, E. Manguelle-Dicoum, C. T. Tabod, and J. M. Tadjou, "Geological models deduced from gravity data in the Garoua basin, Cameroon," *Journal of Geophysics and Engineering*, vol. 2, no. 2, pp. 147–152, 2005.

[5] O. P. N. Eloumala, P. M. Mouzong, and B. Ateba, "Crustal structure and seismogenic zone of cameroon: integrated seismic, geological and geophysical data," *Open Journal of Earthquake Research*, vol. 3, pp. 152–161, 2014.

[6] P. Wessel and W. H. F. Smith, "New version of GMT released," *Transactions of The American Geophysical Union*, vol. 72, no. 441, pp. 445-446, 1995.

[7] W. H. F. Smith and P. Wessel, "Gridding with continuous curvature splines in tension," *Geophysics*, vol. 55, no. 3, pp. 293–305, 1990.

[8] D. G. Krige, *Geostatistics for Oreevaluation*, South African Institute of Mining and Metallurgy, Johannesburg, South Africa, 1978.

[9] N. Cressie, "Spatial prediction and ordinary kriging," *Mathematical Geology*, vol. 20, no. 4, pp. 405–421, 1988.

[10] D. C. Skeels, "What is residual gravity?" *Geophysics*, vol. 32, no. 5, pp. 872–876, 1967.

[11] H. Inoue, "A least-squares smooth fitting for irregularly spaced data: finite-element approach using cubic B-spline basis.," *Geophysics*, vol. 51, no. 11, pp. 2051–2066, 1986.

[12] S. S. Haykin, *Neural Networks and Learning Machines*, MCMaster University, Hamilton, Canada, 3rd edition, 1993.

[13] M. E. Murat and A. J. Rudman, "Automated first arrival picking. A neural network approach," *Geophysical Prospecting*, vol. 40, pp. 587–604, 1992.

[14] M. D. McCormack, D. E. Zaucha, and D. W. Dushek, "First-break refraction event picking and seismic data trace editing using neural networks," *Geophysics*, vol. 58, no. 1, pp. 67–78, 1993.

[15] M. Moustra, M. Avraamides, and C. Christodoulou, "Artificial neural networks for earthquake prediction using time series magnitude data or Seismic Electric Signals," *Expert Systems with Applications*, vol. 38, no. 12, pp. 15032–15039, 2011.

[16] J. Reyes, A. Morales-Esteban, and F. Martínez-Álvarez, "Neural networks to predict earthquakes in Chile," *Applied Soft Computing*, vol. 13, no. 2, pp. 1314–1328, 2013.

[17] M. M. Poulton, B. K. Sternberg, and C. E. Glass, "Location of subsurface targets in geophysical data using neural networks," *Geophysics*, vol. 57, no. 12, pp. 1534–1544, 1992.

[18] Y. Zhang and K. V. Paulson, "Magnetotelluric inversion using regularized Hopfield neural networks," *Geophysical Prospecting*, vol. 45, no. 5, pp. 725–743, 1997.

[19] G. Roth and A. Tarantola, "Neural networks and inversion of seismic data," *Journal of Geophysical Research: Atmospheres*, vol. 99, no. 4, pp. 6753–6768, 1994.

[20] H. Langer, G. Nunnari, and L. Occhipinti, "Estimation of seismic waveform governing parameters with neural networks," *Journal of Geophysical Research: Solid Earth*, vol. 101, no. 9, pp. 20109–20118, 1996.

[21] C. Calderón-Macías, M. K. Sen, and P. L. Stoffa, "Automatic NMO correction and velocity estimation by a feedforward neural network," *Geophysics*, vol. 63, no. 5, pp. 1696–1707, 1998.

[22] H. Dai and C. MacBeth, "Split shear-wave analysis using an artificial neural network?" in *First Break*, vol. 12, pp. 605–613, 1994.

[23] Z. Huang, J. Shimeld, M. Williamson, and J. Katsube, "Permeability prediction with artificial neural network modeling in the Ventura gas field, offshore eastern Canada," *Geophysics*, vol. 61, pp. 422–436, 1996.

[24] L.-X. Wang and J. M. Mendel, "Adaptive minimum prediction-error deconvolution and source wavelet estimation using Hopfield neural networks," *Geophysics*, vol. 57, no. 5, pp. 670–679, 1992.

[25] C. Calderón-Macías, M. K. Sen, and P. L. Stoffa, "Hopfield neural networks, and mean field annealing for seismic deconvolution and multiple attenuation," *Geophysics*, vol. 62, no. 3, pp. 992–1002, 1997.

[26] F. U. Dowla, S. R. Taylor, and R. W. Anderson, "Seismic discrimination with artificial neural networks: Preliminary results with regional spectral data," *Bulletin of the Seismological Society of America*, vol. 80, pp. 1346–1373, 1990.

[27] G. Romeo, "Seismic signals detection and classification using artificial neural networks," *Annals of Geophysics*, vol. 37, pp. 343–353, 1994.

[28] A. A. Grêt, E. E. Klingelé, and H.-G. Kahle, "Application of artificial neural networks for gravity interpretation in two dimensions: A test study," *Bollettino di Geofisica Teorica e Applicata*, vol. 41, no. 1, pp. 1–20, 2000.

[29] M. Ghalambaz, A. R. Noghrehabadi, M. A. Behrang, E. Assareh, A. Ghanbarzadeh, and N. Heayat, "A hybrid neural network and gravitational search algorithm (HNNGSA) method to solve well known Wessinger's equation," *International Journal of Mechanical, Aerospace, Industrial, Mechatronic and Manufacturing Engineering*, vol. 5, no. 1, pp. 147–151, 2011.

[30] J. D. Fairhead and C. S. Okereke, "A regional gravity study of the West African rift system in Nigeria and Cameroon and its tectonic interpretation," *Tectonophysics*, vol. 143, no. 1-3, pp. 141–159, 1987.

[31] I. Ngounouno, B. Déruelle, and D. Demaiffe, "Petrology of the bimodal Cenozoic volcanism of the Kapsiki plateau (northernmost Cameroon, Central Africa)," *Journal of Volcanology and*

Geothermal Research, vol. 102, no. 1-2, pp. 21–44, 2000.

[32] J.-C. Dumort and Y. Péronne, *Carte Géologique de Reconnaissance à l'échelle du 1/500000, République Féderale du Cameroun, feuille Maroua*, avec notice explicative, BRGM et Direction des Mines et de la Géologie, Kribi, Cameroun, 1966.

[33] S. Bofinger and G. Heilscher, "Solar electricity forecast-approaches and first results," in *Proceedings of the in. 21th PV Conference*, Dresden, Germany, 2006.

[34] A. Bosch, A. Miñán, C. Vescina et al., "Fourier Transform Infrared Spectroscopy for rapid identification of nonfermenting gram-negative bacteria isolated from sputum samples from cystic fibrosis patients," *Journal of Clinical Microbiology*, vol. 46, no. 12, pp. 2535–2546, 2008.

[35] J. F. Jodouin, *Les réseaux de neurones: principes et définitions*, Hermès, Paris, France, 1994.

[36] K. Hornik, "Approximation capabilities of multilayer feedforward networks," *Neural Networks*, vol. 4, no. 2, pp. 251–257, 1991.

[37] F. Collignon, *Gravimétrie de reconnaissance de la République Fédérale du Cameroun*, ORSTOM, Paris, France, 1968.

[38] Y. H. Poudjom-Djomani, *Apport de la gravimétrie à l'étude de la lithosphère continentale et implications géodynamiques. Etude d'un bombement intraplaque: le massif de l'Adamaoua (Cameroun) [Thèse de Doctorat]*, Université de Paris-Sud, Orsay, France, 1993.

[39] H. B. Scott, "Multiple Linear Regression Analysis: a matrix approach in MATLAB," *Alabama Journal of Mathematics, Spring/Fall2009*, 3 pages, 2009.

[40] A. B. Reid, J. M. Allsop, H. Granser, A. J. Millett, and I. W. Somerton, "Magnetic interpretation in three dimensions using Euler deconvolution," *Geophysics*, vol. 55, no. 1, pp. 80–91, 1990.

[41] M. P. Mouzong, J. Kamguia, S. Nguiya, Y. Shandini, and E. Manguelle-Dicoum, "Geometrical and structural characterization of Garoua sedimentary basin , Benue Trough, North Cameroon, using gravity data," *Journal of Biology and Earth Sciences*, vol. 4, no. 1, pp. E25–E33, 2014.

2

Variability and Trend of Annual Maximum Daily Rainfall in Northern Algeria

Abderrahmane Nekkache Ghenim and Abdesselam Megnounif

"Eau et Ouvrage dans Leur Environnement" Laboratory, Tlemcen University, BP 230, 13000 Tlemcen, Algeria

Correspondence should be addressed to Abderrahmane Nekkache Ghenim; anghenim@yahoo.fr

Academic Editor: Robert Tenzer

The daily rainfall dataset of 35 weather stations covering the north of Algeria was studied for a period up to 43 years, recorded after 1970s. The variability and trends in annual maximum daily rainfall (AMDR) time series and their contributions in annual rainfall (AR) were investigated. The analysis of the series was based on statistical characteristics, Burn's seasonality procedure, Mann-Kendall test, and linear regression technique. The contribution of the AMDR to AR analysis was subjected to both the Buishand test and the double mass curve technique. The AMDR characteristics reveal a strong temporal irregularity and have a wide frequency of occurrence in the months of November and December while the maximum intensity occurred in October. The observed phenomenon was so irregular that there was no dominant season and the occurrence of extreme event can arrive at any time of the year. The AMDR trends showed that only six of 35 stations have significant trend. For other stations, no clear trend was highlighted. This result was confirmed by the linear regression procedure. On the contrary, the contribution of AMDR in annual totals exhibited a significant increasing trend for 57% of the sites studied with a growth rate of up to 50%.

1. Introduction

Rainfall is a fundamental element of climate which is, for several decades, in perpetual mutations. For most regions around the Mediterranean, these changes resulted in significant rainfall deficits [1, 2] accompanied by an increase of exceptional events such as severe droughts and devastating floods [3, 4]. The Mediterranean environments, typical of semiarid regions that enjoy a rather pleasant climate with sunshine and its fine weather, can suffer hazardous situations since several regions are regularly struck by severe rainstorms. Such events are highly variable in the time and space [5] and often lasted less than one day [6]. Therefore, the critical parameter of these rainstorms is the maximum daily rainfall rather than the total rainfall. The annual maximum daily rainfall is defined as an extreme instance, with critical duration for a watershed, region, or state [7]. In the hydrological year, the daily maximum rainfall is the parameter considered to assess the immediate impact on the hydrological response of streams, flooding cities, soil erosion, dams silting, and agricultural production [8, 9].

Located on the southern shore of the Mediterranean, Algeria suffers a semiarid to arid climate. Despite a drought that has lasted for over three decades, brief, intense, and devastating floods often affect cities. Due to the intensity of rainfall events that typically last less than 24 hours and the vulnerability of urban areas, the floods have caused damage and significant loss of human lives. The only flood that occurred in the city of Algiers in November 2001 caused some 740 casualties [10–12].

The objective of this study is to contribute to the knowledge of the variability of annual maximum daily rainfall and its changes in the north of Algeria. A particular attention is paid to detect possible trends characterizing AMDR series and to evaluate changes in AMDR contribution to annual totals.

2. Study Area/Materials and Methods

The study area is the north of Algeria and covers 15% of the total surface of the country (2.38 million km^2). The length

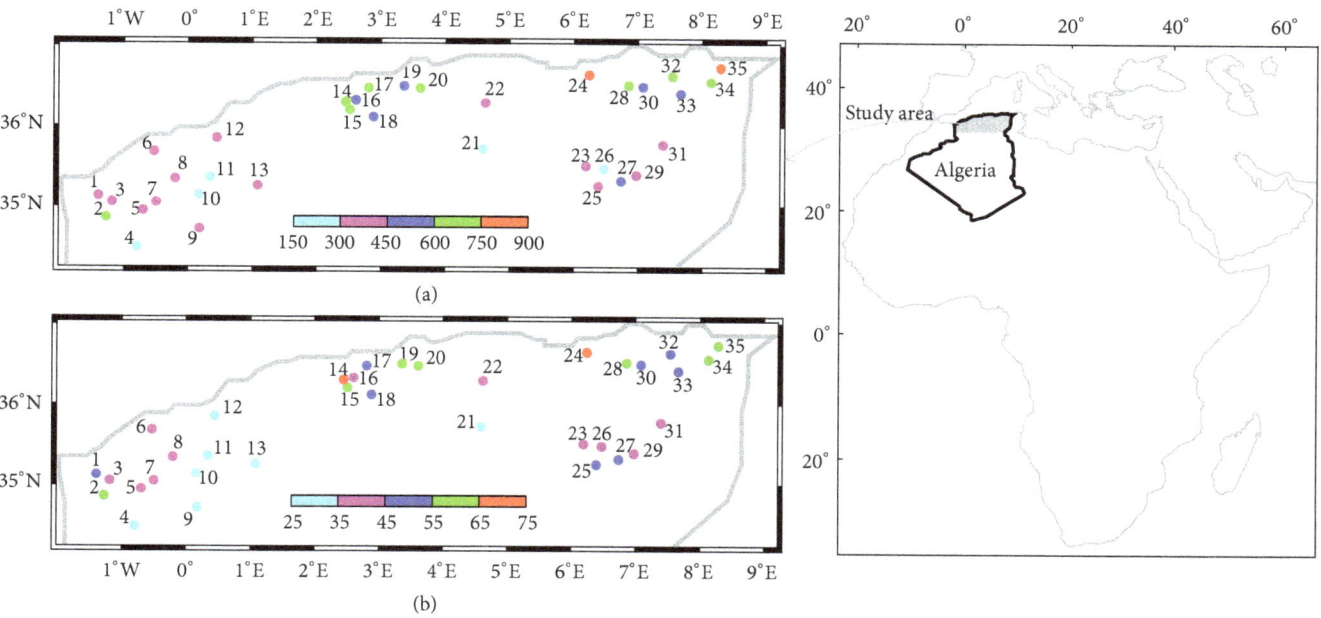

FIGURE 1: Locations of rainfall stations in northern Algeria and mean values intervals of annual rainfall (a) and annual maximum daily rainfall (b).

of the Mediterranean coastline is about 1200 km. Because of its geographic position and its mountainous nature, rainfall shows strongest contrasts between the various areas. This region has a high spatial and temporal climatic variability known as the Mediterranean climate. Intra-annual rainfall is irregular and most rains fall between October and April, while infrequent and local storms occur in the dry season from July to October.

Rainfall data were collected by the National Agency of the Hydraulic Resources (ARNH) (http://www.anrh.dz/). Because there were too many gaps after colonial period in the 1960s and where the number of years of observations was too low for statistical purposes, many station series were discarded from the data set. As a result, only 35 rainfall series were selected (Figure 1).

The data collected are a set of daily rainfall series for the period after 1970. They include less than 5% of gaps. These are replaced by the values of the station which has the best correlation. The data were tested for their quality control. The suspect data were cross-checked with those of nearby stations (some are not used in this study). The arithmetic sum of daily heights recorded is the annual rainfall value (AR) while the daily maximum observed during the year represents the annual maximum daily rainfall (AMDR). Data were aggregated according to hydrological year from September to August. The selected stations are described by the geographical coordinates, number of years of observation, and statistical characteristics (Table 1).

In the Mediterranean region, it is unlikely that rainstorm lasts more than 24 hours [6]. So, the annual maximum daily rainfall (AMDR) series may be introduced to study

the distribution of extreme precipitation occurrences within a year. To examine the regularity for such series, we applied the seasonality as described by Burn [13]. This method used to estimate the timing and regularity of floods was then transposed to extreme rainfall in particular AMDR [14]. It assesses the degree of similarity of watersheds in relation to the hydrological response or the occurrence of extreme rainfall. Burn's vector defined by $(\bar{\theta}, \bar{r})$ represents the variability of the date of occurrence of all extremes events. Its direction is the mean date of the occurrence of the extremes events and the modulus is the variability around the mean value. The dates of the AMDR occurrence are based on the calendar year. January 1 is the 1st day and December 31 is the 365th day.

Each date j_i was replaced by an angle:

$$\theta_i = j_i \frac{2\pi}{365.25}, \quad \theta_i \in \{0, 75°, \ldots, 360°\}. \quad (1)$$

The obtained series covers the unit circle where each term can be described in polar coordinates as a vector $(\cos\theta_i, \sin\theta_i)$, where θ_i indicates the direction expressed in radians. Following the Burn approach, the Cartesian coordinates, $x_i = \cos\theta_i$ and $y_i = \sin\theta_i$, are used to evaluate Burn's vector defined by $(\bar{\theta}, \bar{r})$ the mean direction and mean modulus given by the following equations:

$$\bar{\theta} = \text{arctg}\left(\frac{\bar{y}}{\bar{x}}\right), \quad \bar{\theta} \in [0°; 360°],$$

$$\bar{r} = \sqrt{\bar{x}^2 + \bar{y}^2}, \quad \bar{r} \in [0; 1], \quad (2)$$

TABLE 1: Descriptive statistic of AMDR in northern Algeria.

Number	Station	N	Coordinates		Average AR (mm)	AMDR (mm)			
			Latitude	Longitude		Average	Max	Cv	Cs
1	Pierre du chat	41	1°26′52″W	35°08′37″N	330	46	255.4	0.81	4.60
2	Meffrouche	41	1°17′31″W	34°51′19″N	608	64	158.1	0.50	1.30
3	Bensekrane	37	1°13′26″W	35°04′28″N	377	46	82.2	0.36	0.46
4	Ras El Ma	39	0°48′34″W	34°29′41″N	194	25	52.2	0.50	0.51
5	S. A. Benyoub	42	0°44′04″W	34°58′38″N	351	35	77.2	0.42	0.85
6	Sarno	41	0°35′52″W	35°44′59″N	357	40	88.7	0.45	1.48
7	Hassi Daho	40	0°32′26″W	35°05′28″N	313	38	103	0.40	2.10
8	Cheurfas	35	0°15′06″W	35°24′15″N	328	38	114.4	0.56	1.93
9	Ain El Hadjer	35	0°08′56″E	34°45′25″N	330	29	61.6	0.38	1.14
10	Ghriss	38	0°09′59″E	35°14′46″N	291	31	52	0.31	0.74
11	Maoussa	35	0°14′53″E	35°22′41″N	292	34	84.9	0.44	1.16
12	Oued El Kheir	41	0°22′50″E	35°57′08″N	301	31	94.4	0.50	2.16
13	Bekhedda	33	1°02′15″E	35°20′32″N	315	30	64.4	0.41	0.89
14	Meured	41	2°24′27″E	36°26′58″N	615	71	176.8	0.51	1.62
15	Djebabra	41	2°26′06″E	36°23′43″N	647	59	176.8	0.54	1.79
16	Ameur El Ain	41	2°34′01″E	36°28′31″N	586	40	83	0.39	0.99
17	Kolea	41	2°46′19″E	36°38′04″N	606	54	105.5	0.34	0.93
18	Ouzera	41	2°50′50″E	36°15′21″N	571	52	134	0.46	1.38
19	Hamiz-D9	38	3°19′49″E	36°39′26″N	598	57	102.9	0.38	0.56
20	Lakhdaria	39	3°35′12″E	36°37′40″N	701	61	117.4	0.40	1.02
21	K'sob	34	4°34′03″E	35°49′32″N	220	25	53.5	0.38	1.20
22	Sidi Yahia	40	4°37′11″E	36°25′19″N	377	35	70.1	0.33	1.00
23	Batna	40	6°10′23″E	35°33′53″N	374	36	64.4	0.35	0.40
24	El Milia	40	6°16′38″E	36°45′21″N	874	70	210.3	0.52	1.70
25	Ain Tinn	40	6°26′21″E	35°22′39″N	430	47	143	0.46	2.44
26	Timgad	40	6°28′07″E	35°29′51″N	289	33	93.7	0.44	1.93
27	Chelia	43	6°39′03″E	35°22′02″N	481	52	177.5	0.56	2.65
28	Zardasas	41	6°53′48″E	36°35′58″N	658	58	137	0.54	1.24
29	Ain Mimoun	40	6°57′22″E	35°24′55″N	434	43	116.1	0.41	2.32
30	Helioplolis	41	7°26′44″E	36°30′32″N	596	54	103.7	0.43	0.79
31	Ain Beida	43	7°27′03″E	35°47′50″N	408	40	104	0.36	2.24
32	Ain Berda	42	7°35′47″E	36°41′30″N	631	53	111	0.43	0.94
33	Bouchegouf	41	7°42′35″E	36°27′33″N	545	32	120	0.41	1.12
34	Ain Kerma	41	8°11′46″E	36°35′23″N	712	57	155.1	0.47	1.41
35	Ain Assel	41	8°21′57″E	36°46′03″N	821	60	142.7	0.38	1.31

N: years of record; Cv: coefficient of variation; Cs: coefficient of skewness.

where

$$\overline{x} = \frac{1}{n}\sum_{i=1}^{n} \cos\theta_i,$$

$$\overline{y} = \frac{1}{n}\sum_{i=1}^{n} \sin\theta_i. \tag{3}$$

When $\overline{r}$ decrease to zero, there is no single dominant season and the occurrence of extreme event can arrive at any time of the year, while $\overline{r} = 0$ is a virtual value indicating that extreme events occur on the same day. The date of occurrence of extreme events is regular, as the modulus approaches the unit more.

The annual trend of AMDR series was analyzed by using two statistical methods: the Mann-Kendall test and the linear regression. The nonparametric Mann-Kendall test [15, 16] detects the direction of trend patterns in hydrological variables. For a time series (x_i) of n values, each value x_i is compared with all corresponding x_j to compute the sign, and the indices i and j take the respective values $i = 1, 2, \ldots, n-1$ and $j = i+1, i+2, i+3, \ldots, n$. Kendall's S-statistics is based on the sum and variance computation. In this study, an error risk of 5% is accepted; that means a probability threshold below which the null hypothesis, the trend series, is monotonic, will be rejected. Thereby, the Mann-Kendall test is expected to be less affected by the outliers because its statistic is based on the sign of differences rather than on the values of the random variable [17].

The linear regression procedure is a statistical technique for estimating the relationships among variables (e.g., X and Y). The straight line $Y = aX + b$ is obtained by the least square regression method. The slope (a) indicates the average rate of change in the variable used. If the change is significantly different from zero, then a real change occurs. Positive slope defines increasing trend while the negative one indicates a decreasing trend.

The temporal series of the rate contribution of the AMDR to annual rainfall was subjected to the Buishand test to assess trends and date of departure of homogeneity [18]. The test is a graphical method based on the evolution of the following equation:

$$CS_k(X) = \sum_{i \le k} S_i = \sum_{i \le k} \left(\frac{X_i - \overline{X}}{s} \right), \tag{4}$$

where X_i is the variable, $\overline{X}$ is the mean value of the series, and s is the standard deviation.

The statistical parameter $\max_k |CS_k|$ is a good indicator of the departure of homogeneity [19]. The increasing or decreasing limbs of both sides of the extremum $\max_k |CS_k|$ correspond to surplus or deficit periods, respectively. So when significant change is confirmed, we applied the double mass curve to quantify the surplus or deficit [20]. The curve is a straight line whose slope is the proportionality constant. A break in slope indicates a change of the proportionality [21]. The break in slope and the angle formed by two straight lines indicate the date and the degree of change in the behavior of the phenomenon.

3. Results

In the study sites located in the north of Algeria where a division in increments of 150 mm is shown in Figure 1(a), the average annual rainfall varies from 194 mm (Ras El Ma station, code number 4) to 874 mm (El Milia station code number 24) (Table 1). Spatially, rainfall is distributed in four areas [22]:

(i) The central highlands, Sersou and Ras El Ma regions, though situated at high altitude, are sheltered compared to wet currents, where rainfall is less than 300 mm.

(ii) The western region of the country is characterized by a relative sheltered position compared to maritime influences and the low volume of the relief. Annual averages are generally less than 450 mm.

(iii) The mountainous areas and high interior plains (mountains of Tlemcen in the west and the mountains of Zaccar and Dahras in the east) are characterized by the relative importance of total rainfall, with annual average exceeding 600 mm.

(iv) The Atlas Tellien region exposed to the north and northwest records rainfall amounts that can exceed 800 mm.

For AMDR values, a division in increments of 10 mm and the main statistical parameters estimated for all the stations are shown, respectively, in Figure 1(b) and Table 1. Spatial variability of intra-annual averages of AMDR ranges from a minimum of 25 mm (station Ras El Ma, code number 4) to a maximum of 71 mm (Station Meured, code number 14). Low values are concentrated on the west side while the high values are in the east and center of the country.

Through the 35 studied stations, the temporal distribution of the annual maximum daily rainfall is irregular. The coefficients of variation, Cv, vary between 31% and 81%, with a spatial average of 44%. The most irregular series, Cv above 0.5, are found near the reliefs. The stations with the code numbers 1 and 24 are positioned facing the north and the stations with code numbers 2, 4, 8, 12, 14, 15, 24, 27, and 28 are positioned facing the south (Figure 1). The largest values of 255 and 210 mm were recorded, respectively, in the west region in station code number 1 in 1999-2000 and at the east in station code number 24 in 1990-1991. The spatial average of maximum daily rainfall is about 114 mm, which represents 24% of the spatial average of mean inter annual rainfall.

For stations under consideration, the skewness coefficient (Cs) has positive values fluctuating from 0.4 to 4.6 with a spatial mean of 1.43. Thereby, at level confidence of 95%, there is rejection for normal distribution for a majority of stations since the values of skewness coefficient are above 0.62 for 31 of 35 stations. That means low values are more frequent while high values are still rare but excessive. The pronounced skewness coefficients occurred at the station Pierre du chat (code number 1).

The difference in positioning stations (near the coastline, facing the sea, at high altitude or between the mountains) prevents a relatively good correlation to the altitude. The spatial variability of rainfall (AR and AMDR) is mainly influenced by latitude (Figure 2). The average increase for AR and AMDR, respectively, is about 140 mm and 25 mm to 100 km latitude.

The occurrence of AMDR was more frequent during November, followed by December and January (Figure 3(a)), while the peaks in order of importance occurred in October, December, and November (Figure 3(b)). In addition, the seasonal concentration index $\overline{r}$ varies from 0.03 to 0.32 with an average of 0.15 (Figure 4). So, there is no single dominant season and the occurrence of an extreme event can arrive at any time of the year. On the other hand, Burn's method shows that AMDR were more concentrated in February through 40% of the stations, against 20% in October (Figure 4).

The results of the Mann-Kendall test are summarized in Table 2 and Figure 5(a). At the 95% confidence level, only 6 of 35 stations show significant trend. The stations Ras El Ma, Ain Berda, and Chelia show an increasing trend, whereas the trend is decreasing in Sarno, Ameur El Ain, and Djebabra, while in the remaining stations no significant trend is observed. This finding corroborates the results obtained by the linear regression analysis (Table 2 and Figure 5(b)).

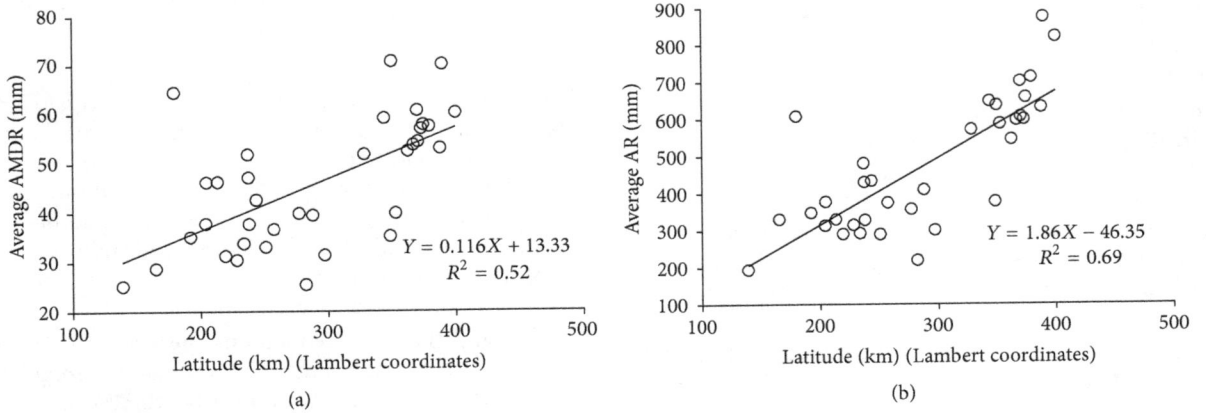

FIGURE 2: Relationship of AMDR (a) and AR (b) with latitude in northern Algeria.

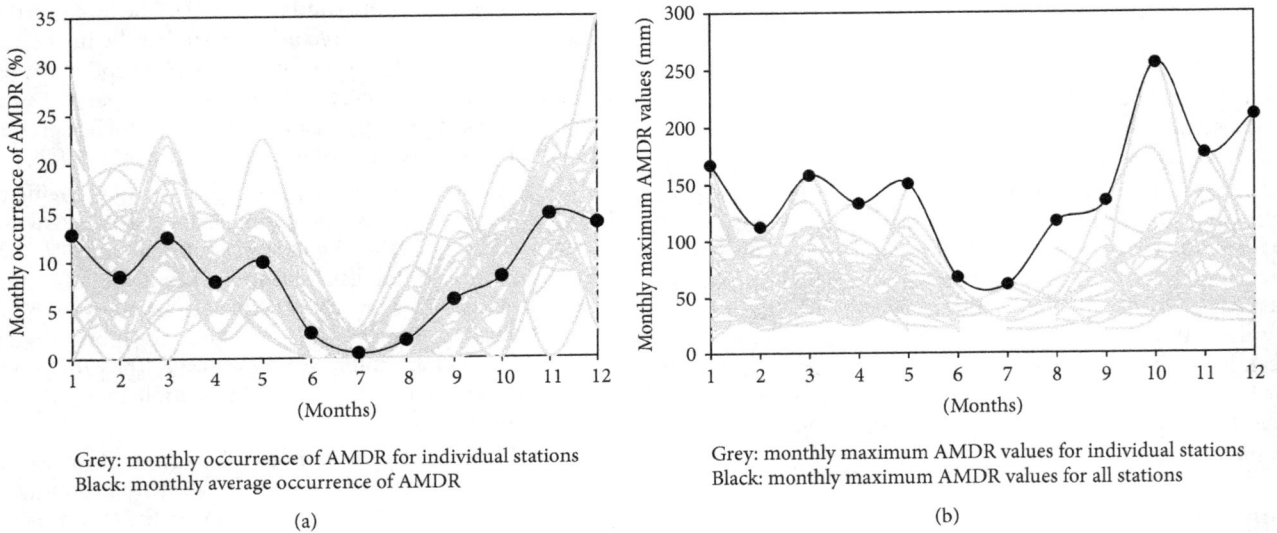

Grey: monthly occurrence of AMDR for individual stations
Black: monthly average occurrence of AMDR

(a)

Grey: monthly maximum AMDR values for individual stations
Black: monthly maximum AMDR values for all stations

(b)

FIGURE 3: Annual maximum daily rainfall (AMDR) frequency (a) and maximum values (b) at monthly scales.

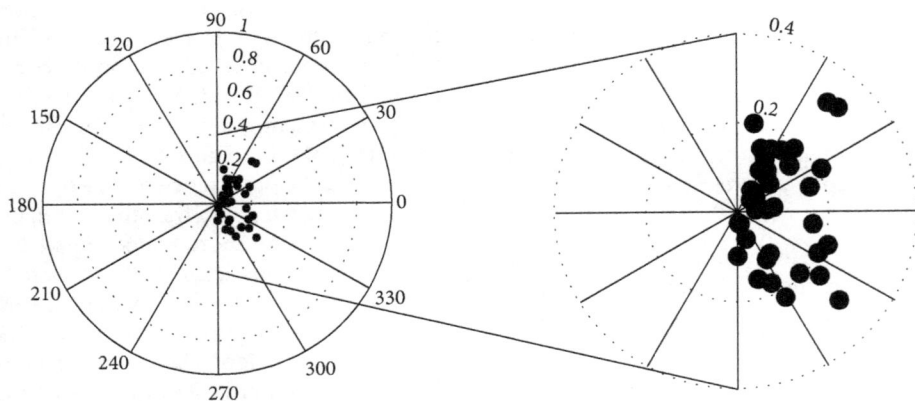

FIGURE 4: Location of seasonal concentration index $\bar{r}$ for AMDR. The angles: 0° represents January 1; 90° represents April 1; 180° represents July 1; and 270° represents October 1.

TABLE 2: P values and Tau of Mann-Kendall test, the slope (a) of linear regression analysis, and AMDR contribution to AR over the period after 1970 in the north of Algeria.

Number	Station	Mann-Kendall test		Linear regression	AMDR contribution to AR	
		P value	Tau	Slope (a)	Date of change	% deficit (−) or surplus (+)
1	Pierre du chat	0.167	−0.151	+0.118	2001	+34.9
2	Meffrouche	0.078	−0.193	−0.872	Complex	
3	Bensekrane	1.000	+0.002	+0.024	1987	+41.2
4	Ras El Ma	0.002*	+0.022	+0.353	1999	+22.9
5	S. A. Benyoub	0.374	+0.104	+0.286	1987	+25.3
6	Sarno	0.047*	−0.217	−0.427	Complex	
7	Hassi Daho	0.305	−0.114	−0.027	1997	+27.8
8	Cheurfas	0.787	−0.034	+0.158	1986	+29.6
9	Ain El Hadjer	0.822	−0.029	−0.140	1988	−17.4
10	Ghriss	0.874	−0.024	−0.089	1983	+42.6
11	Maoussa	0.287	+0.128	+0.335	Complex	
12	Oued El Kheir	0.928	+0.011	+0.032	1997	+18.9
13	Bekhedda	0.687	+0.051	+0.061	1991	+18.3
14	Meured	0.240	−0.129	−0.437	Complex	
15	Djebabra	0.001*	−0.368	−1.223	1985	−22.7
16	Ameur El Ain	$<10^{-3*}$	−0.431	−0.802	Complex	
17	Kolea	0.574	−0.062	−0.045	1990	+07.9
18	Ouzera	0.084	−0.189	−0.588	No trend	
19	Hamiz-D9	0.372	−0.102	−0.349	Complex	
20	Lakhdaria	0.200	−0.145	−0.301	1990	+8.5
21	K'sob	0.778	−0.036	−0.057	1981	+18.4
22	Sidi Yahia	0.577	+0.066	+0.034	1987	−18.8
23	Batna	0.576	+0.063	+0.093	No trend	
24	El Milia	0.363	+0.101	+0.285	1981	+49.9
25	Ain Tinn	0.651	+0.111	−0.166	1985	+12.2
26	Timgad	0.402	+0.094	+0.275	1992	+10.0
27	Chelia	0.013*	+0.265	+0.664	No trend	
28	Zardasas	0.357	+0.101	+0.293	1979	+29.6
29	Ain Mimoun	0.807	−0.028	+0.042	1986	−11.01
30	Helioplolis	0.200	+0.140	+0.425	1991	+28.1
31	Ain Beida	0.944	−0.009	−0.151	1989	−12.1
32	Ain Berda	0.032*	+0.232	+0.601	1987	+14.2
33	Bouchegouf	0.094	+0.183	+0.411	1998	+33.2
34	Ain Kerma	0.551	−0.066	−0.255	Complex	
35	Ain Assel	0.142	+0.161	+0.345	1990	+14.2

* Trend statistically significant at 5% level.

Indeed, for 34 stations the slope (a) is less than |1| leading to stationarity of the series. The only Djebabra station shows a moderate decreasing trend.

The curves resulting from Buishand procedure were classified into four distinct cases. The first case concerns five rainfall stations (Ain Beida, Ain El Hadjer, Sidi Yahia, Aïn Mimoun, and Djebabra) where Cs_k curve shows ascending and descending lambs, and the maximum of the Cs_k curve corresponds to the date of change (Figure 6(a)). The decline of the AMDR contribution to annual rainfall, manifested by the falling lamb, was quantified by the double mass curve and varies between 12.1 and 22.7%. The dates of change were observed during the 1980s (Table 2). The second case grouped 20 rainfall series exhibiting opposite behavior (Figure 6(b)). The AMDR contribution has increased significantly (e.g., at El Milia station the increase reaches about 50%). The dates of change occurred from the end of 1970s to the end of 1990s. The third case represents seven of 35 rainfall series and is manifested by a complex form showing more than one date of the change. However, lambs of the Cs_k curves were not sufficiently long to apply the double mass method (Figure 6(c)). For the last case, three of 35 rainfall series, no clear trend can be detected for any rainfall sequence (e.g., Ouzera, Batna, and Chelia) (Figure 6(d)).

(a) Mann-Kendall trend

(b) Linear regression trend

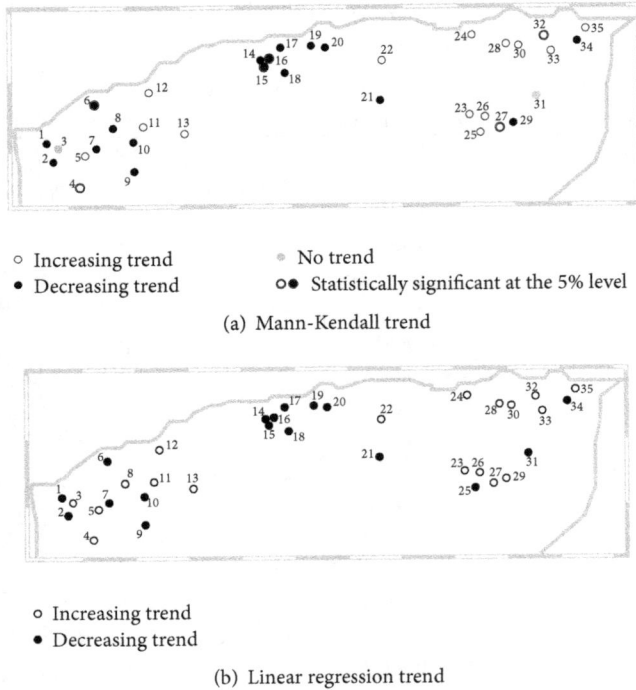

FIGURE 5: Mann-Kendall (a) and linear regression (b) trend of AMDR over the period after 1970 in northern Algeria.

4. Discussion and Conclusion

The data of 35 rainfall stations recorded after 1970 and covering the north of Algeria reveal that rainfall undergoes an overall downward trend. Deficit rate between 20 and 40% was estimated in several parts of the country [23–27]. This decrease in annual rainfall was accompanied by a large temporal irregularity like the Mediterranean region [28, 29]. In terms of AMDR values, coefficients of variation for the different sites vary at ±12% around the average of 44% showing a more pronounced irregularity occurrence, but in the same proportions as in other regions [30–32]. Indeed, the skewness values of the time series, which show if the empirical distribution of the data follows a normal distribution, are more important in the extreme east of the country than in the west and the center (a ratio of 1.66). All positive values of Cs found indicating that distribution shifted to the left of the median (Table 1) are of the same order as those of Chott-Chergui basin in western Algeria [33]. In this analysis, we exclude Pierre du chat station for which an exceptional value of 255.4 mm has increased all coefficients.

For a better understanding of water availability, it is very interesting to demarcate the quasi-homogeneous climatic zones and identify the climatic subregions in an observation network [34]. Nevertheless, in the north of Algeria, the proximity of the Mediterranean and the variety surrounding reliefs make it difficult to delimiting homogeneous areas. Moreover, the increased baroclinic instability in saturated air

is closely related to latent heat release and thus to the development of convective phenomena. During the rainy season, northern Algeria is affected by the polar front, especially the east of the country. Further south, the highlands are generally affected by western disturbances following the orographic forcing that causes thunderstorms with heavy rainfall [35]. Then, the correlation between rainfall and altitude is complex. It may be valid only for limited areas. Contrary to that, the rainfall is positively correlated with latitude and the east of the country is much wetter than the west (Figure 2).

For almost all stations, mean values of AMDR as well as AR are two to five times higher for particular years. This disparity between the average values and peaks combined with statistical parameters cited above is illustrated in Figure 4 according to Burn [13]. The seasonal concentration index $\bar{r}$ varies between 0.03 and 0.32 showing that there is no single dominant season, and the time of the occurrence of an extreme event is distributed around the year. This demonstrates the extent of rainfall irregularity in the south of the Mediterranean. This is not always the case through other regions. In central Slovakia, for example, the phenomenon has a spatiotemporal occurrence more regular [14]. Despite this, October and November are the months when the floods are the deadliest. It should be noted that Algeria is the country where the number of flood victims is the highest of the Mediterranean countries [36].

The examination of AMDR trends using the Mann-Kendall test and linear regression procedure shows that, unlike the annual rainfall undergoing a downward trend in most of the rainfall stations in Algeria [28, 37–39], there is no clear trend to rise or fall in the series of AMDR for the majority of the sites studied. The region of study is characterized by a great number of complexities related to geography and topography where the combination of the effects produced makes weather forecasting exceedingly difficult causing high spatiotemporal variability. So, like many other parts of the Mediterranean area, a majority of AMDR series recorded in the north of Algeria show nonsignificant trends [38, 40, 41]. However, globally, AMDR series show an increase in tendency in the east of the country and a decrease in the center while in the west, the system is more complicated and there is no majority of stations that differs from others. In this context, the IPCC report confirmed that the behavior of extreme rainfall differs considerably from the annual totals by some considerable geographical differences in the frequency, timing, and magnitude of events [42].

The present study does not converge with the results developed by some authors and in the occurrence of extreme rainfall forecasts which predict an increase in extreme rainfall in many parts of the world, even in areas where the average annual rainfall has a downward trend [42–45]. For some authors, this finding is apparently valid for Mediterranean countries that record rainfall deficits since 3-4 decades [3, 28]. However, our study reveals that not only does the trend of the AMDR series depend on global and regional settings, but also, it is highly influenced by local geographical characteristics. This result corroborates the finding of Tramblay et al. [38] and Jones et al. [46] who cited the orography influence as a main cause generating local climatic processes. Other

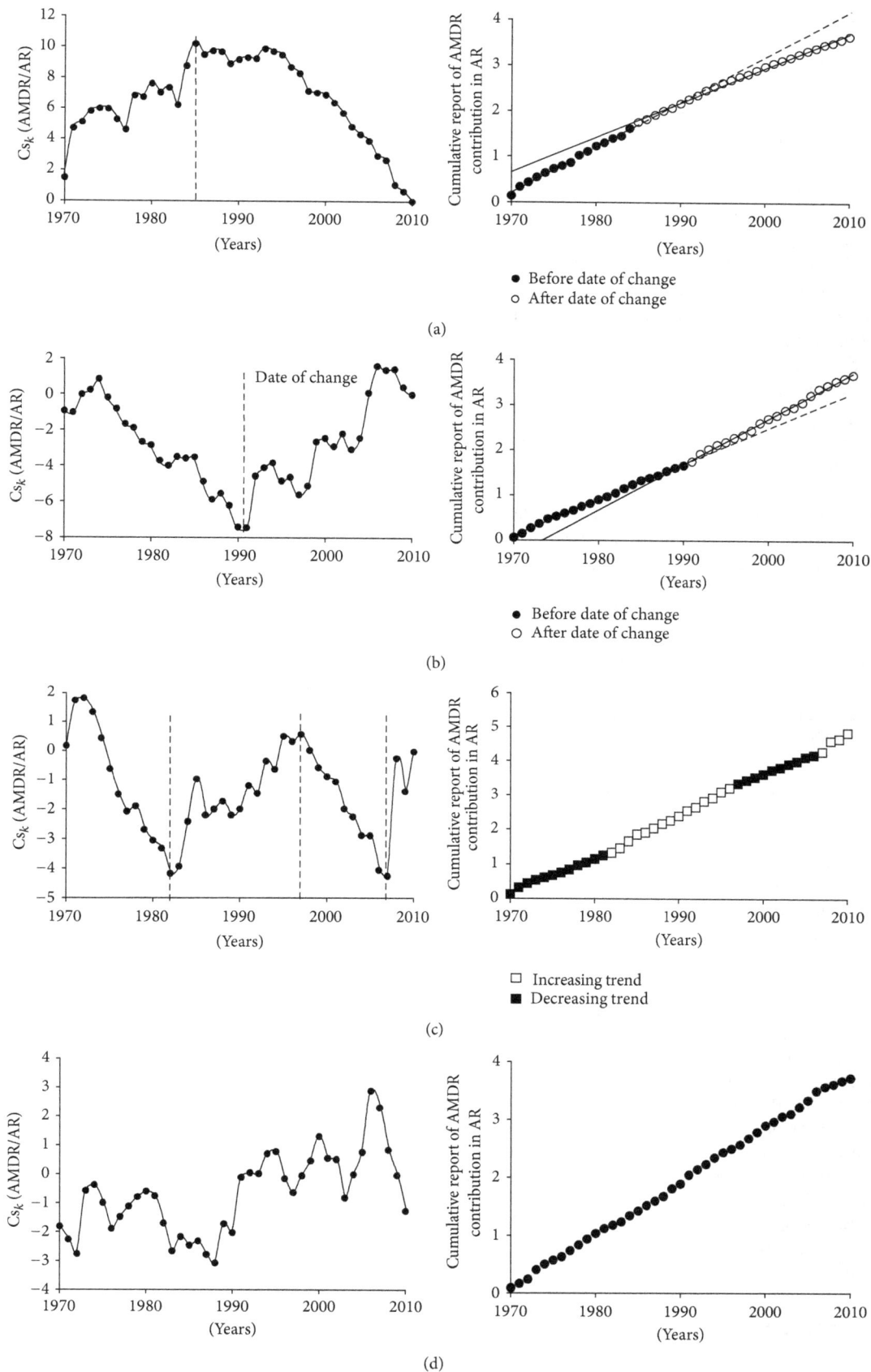

FIGURE 6: Buishand and double mass methods applied to AMDR contribution to AR.

investigations of extreme rainfall across many countries of the globe show that extreme rainfall trend is complex. For example, it had increased in the USA, China, Australia, Canada, Norway, Mexico, and Poland [47], where no clear trend was detected in other regions such as in Brazil and Ethiopia [7, 48]. Although a general trend for drought and a reduction in rainfall intensity was predicted for eastern Mediterranean [49], the extreme precipitation trend is more misunderstood as in Jordan [32].

Without having a provided dominant season, the AMDR occurred mainly between October and March (Figure 4) and the highest values were recorded in autumn (Figure 3). Similar results were observed in other Mediterranean sites where a slight increase in autumn rainfall was mainly felt in October [50, 51], leading to more concentrated rainfall during the hydrological year [52].

The relationship between the values of the AMDR and AR in the north of Algeria has experienced significant temporal changes in most of the studied stations. However, four configurations of the relationship between AMDR and AR (Figure 6) are distinguished and seem to be related to the climate change which operated in the Mediterranean region. Despite the lack of trend in AMDR series, their contributions in the annual rainfall are on the rise. This occurred in 80% of stations spread from the east to the west (not taking into account the stations that submitted a complex or nonsignificant trend behavior). The rate of increase of this contribution will be from 8 to about 50%. Although the rates of reduction of annual precipitation in the post-1970 period are higher than 20% for most stations in Algeria, some very high rates of increase in the contribution of the AMDR in annual totals are amplified by extremely high values of the recorded AMDR (example of El Milia Station). In this context and by analyzing the contribution of events in annual precipitation in the north of the Mediterranean, De Luis et al. [53] reported that, in all the studied stations, more than 37% experienced an increasing contribution against 15% with decreasing contribution.

The start of the change leading to the increasing contribution of the AMDR to the annual rainfall occurred mainly during the 1980s and 1990s and the AMDR occurred principally in the autumn season. However, this observation is misleading because the decrease of annual rainfall is mainly due to the sharp diminution of the winter rainfall. Indeed, the values of the AMDR that occurred during autumn are in the same proportions observed in winter (Figure 3). In the Mediterranean region, Giorgi and Lionello [50] observed a sharp falling in the winter rainfall against a small increase in autumn. They mainly attributed the increase in extreme daily intensities to the reduction of annual precipitation. The rate of this reduction was estimated to be more than 30% in the northwest of Algeria [54].

Competing Interests

The authors declare that there is no conflict of interests regarding the publication of this paper.

References

[1] A. Longobardi and P. Villani, "Trend analysis of annual and seasonal rainfall time series in the Mediterranean area," *International Journal of Climatology*, vol. 30, no. 10, pp. 1538–1546, 2010.

[2] P. T. Nastos, "Trends and variability of precipitation within the Mediterranean region, based on Global Precipitation Climatology Project (GPCP) and ground based datasets," in *Advances in the Research of Aquatic Environment*, vol. 1, pp. 67–74, Springer, 2011.

[3] P. Alpert, T. Ben-Gai, A. Baharad et al., "The paradoxical increase of Mediterranean extreme daily rainfall in spite of decrease in total values," *Geophysical Research Letters*, vol. 29, no. 11, pp. 1–31, 2002.

[4] A. Bodini and Q. A. Cossu, "Vulnerability assessment of Central-East Sardinia (Italy) to extreme rainfall events," *Natural Hazards and Earth System Science*, vol. 10, no. 1, pp. 61–72, 2010.

[5] S. Beguería, S. M. Vicente-Serrano, J. I. López-Moreno, and J. M. García-Ruiz, "Annual and seasonal mapping of peak intensity, magnitude and duration of extreme precipitation events across a climatic gradient, northeast Spain," *International Journal of Climatology*, vol. 29, no. 12, pp. 1759–1779, 2009.

[6] T. Haktanir, S. Bajabaa, and M. Masoud, "Stochastic analyses of maximum daily rainfall series recorded at two stations across the Mediterranean Sea," *Arabian Journal of Geosciences*, vol. 6, no. 10, pp. 3943–3958, 2013.

[7] J. R. Porto de Carvalho, E. D. Assad, A. F. de Oliveira, and H. Silveira Pinto, "Annual maximum daily rainfall trends in the midwest, southeast and southern Brazil in the last 71 years," *Weather and Climate Extremes*, vol. 5, no. 1, pp. 7–15, 2014.

[8] M. J. M. Römkens, K. Helming, and S. N. Prasad, "Soil erosion under different rainfall intensities, surface roughness, and soil water regimes," *Catena*, vol. 46, no. 2-3, pp. 103–123, 2002.

[9] X.-C. Zhang and W.-Z. Liu, "Simulating potential response of hydrology, soil erosion, and crop productivity to climate change in Changwu tableland region on the Loess Plateau of China," *Agricultural and Forest Meteorology*, vol. 131, no. 3-4, pp. 127–142, 2005.

[10] G. J. Tripoli, C. M. Medaglia, S. Dietrich et al., "The 9-10 November 2001 Algerian flood: a numerical study," *Bulletin of the American Meteorological Society*, vol. 86, no. 9, pp. 1229–1235, 2005.

[11] S. Argence, D. Lambert, E. Richard et al., "High resolution numerical study of the Algiers 2001 flash flood: sensitivity to the upper-level potential vorticity anomaly," *Advances in Geosciences*, vol. 7, pp. 251–257, 2006.

[12] N. Söhne, J.-P. Chaboureau, S. Argence, D. Lambert, and E. Richard, "Objective evaluation of mesoscale simulations of the Algiers 2001 flash flood by the model-to-satellite approach," *Advances in Geosciences*, vol. 7, pp. 247–250, 2006.

[13] D. H. Burn, "Catchment similarity for regional flood frequency analysis using seasonality measures," *Journal of Hydrology*, vol. 202, no. 1-4, pp. 212–230, 1997.

[14] J. Szolgay, J. Parajka, S. Kohnová, and K. Hlavčová, "Comparison of mapping approaches of design annual maximum daily precipitation," *Atmospheric Research*, vol. 92, no. 3, pp. 289–307, 2009.

[15] H. B. Mann, "Nonparametric tests against trend," *Econometrica*, vol. 13, pp. 245–259, 1945.

[16] M. G. Kendall, *Rank Correlation Methods*, Charles Griffin, London, UK, 4th edition, 1975.

[17] D. R. Helsel and R. M. Hirsch, *Statistical Methods in Water Resources*, vol. 529, Elsevier, Amsterdam, The Netherlands, 1992.

[18] T. A. Buishand, "Some methods for testing the homogeneity of rainfall records," *Journal of Hydrology*, vol. 58, no. 1-2, pp. 11–27, 1982.

[19] D. Raes, D. Mallants, and Z. Song, "RAINBOW: a software package for analysing hydrologic data," in *Hydraulic Engineering Software VI*, W. R. Blain, Ed., pp. 525–534, Computational Mechanics Publication, Boston, Mass, USA, 1996.

[20] C. F. Merriam, "A comprehensive study of the rainfall on the Susquehanna Valley," *Transactions American Geophysical Union*, vol. 18, no. 2, pp. 471–476, 1937.

[21] W.-W. Zhao, B.-J. Fu, Q.-H. Meng, Q.-J. Zhang, and Y.-H. Zhang, "Effects of land-use pattern change on rainfall-runoff and runoff-sediment relations: a case study in Zichang watershed of the Loess Plateau of China," *Journal of Environmental Sciences*, vol. 16, no. 3, pp. 436–442, 2004.

[22] A. Medjerab and L. Henia, "Régionalisation des pluies annuelles dans l'Algérie nord-occidentale," *Revue Géographique de l'Est*, vol. 45, no. 2, pp. 1–12, 2005.

[23] H. Meddi and M. Meddi, "Variabilité des précipitations annuelles du Nord-Ouest de l'Algérie," *Sécheresse*, vol. 20, no. 1, pp. 57–65, 2009.

[24] A. N. Ghenim, A. Megnounif, A. Seddini, and A. Terfous, "Fluctuations hydropluviométriques du bassin versant de l'Oued Tafna à Béni Bahdel (Nord Ouest Algérien)," *Sécheresse*, vol. 21, no. 2, pp. 115–120, 2010.

[25] A. N. Ghenim and A. Megnounif, "Analyse des précipitations dans le Nord-Ouest Algérien," *Sécheresse*, vol. 24, no. 2, pp. 107–114, 2013.

[26] A. N. Ghenim and A. Megnounif, "Ampleur de la sécheresse dans le bassin d'alimentation du barrage Meffrouche (Nord-Ouest de l'Algérie)," *Géographie Physique et Environnement (Physio-Géo)*, vol. 7, pp. 35–49, 2013.

[27] A. Dahmani and M. Meddi, "Impact of rainfall deficiency on water resources in the plain Ghriss Wilaya of Mascara (West of Algeria)," *American Journal of Scientific and Industrial Research*, vol. 2, no. 5, pp. 755–760, 2011.

[28] C. M. Philandras, P. T. Nastos, J. Kapsomenakis, K. C. Douvis, G. Tselioudis, and C. S. Zerefos, "Long term precipitation trends and variability within the Mediterranean region," *Natural Hazards and Earth System Sciences*, vol. 11, no. 12, pp. 3235–3250, 2011.

[29] H. Reiser and H. Kutiel, "Rainfall uncertainty in the Mediterranean: time series, uncertainty, and extreme events," *Theoretical and Applied Climatology*, vol. 104, no. 3-4, pp. 357–375, 2011.

[30] C. Maciel Vaz, "Trend analysis in annual maximum daily rainfall series," Tech. Rep., Universidade Técnica de Lisboa, Lisbon, Portugal, 2008.

[31] S. Deka, M. Borah, and SC. Kakaty, "Distributions of annual maximum rainfall series of north-east India," *European Water Publications*, vol. 27-28, pp. 3–14, 2009.

[32] K. A. Al-Qudah and A. A. Smadi, "Trends in maximum daily rainfall in marginal desert environment: signs of climate change," *American Journal of Environmental Sciences*, vol. 7, no. 4, pp. 331–337, 2011.

[33] B. Habibi, M. Meddi, and A. Boucefiane, "Analyse fréquentielle des pluies journalières maximales," *Cas du Bassin Chott-Chergui. Nature & Technologie*, vol. 8, pp. 41–48, 2013.

[34] J. Guiot, "Sur la détermination des régions climatiques quasi homogènes," *Revue de Statistique Appliquée*, vol. 34, no. 2, pp. 15–34, 1986.

[35] A. Benhamrouche, D. Boucherf, R. Hamadache, L. Bendahmane, J. Martin-Vide, and J. Teixeira Nery, "Spatial distribution of the daily precipitation concentration index in Algeria," *Natural Hazards and Earth System Sciences*, vol. 15, no. 3, pp. 617–625, 2015.

[36] M. C. Llasat, M. Llasat-Botija, M. A. Prat et al., "High-impact floods and flash floods in Mediterranean countries: the FLASH preliminary database," *Advances in Geosciences*, vol. 23, pp. 47–55, 2010.

[37] A. Bakreti, I. Braud, E. Leblois, and A. Benali, "Analyse conjointe des régimes pluviométriques et hydrologiques dans le bassin de la Tafna (Algérie Occidentale)," *Hydrological Sciences Journal*, vol. 58, no. 1, pp. 133–151, 2013.

[38] Y. Tramblay, S. El Adlouni, and E. Servat, "Trends and variability in extreme precipitation indices over maghreb countries," *Natural Hazards and Earth System Sciences*, vol. 13, no. 12, pp. 3235–3248, 2013.

[39] M. Lazri and S. Ameur, "Analysis of the time trends of precipitation over mediterranean region," *International Journal of Information Engineering and Electronic Business*, vol. 6, no. 4, pp. 38–44, 2014.

[40] G. Villarini, "Analyses of annual and seasonal maximum daily rainfall accumulations for Ukraine, Moldova, and Romania," *International Journal of Climatology*, vol. 32, no. 14, pp. 2213–2226, 2012.

[41] T. S. Stephenson, L. A. Vincent, T. Allen et al., "Changes in extreme temperature and precipitation in the Caribbean region, 1961–2010," *International Journal of Climatology*, vol. 34, no. 9, pp. 2957–2971, 2014.

[42] IPCC, "Special Report on Managing the risks of extreme events and disasters to advance climate change adaptation(SREX)," in *A Special Report of Working Group I and Working Group II of the Intergovernmental Panel on Climate Change*, C. B. Field, V. Barros, T. F. Stocker et al., Eds., p. 582, Cambridge University Press, New York, NY, USA, 2012.

[43] D. H. Burn, R. Mansour, K. Zhang, and P. H. Whitfield, "Trends and variability in extreme rainfall events in British Columbia," *Canadian Water Resources Journal*, vol. 36, no. 1, pp. 67–82, 2011.

[44] E. M. Douglas and C. A. Fairbank, "Is precipitation in Northern New England becoming more extreme? statistical analysis of extreme rainfall in Massachusetts, New Hampshire, and Maine and updated estimates of the 100-year storm," *Journal of Hydrologic Engineering ASCE*, vol. 16, no. 3, pp. 203–217, 2011.

[45] S. Westra, L. V. Alexander, and F. W. Zwiers, "Global increasing trends in annual maximum daily precipitation," *Journal of Climate*, vol. 26, no. 11, pp. 3904–3918, 2013.

[46] M. R. Jones, S. Blenkinsop, H. J. Fowler, and C. G. Kilsby, "Objective classification of extreme rainfall regions for the UK and updated estimates of trends in regional extreme rainfall," *International Journal of Climatology*, vol. 34, no. 3, pp. 751–765, 2014.

[47] P. Y. Groisman, T. R. Karl, D. R. Easterling et al., "Changes in the probability of heavy precipitation: important indicators of climatic change," *Climatic Change*, vol. 42, no. 1, pp. 243–283, 1999.

[48] M. A. Degefu and W. Bewket, "Variability and trends in rainfall amount and extreme event indices in the Omo-Ghibe River Basin, Ethiopia," *Regional Environmental Change*, vol. 14, no. 2, pp. 799–810, 2014.

[49] C. Oikonomou, H. A. Flocas, M. Hatzaki, D. N. Asimakopoulos, and C. Giannakopoulos, "Future changes in the occurrence of

extreme precipitation events in eastern Mediterranean," *Global Nest Journal*, vol. 10, no. 2, pp. 255–262, 2008.

[50] F. Giorgi and P. Lionello, "Climate change projections for the Mediterranean region," *Global and Planetary Change*, vol. 63, no. 2-3, pp. 90–104, 2008.

[51] C. Norrant and A. Douguédroit, "Tendances des précipitations mensuelles et quotidiennes dans le sud-est méditerranéen français (1950-51/1999-2000)," *Climatologie*, vol. 1, pp. 45–64, 2004.

[52] A. Megnounif and A. N. Ghenim, "Rainfall irregularity and its impact on the sediment yield in Wadi Sebdou watershed, Algeria," *Arabian Journal of Geosciences*, vol. 9, no. 4, pp. 1–15, 2016.

[53] M. De Luis, J. Raventos, J. R. Sanchez, J. C. Gonzalez, J. Cortina, and M. F. Garcia-Cano, "Event contribution to annual precipitation: a trend analysis," in *Proceedings of the 2nd International Conference on Climate and Water*, pp. 290–300, Espoo, Finland, 1998.

[54] A. N. Ghenim, *Ecoulements et Transports Solides dans les Régions Semi-Arides*, Editions Universitaires Européennes, 2012.

3

Integration of Earth Observation Data and Spatial Approach to Delineate and Manage Aeolian Sand-Affected Wasteland in Highly Productive Lands of Haryana, India

Kishan Singh Rawat [ID],[1] **Shashi Vind Mishra,**[2] **and Sudhir Kumar Singh**[3]

[1]*Centre for Remote Sensing and Geoinformatics, Sathyabama Institute of Science and Technology, Chennai 600119, India*
[2]*Division of Environmental Sciences, Indian Agricultural Research Institute, New Delhi 110012, India*
[3]*K. Banerjee Centre of Atmospheric and Ocean Studies, IIDS, Nehru Science Centre, University of Allahabad, Allahabad, India*

Correspondence should be addressed to Kishan Singh Rawat; ksr.kishan@gmail.com

Academic Editor: Akhilesh Mishra

The western part of the country India is surrounded by Thar desert. Due to climate change, many regions in the world are facing different challenges. The objective of the study was to quantify the aeolian sand-affected land through integrated approach. The LANDSAT-ETM+ satellite image of 2009 has been used to distinguish recently affected areas by aeolian sand. A combined approach of digital classification backed with visual interpretation and ground verification was adopted. In addition to classification accuracy assessment was performed using field observations. Evidence based results of aeolian sand-affected areas have suggested that wasteland area has increased up to 4,427.55 ha (6.79%) of total geographical area. Two types of aeolian sands areas have been detected, namely, moderately affected (3,881.77 ha) and severely affected (545.79 ha). Moderately and severely affected aeolian soil lands have been more accurately mapped with reasonably good accuracy whereas smaller aeolian affected areas within croplands are mapped with low accuracy. The present study provides easy methodology for delineation, classification, and characterization of aeolian affected sands.

1. Introduction

Land resources are a valuable natural resource and it acts as a key for sustenance of mankind [1, 2]. Overexploitation of land resources causes a significant change to the landforms, which has adverse effect to the environment [3]. The high population pressure, fast urbanization, rapid industrialization, and extensive agriculture have put great stresses on land resources, resulting into the substantial reduction in agricultural area and natural resources [4]. Tremendous population pressure is also leading to deforestation and resource degradation that has disturbed ecological balance of terrestrial ecosystems [5, 6]. Hence, quantitative information about the nature, degree of extent, and spatiotemporal distribution of affected soils of India and the world is needed. It is essential for improved planning and need to implement strategic reclamation programs in proper time and cost-effective manner for huge crops production. In India at national level wasteland mapping was conducted by conventional surveys and integrated (remote sensing and GIS) approaches from last four decades.

Management of land and water resources is essential to meet the economic growth of people in any country [7, 8]. Land degradation is a serious problem; it can be controlled by afforestation practices on available wastelands in more scientific way [9, 10] and keep natural ecosystem in harmony and maintain ecological balance. Thus, the up-to-date and appropriate information about location and spatial extent of vacant/wastelands has played very important role for better planning of afforestation and treatment to eradicate the negative effects of land degradation [11, 12]. Hence, there is great demand to identify and reclaim these degraded lands in many countries [6] and in district of Sirsa of Haryana [13, 14].

Recent developments in geographical mapping allow the researchers to exercise the spatiotemporal distribution pattern and location aspects of land use/land cover (LULC)

that can be studied more accurately using geospatial techniques (Saha 1990, [15–17]). Many studies have proven the applications of Remote Sensing and GIS in monitoring and management of natural resources [6, 18–20]. Remote sensed data sets are used widely in studies, namely, groundwater [21], lake and wetlands [22, 23], land use/land cover mapping [24, 25], land use/land cover modeling [26, 27], crop suitability (Mustak et al. 2013), urban land use dynamics, forest mapping [28], soil characterization [29], slope estimation [30], landscape ecology [31], and watershed management [32]. Further, RS technology has proven its application in assessment of wasteland and its temporal monitoring [33, 34]. With the advent of satellite remote sensing mapping of degraded land has been started more efficiently at much finer scales [35]. The raw data which are affected by panoramic distortion, earth curvature, sensor detector failed, and detector line losses which are primarily corrected by data providers [36], hence, provide better mapping possibilities. Satellite remote sensing provides unbiased information about the objects. Satellite remote sensing has advantage over field based method in the forms of multispectral, synoptic coverage, very high temporal resolution, and cost effectiveness [37, 38]. Grunwald [4] has identified the spatial patterns and variations in landforms, land use, wasteland, and demographic characteristics and their spatial associations in satellite data. Genesis of wasteland and its typologies into degraded forests, undulating land, gullied and ravinous land, and degraded pastures, waterlogged, salt-affected, and sand-affected areas can be mapped and analyzed accurately with special reference to their relationship to natural environment [7]. Satellite data are of immense use for fine scale mapping of wastelands and may be helpful in implementing the schemes for wastelands development in given time [9, 39, 40].

Saha et al. [41] integrated LANDSAT-TM dataset into GIS environment to map salt-affected and surface waterlogged/marshy lands with an accuracy of about 96% in parts of Aligarh district, Uttar Pradesh, India. Jain et al. [33] have identified the highly degraded scrub and sandy land on fringes of town and provided recommendations for future urban planning of the town where further development can be planned to avoid encroachment on good agricultural lands. Wasteland mappings of Karnal district (Haryana, India) using visual, monoscopic interpretations of MS LANDSAT-TM dataset of 1986 and ground truth were attempted [42, 43]. Pramila [18] used LANDSAT-TM dataset of 1988 and found that a major reason for formation of wasteland was severe wind erosion (Jalor and Ahor Tehsil of Jalor district of Western Rajasthan). The reflection of sand particles is being high; hence the sandy soil could be easily identified by false color composition (FCC) in LANDSAT-TM dataset. Sugumaran et al. [15] used IRS-1A, LISS-II data sets to delineate more accurately the wastelands at microlevel in *Matar taluka* of *Kheda* district (states of Gujarat, India) and generated area statistics of wastelands. Status of desertification was mapped with the application of satellite data in dry subhumid region of Panchkula district of Haryana by Arya et al. [44]. They found that nearly 47.5% of the total geographical area faces desertification (formation of wasteland states). Major outcomes from the study were that near about 32%

of the total area of district's scrublands converted into desert mainly due to the unchecked desertification process. The change status of sodic lands has been prepared using satellite remote sensing data by Singh [16]. The desertification process is going on these regions; hence, we formulated our study with objectives as (i) identification and delineation of wastelands (at 1 : 50,000 scale) using LANDSAT-ETM+ (2009) data set and (ii) creation of digital database in GIS environment.

2. Materials and Methods

2.1. Study Area. Sirsa district has an area of $4,276 \, \text{km}^2$ ($29.5400°N$, $75.0300°E$) with its headquarters being situated in Sirsa town (Figure 1). It is situated at nearly 255 km west of Delhi and 280 km from Southwest of Chandigarh. The Ghaggar River is flowing through central part of study area and Bhakra Canal is prime source of surface water for irrigation and drinking uses. The climate of region is tropical in nature with intensive hot summer and cool winter with a temperature variation of $47°C$ in June and $3°C$ in December and January. The average rainfall ranges between 200 and 300 mm and 80% precipitation is received during the four months (July to September; [45]). Consequently the agriculture in Sirsa district is threatened with nonavailability of water and the land related constraints too which is rendering a sizeable area into unproductive land due to aeolian spread of sands from the adjoining lands in the summers of each year. The terrain is broadly classified into moderately slopy to steep slope and highly undulating [46]. The terrain is divided into three major types, that is, Haryana plain, alluvial bed (Ghaggar or *"nali"*), and sand dune tract. The general slope varies from North to South.

2.2. Data Used and Analytical Procedures. In study the wastelands of Sirsa district, Haryana, have been delineated and mapped on 1 : 50,000 scale through digital image processing of LANDSAT-ETM+ image of 2009. Geocoded data was processed using image processing (ERDAS IMAGINE 8.7) software supported with the visual interpretation and ground truth. ArcGIS 9.3 was used for creation of digital database. LANDSAT-ETM+ digital data of 2 October 2009 was downloaded from web site (https://glovis.usgs.gov), and gap filling was applied using ENV 4.8ver.

Collateral Data. For reference purposes, latest published reports, papers, and maps were used. For identification of village location, major transport networks, cultural features, and annotation of major towns and cities topo-sheets were obtained from different agencies and used. The groundwater report was used from Central Groundwater Board Report [46], *agriculture information from* [45], and information about the forestry from [45].

3. Methodology

Geometric correction was performed and later overlay was carried out using administrative boundary. False color composition (FCC) was generated using bands 2, 3, and 4 in blue, red, and green filters (Figure 2). Based on the standard

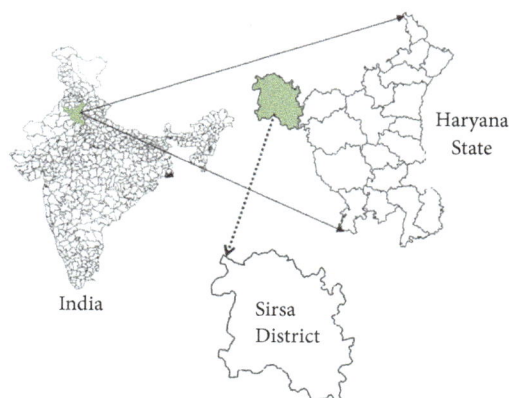

FIGURE 1: Location map of study area (Sirsa district of Haryana, India).

TABLE 1: Accuracy assessment report.

Class	Severely affected	Moderately affected	Settlement	Row total
Severely affected	70	5	0	75
Moderately affected	3	55	0	58
Settlement	0	0	121	121
Column total	**73**	**60**	**121**	**254**
Error matrix				
Agreement/accuracy (in %)	95.89	91.67	100	
Omission error (in %)	4.11	8.33	100	
Commission error (in %)	6.67			
Overall accuracy (in %)	96.85			

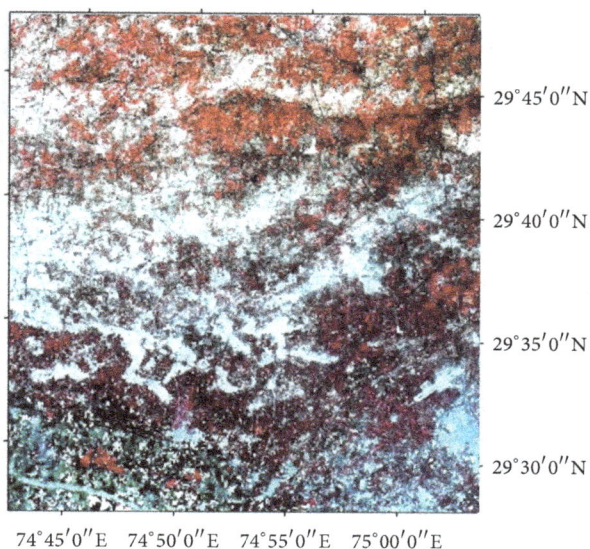

FIGURE 2: LANDSAT-ETM5+ false color composite map showing area affected by aeolian sand (in bright color, tones, and textures).

image interpretation key such as tone, texture, pattern, shape, size, location, and association, image was classified using Maximum Likelihood Classification (MLC) and visual interpretation. The readers can find the details of MLC in any standard remote sensing book (e.g., [48]). Following a standard legend prepared by Department of Space, Government of India (DOS, GOI) to delineate different wastelands categories was performed. A separate layer of settlement areas along with their names and major roads was also prepared. These maps were kept in GIS format further to create the database.

Training Signatures and Object Identification. By operating the classifier panel of the ERDAS IMAGINE 8.7, training signatures of the target (aeolian sand soil) were identified in two steps as (i) marking of training windows for various features by locating ground truth sites in the images and (ii) generation of signatures for training windows. Prior to training signature generation, identification of target is important. It was found that different objects had unique spectral signatures. Differences in photographic tone or texture or both are the basis for target identification coupled with knowledge of the target. They depend on the reflectance power of leaves and also on the soil cover ratio, which in turn depends on the stage of nutrient status and moisture status of the soil. Photographic texture depends on nature of soil and spacing between soil particles and scale of the photograph. The spectral signature of soil is influenced by the underlying soil and residue. The signature is subsequently modified by the appearance and gradual increase in surface coverage of particular soil. Knowing the manner in which soil characteristics is recorded photographically and the ability to associate these with the knowledge of soil identification was done successfully. The exact locations of ground truth sites were recorded with the help of Global Positioning System (GPS). Before freezing the training signatures, trial classification was run on respective sampled images. As per the requirement some training sites which were not suitable were hence discarded while some were modified to proceed with classification process. MLC method was used to classify LANDSAT-ETM+ (Path/Row 148/40) imagery using ground truth data which widely used parametric classifier for satellite data analysis and it relies on the second-order statistics of Gaussian probability density functions (PDFs) used to classify unidentified pixel belonging to each category.

4. Results and Discussions

The accuracy assessment in Table 1 shows that the overall accuracy of classified image is 96.85. The delineated wasteland areas are illustrated (Figure 3). The district is comprised of 4.27 lac hectares (ha), which is 9.66% of area of Haryana state. The area taken under study was 65,176.93 ha which constitutes nearly 15.24% of the total geographical area of the district. It was found that the wastelands in the study area covered approximately 4,427.55 ha which constitutes nearly 6.79% of the total geographical extents. The main wasteland category is aeolian sand. This analysis has completely upgraded the land use statistics of district which can be used for effective planning of its proper development and management of resources in a more sustainable manner.

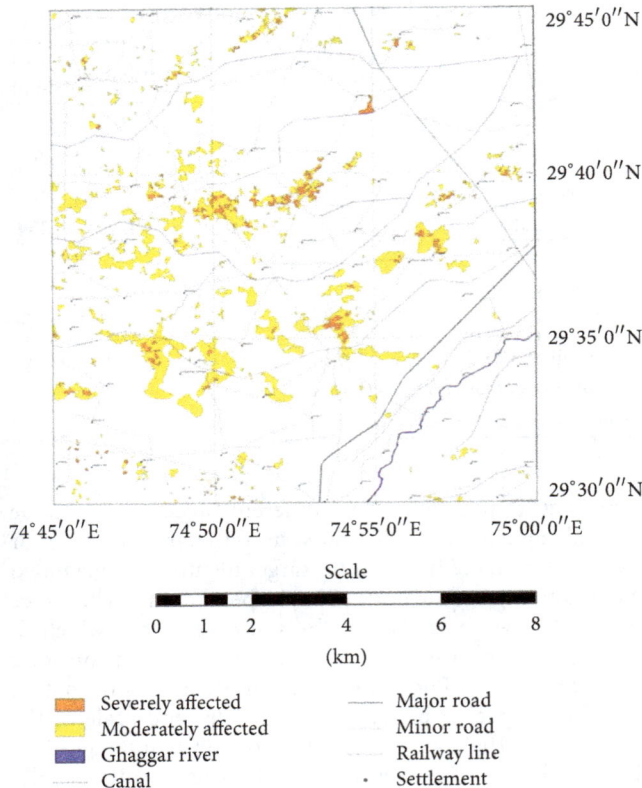

FIGURE 3: Map showing area affected by aeolian sand in Sirsa District of Haryana, India.

4.1. Aeolian Sand. The majority of land areas is sandy and is characterized by accumulation of sand in the form of varying size of sand dunes having variable shapes and sizes that have developed as a result of transportation of soil through wind processes. The main cause of aeolian sand is due to adjacency to the Rajasthan State that lies under arid climatic condition and resulted into soil erosion due to wind. The directional flow of sand is more in the Sirsa region. The area encompasses sandy soils with low to very low soil moisture and low organic matter. The productivity of such types of soil is low. Hence soil loss in the form of deposits of sand dune and sand transportation is very frequent by wind erosion. Wind erosion generally forms two types of wasteland, which are discussed below.

4.1.1. Lands Moderately Affected by Aeolian Process. Areas that come under this category were found to be approximately 3,881.77 ha (5.96%) of study area. The majority of land is moderately affected by aeolian sands and has good amount of organic matter content and higher soil moisture retention compared to severely affected lands. The slightly better soils have good agricultural activities in this region and have better productive yield. However, it still needs to be managed through appropriate agricultural practices in proper manner. In condition of extreme utilization without proper conservation and management practices, it enhances severe soil loss condition and complete degradation. Moderately

affected soils may convert into fertile soil by adopting proper agricultural practices and thus can be optimized for better agriculture utilities.

4.1.2. Lands Severely Affected by Aeolian Process. The severely affected soils have covered approximately 545.79 ha (0.84%); hence urgent need is required to take proper care and management. The moderately affected land is also getting converted into severely affected land because of deforestation and nonscientific agriculture practices. It is very difficult to ameliorate severely affected soils because of their bad physical and biological conditions, although they can be used for horticultural plantations, agri-silvi or agri-silvi-pastoral purposes, and for conservation agriculture to some extent with much effort on soil conservation measures.

4.2. The Piedmont Plains of Ghaggar River. The Ghaggar, an important rain fed river of length 85 km in district, has major drainage channel. The area occupied by the *piedmont plains of Ghaggar River* is approximately 0.14%. The river flooded in South West monsoon and causes extensive damage to crops and property. Soils around the river region are sandy to loamy sand at places underlined by lime concretion and gypsiferous substrata. Some places are also covered with sand hummocks and sand dunes. The soil is low in nitrogen, organic carbon, and phosphorus contents.

Most of the area is covered by aeolian sand; it has fine particles and is continuously varying. Because wastelands are an important aspect in land use planning and developmental activities of any area, so all the database on wastelands was created and put in GIS.

5. Recommendations and Reclamation Measures, Management, and Adaptation

Wind erosion is a major problem. The top soil erosion leads to loss of organic matter, damage to crops, and burial of productive agricultural lands. Farmers are protecting and managing their agricultural fields especially through crop residue and fencing during critical periods. Two major control activities are recommended as aeolian sand stabilization and shelter belt plantation (Table 2) and some improved methods are also given below for aeolian sand stabilization being followed in similar conditions elsewhere in the world [8, 49–53].

Gap Filling/Regeneration of Tree Species on Degraded Forest Land. Afforestation can be used in gap filling with transplanting. Suitable plant species like *Acacia nilotica, Acacia tortilis, Dalbergia sissoo, Azadirachta indica, Eucalyptus camaldulensis, Popular, Tamarix* species, and so forth can be used for the transplanting. The ground flora should be enriched by growing suitable shrub and grass species in contour furrows. However, uncontrolled grazing should be restricted and properly regulated for forest regeneration.

Brushes and Mulches. Brushes or mulches act as surface covers and wind barriers against erosion from the wind. They prevent loss of top soil, retain moisture and provide shelter for developing seedlings, and enhance sand accretion. Accreting

TABLE 2: Plant species suitable for aeolian sand stabilization.

Annual rainfall zone (mm)	Trees	Shrubs	Grasses
150–300	*Prosopis juliflora, Acacia tortilis, A. senegal*	*Calligonum polygonoides, Ziziphus nummularia, Citrullus colocynthis, Ziziphus mauritiana*	*Lasiurus sindicus*
300–400	*A. tortilis, A. senegal, P. juliflora, P. cineraria, Tecomella undulata, Parkinsonia aculeata, Acacia nubica, Dichrostachys glomerata, Colophospermum mopane, Cordia rothii*	*Z. nummularia, C. polygonoides, Citrullus colocynthis*	*C. setigerus, L. sindicus, Saccharum munja, C. ciliaris*
400–550	*A. tortilis, P. cineraria, P. juliflora, A. senegal, Dalbergia sissoo, Ailanthus excelsa, Albizia lebbeck, P. aculeata, T. undulata, D. glomerata, C. mopane*	*Z. mauritiana, Cassia auriculata*	*Cenchrus ciliaris, C. setigerus, S. munja, Panicum antidotale*

Source. Adopted from [47].

sand stimulates the growth of primary colonizing grasses such as *Spinifex.*

Liquid Sprays. Liquid sprays such as emulsified bitumen or dispersed organic polymers provide temporary stabilization by aggregating or cementing sand grains so that they cannot be moved by the wind. Unlike brushes and mulches, these products do not enhance sand accretion; mobile sand simply passes over the surface and continues until intercepted by other obstacles. Recommended sprays do not interfere with the germination or with the growth of seedlings, transplanted culms, or runners. Most allow reasonable air and water exchange between the atmosphere and stabilized sand [54].

Cover Crops. Cover crops are generally intolerant of strong winds and significant sand burial. Thus, where there is no migrating sand to accrete and it is a matter of holding the sand surface in place while secondary or tertiary species are established, then a living plant mulch or cover crop is appropriate. By necessity, cover crops must germinate rapidly and grow vigorously to provide a dense vegetative cover capable of reducing the wind velocity at the sand surface. Cover crops may be used to provide sand surface stability and to protect emerging secondary species or they may be used in their own right as a longer term surface stabilizer. The length of time they persist can be controlled by species selection and by management of fertilizer inputs.

6. Conclusions

In this work, wasteland mapping has been carried out of year 2009. The satellite image was classified using MLC based supervised classification, and total 65,176.93 ha of land area is classified as wasteland, in which 3,881.77 ha area was mapped as moderately and 545.79 ha as severely affected by aeolian sand. The more accurate mapping of aeolian soil encrusted lands with large contiguous areas whereas slightly affected land having less affected areas within croplands was mapped less accurately. The previous and current national scenarios of aeolian affected soils using traditional and RS

approaches can be refined and all other national estimates of wasteland (especially aeolian sand-affected soil) mapping can be subjected to the present methodology and approach can be subjected to the reconciliation. It can be stated that fine resolution satellite data sets facilitate interpretation to further differentiate a wasteland class delineated on moderate spatial resolution satellite data. Moreover, these data are found to be useful in achieving high mapping accuracy in delineation of wasteland classes.

Acknowledgments

The authors are grateful to the authorities of their respective organization for having allowed them to undertake parts of the whole analysis/ground truth/collection of secondary information and so forth. Thanks are due to the Chairman and members of the advisory of the fourth author. All help and information received from known and unknown sources are also duly acknowledged.

References

[1] National Commission on Agriculture, *Report of the National Commission on Agriculture*, Ministry of Agriculture and Irrigation, Government of India, New Delhi, India, 1976.

[2] T. V. Ramachandra, "Comparative assessment of techniques for bio-resource monitoring using GIS and remote sensing," *The ICFAI; Journal of Environmental Sciences*, vol. 1, no. 2, pp. 2–8, 2007.

[3] V. K. Verma, *Assessment of land degradation by integrated analysis of spectrally based information and terrain attributes in semi-arid region [M.Sc. thesis]*, IIRS and ITC, Dehradun, India, 2005.

[4] S. Grunwald, "The current state of digital soil mapping and what is next," in *Digital Soil Mapping: Bridging Research, Production and Environmental Applications*, J. L. Boetinger, D. W. Howell,

A. C. Moore, A. E. Hartemink, and S. Kienst-Brown, Eds., pp. 3–12, Springer, Heidelberg, Germany, 2013.

[5] A. Sharma, T. Moorti, and S. K. Chauhan, "A study on the estimation of wasteland and proposed strategy for their regeneration in Western Himalayas; agricultural situation in India 47 CSSRI–2007," Annual Report 2006-07, 1992.

[6] T. Chandramohan and D. G. Durbude, "Estimation of soil erosion potential using universal soil loss equation," Journal of the Indian Society of Remote Sensing, vol. 30, no. 4, pp. 181–190, 2002.

[7] G. Francis, R. Edinger, and K. Becker, "A concept for simultaneous wasteland reclamation, fuel production, and socioeconomic development in degraded areas in India: Need, potential and perspectives of Jatropha plantations," Natural Resources Forum, vol. 29, no. 1, pp. 12–24, 2005.

[8] S. Diniega, C. J. Hansen, J. N. McElwaine et al., "A new dry hypothesis for the formation of martian linear gullies," Icarus, vol. 225, no. 1, pp. 526–537, 2013.

[9] R. S. Dwivedi, T. Ravi Sankar, L. Venkataratnam et al., "The inventory and monitoring of eroded lands using remote sensing data," International Journal of Remote Sensing, vol. 18, no. 1, pp. 107–119, 1997.

[10] Haryana Space Applications Centre, Wastelands Atlas of Haryana, Haryana State Remote Sensing Application Centre (HARSAC), Department of Science and Technology, Government of Haryana, 2006.

[11] J. F. L. Contador, S. Schnabel, A. Gómez Gutiérrez, and M. P. Fernández, "Mapping sensitivity to land degradation in Extremadura, SW Spain," Boletín de la Asociación de Geógrafos Españoles, vol. 53, pp. 387–390, 2008.

[12] S. Nawar, H. Buddenbaum, and J. Hill, "Digital mapping of soil properties using multivariate statistical analysis and ASTER data in an Arid region," Remote Sensing, vol. 7, no. 2, pp. 1181–1205, 2015.

[13] V. P. Goyal, R. L. Ahuja, B. S. Sangwan, and M. L. Manchanda, "Application of remote sensing technique in wasteland mapping and their landuse planning in Karnal District of Haryana State (India)," in Proceedings of the 13th Annual International Geoscience and Remote Sensing Symposium, vol. 2, pp. 932–934, Tokyo, Japan, August 1993.

[14] Agriculture Situations in District Sirsa, Haryana, India, 2014, https://sirsa.gov.in/.

[15] R. Sugumaran, G. Sandhya, K. S. Rao, R. N. Jadhav, and M. M. Kimothi, "Potential of satellite data in delineation of wastelands and correlation with ground information," Journal of the Indian Society of Remote Sensing, vol. 22, no. 2, pp. 113–118, 1994.

[16] A. N. Singh, "Geospatial database for sustainable reclamation of degraded lands," in Proceedings of the Abstracts of the ISPRS TC-IV International Symposium on Geo Spatial Databases for Sustainable Development, pp. 151-152, Goa, India, 2006.

[17] D. P. Rao, "emote sensing application for land use and urban planning: retrospective and perspective," in Proceedings of the ISRS National Symposium on Remote Sensing Application for Natural Resources Retrospective and Perspective, pp. 287–297, Bangalore, India, January 1999.

[18] R. Pramila, "Assessment of soil degradation Hazards in Jalor and Ahor Tehsil of Jalor district (Western Rajasthan) by remote sensing," Journal of the Indian Society of Remote Sensing, vol. 22, no. 3, pp. 169–181, 1994.

[19] G. Metternicht I and J. Zinck A, Remote Sensing of Soil Salinization: Impact on Land Management, CRC Press, Taylor & Francis Group, Boca Raton, Fla, USA, 2008.

[20] V. L. Mulder, S. de Bruin, and M. E. Schaepman, "Representing major soil variability at regional scale by constrained Latin Hypercube Sampling of remote sensing data," International Journal of Applied Earth Observation and Geoinformation, vol. 21, no. 1, pp. 301–310, 2013.

[21] S. K. Singh, C. K. Singh, and S. Mukherjee, "Impact of land-use and land-cover change on groundwater quality in the Lower Shiwalik hills: A remote sensing and GIS based approach," Central European Journal of Geosciences, vol. 2, no. 2, pp. 124–131, 2010.

[22] J. K. Thakur, P. K. Srivastava, A. K. Pratihast, and S. K. Singh, "Estimation of evapotranspiration from wetlands using geospatial and hydrometeorological data," in Geospatial Techniques for Managing Natural Resources, J. K. Thakur, S. K. Singh, A. Ramanathan, M. B. K. Prasad, and W. Gossel, Eds., pp. 53–67, Springer and Capital, 2012.

[23] J. K. Thakur, P. K. Srivastava, and S. K. Singh, "Ecological monitoring of wetlands in semi-arid Konya closed basin, Turkey," Regional Environmental Change, vol. 12, no. 1, pp. 133–144, 2012.

[24] S. K. Singh, A. C. Pandey, and D. Singh, "Land use fragmentation analysis using remote sensing and Fragstats," in Remote Sensing Applications in Environmental Research, P. K. Srivastava, S. Mukherjee, T. Islam, and M. Gupta, Eds., chapter 9, pp. 151–176, Springer International Publishing, Cham, Switzerland, 2014.

[25] S. K. Singh, P. K. Srivastava, M. Gupta, J. K. Thakur, and S. Mukherjee, "Appraisal of land use/land cover of mangrove forest ecosystem using support vector machine," Environmental Earth Sciences, vol. 71, no. 5, pp. 2245–2255, 2014.

[26] S. K. Singh, S. Mustak, P. K. Srivastava, S. Szabó, and T. Islam, "Predicting spatial and decadal LULC changes through cellular automata markov chain models using earth observation datasets and geo-information," Environmental Processes, vol. 2, no. 1, pp. 61–78, 2015.

[27] S. Mustak, N. K. Baghmar, and S. K. Singh, "Prediction of industrial land use using linear regression and MOLA techniques: a case study of Siltara industrial belt," Landscape & Environment, vol. 9, no. 2, pp. 59–70, 2015.

[28] S. K. Singh, S. K. Kewat, B. Aier, V. P. Kanduri, and S. Ahirwar, "Plant community characteristics and soil status in different land use systems at Dimapur, Nagaland, India," Forest Research Papers, vol. 73, no. 4, pp. 305–312, 2012.

[29] D. Paudel, J. K. Thakur, S. K. Singh, and P. K. Srivastava, "Soil characterization based on land cover heterogeneity over a tropical landscape: an integrated approach using earth observation data-sets," Geocarto International, vol. 30, no. 2, pp. 218–241, 2015.

[30] G. Szabó, S. K. Singh, and S. Szabó, "Slope angle and aspect as influencing factors on the accuracy of the SRTM and the ASTER GDEM databases," Physics and Chemistry of the Earth, vol. 83-84, pp. 137–145, 2015.

[31] S. K. Singh, P. K. Srivastava, S. Szabó, G. P. Petropoulos, M. Gupta, and T. Islam, "Landscape transform and spatial metrics for mapping spatiotemporal land cover dynamics using Earth Observation data-sets," Geocarto International, vol. 32, no. 2, pp. 113–127, 2017.

[32] S. K. Yadav, S. K. Singh, M. Gupta, and P. K. Srivastava, "Morphometric analysis of Upper Tons basin from Northern Foreland of Peninsular India using CARTOSAT satellite and GIS," Geocarto International, vol. 29, no. 8, pp. 895–914, 2014.

[33] A. K. Jain, R. S. Hooda, J. Nath, and M. L. Manchanda, "Mapping and monitoring of urban landuse of Hisar Town, Haryana using

remote sensing techniques," *Journal of the Indian Society of Remote Sensing*, vol. 19, no. 2, pp. 125–134, 1991.

[34] T. E. Barchyn, R. L. Martin, J. F. Kok, and C. H. Hugenholtz, "Fundamental mismatches between measurements and models in aeolian sediment transport prediction: The role of small-scale variability," *Aeolian Research*, vol. 15, pp. 245–251, 2014.

[35] F. M. Breunig, L. S. Galvão, and A. R. Formaggio, "Detection of sandy soil surfaces using ASTER-derived reflectance, emissivity and elevation data: Potential for the identification of land degradation," *International Journal of Remote Sensing*, vol. 29, no. 6, pp. 1833–1840, 2008.

[36] S. Mustak, N. K. Baghmar, and S. K. Singh, "Correction of atmospheric haze of IRS-1C LISS-III multispectral satellite imagery: an empirical and semi-empirical based approach," *Landscape & Environment*, vol. 10, no. 2, pp. 63–74, 2016.

[37] L. Xiaoyan, Z. Wang, S. Kaishan, Z. Bai, L. Dianwei, and G. Zhixing, "Assessment for salinized wasteland expansion and land use change using GIS and remote sensing in the west part of Northeast China," *Environmental Modeling & Assessment*, vol. 131, no. 1-3, pp. 421–437, 2007.

[38] K. Whitehead, C. H. Hugenholtz, S. Myshak et al., "Remote sensing of the environment with small unmanned aircraft systems (UASs), part 2: scientific and commercial applications," *Journal of Unmanned Vehicle Systems*, vol. 2, no. 3, pp. 86–102, 2014.

[39] S. Grunwald, J. A. Thompson, and J. L. Boettinger, "Digital soil mapping and modeling at continental scales: Finding solutions for global issues," *Soil Science Society of America Journal*, vol. 75, no. 4, pp. 1201–1213, 2011.

[40] I. Delgado-Fernandez, R. Davidson-Arnott, B. O. Bauer, I. J. Walker, J. Ollerhead, and H. Rhew, "Assessing aeolian beach-surface dynamics using a remote sensing approach," *Earth Surface Processes and Landforms*, vol. 37, no. 15, pp. 1651–1660, 2012.

[41] S. K. Saha, M. Kudrat, and S. K. Bhan, "Digital processing of Landsat TM data for wasteland mapping in parts of Aligarh District (Uttar Pradesh), India," *International Journal of Remote Sensing*, vol. 11, no. 3, pp. 485–492, 1990.

[42] K. Anderson and H. Croft, "Remote sensing of soil surface properties," *Progress in Physical Geography*, vol. 33, no. 4, pp. 457–473, 2009.

[43] V. L. Mulder, S. de Bruin, M. E. Schaepman, and T. R. Mayr, "The use of remote sensing in soil and terrain mapping—a review," *Geoderma*, vol. 162, no. 1-2, pp. 1–19, 2011.

[44] V. S. Arya, H. Singh, R. S. Hooda, and A. S. Arya, "Desertification change analysis in siwalik hills of Haryana using geoinformatics," in *International Archives of the Photogrammetry, Remote Sensing and Spatial Information Sciences*, vol. XL-8 of *2014 ISPRS Technical Commission VIII Symposium, December 2014, Hyderabad, India*, 2014.

[45] https://sirsa.gov.in/.

[46] http://cgwb.gov.in/District_Profile/Haryana/Sirsa.pdf.

[47] P. C. Moharana, P. Santra, D. V. Singh et al., "ICAR-Central Arid Zone Research Institute, Jodhpur: Erosion processes and desertification in the Thar Desert of India," *Proceedings of the Indian Academy of Science*, vol. 82, no. 3, pp. 1117–1140, 2016.

[48] T. M. Lillesand and R. W. Kiefer, *Remote Sensing and Image Interpretation*, John Wiley, New York, NY, USA, 5th edition, 2004.

[49] A. S. Hickin, B. Kerr, T. E. Barchyn, and R. C. Paulen, "Using ground-penetrating radar and capacπtvely coupled resistivity to investigate 3-D fluvial architecture and grain-size distribution of a gravel floodplain in northeast British Columbia, Canada," *Journal of Sedimentary Research*, vol. 79, no. 6, pp. 457–477, 2009.

[50] C. H. Hugenholtz and T. E. Barchyn, "Spatial analysis of sand dunes with a new global topographic dataset: New approaches and opportunities," *Earth Surface Processes and Landforms*, vol. 35, no. 8, pp. 986–992, 2010.

[51] C. H. Hugenholtz, N. Levin, T. E. Barchyn, and M. C. Baddock, "Remote sensing and spatial analysis of aeolian sand dunes: A review and outlook," *Earth-Science Reviews*, vol. 111, no. 3-4, pp. 319–334, 2012.

[52] C. H. Hugenholtz, O. W. Brown, and T. E. Barchyn, "Estimating aerodynamic roughness (z_0) from terrestrial laser scanning point cloud data over un-vegetated surfaces," *Aeolian Research*, vol. 10, pp. 161–169, 2013.

[53] C. H. Hugenholtz, K. Whitehead, O. W. Brown et al., "Geomorphological mapping with a small unmanned aircraft system (sUAS): Feature detection and accuracy assessment of a photogrammetrically-derived digital terrain model," *Geomorphology*, vol. 194, pp. 16–24, 2013.

[54] A. S. Hickin, B. Kerr, D. G. Turner, and T. E. Barchyn, "Mapping Quaternary paleovalleys and drift thickness using petrophysical logs, northeast British Columbia, Fontas map sheet, NTS 941," *Canadian Journal of Earth Sciences*, vol. 45, no. 5, pp. 577–591, 2008.

4

Machine Learning: A Novel Approach to Predicting Slope Instabilities

Upasna Chandarana Kothari⦿ **and Moe Momayez**⦿

Mining & Geological Engineering, University of Arizona, Tucson, AZ, USA

Correspondence should be addressed to Upasna Chandarana Kothari; upasnap@gmail.com

Academic Editor: Yun-tai Chen

Geomechanical analysis plays a major role in providing a safe working environment in an active mine. Geomechanical analysis includes but is not limited to providing active monitoring of pit walls and predicting slope failures. During the analysis of a slope failure, it is essential to provide a safe prediction, that is, a predicted time of failure prior to the actual failure. Modern-day monitoring technology is a powerful tool used to obtain the time and deformation data used to predict the time of slope failure. This research aims to demonstrate the use of machine learning (ML) to predict the time of slope failures. Twenty-two datasets of past failures collected from radar monitoring systems were utilized in this study. A two-layer feed-forward prediction network was used to make multistep predictions into the future. The results show an 86% improvement in the predicted values compared to the inverse velocity (IV) method. Eighty-two percent of the failure predictions made using ML method fell in the safe zone. While 18% of the predictions were in the unsafe zone, all the unsafe predictions were within five minutes of the actual failure time, all practical purposes making the entire set of predictions safe and reliable.

1. Introduction

Monitoring slope stability is an essential requirement in the field of geomechanics due to the potential threat a moving slope can cause to the workers or the business. Slope stability is an important concern for mining and civil engineers that deal with man-made slopes such as open-pit walls, dams, embankments of highways and railways, and hills. The causes of instability are often complex and creep theory is used in the design of rock slopes. The complexity of the causes of slope movement makes the time of slope failure prediction challenging. In recent years, the use of modern monitoring technologies has helped engineers better prepare for the outcomes of slope failures in open-pit mines [1].

Many attempts have been made to develop a method to predict the time of failure. Factors affecting slope instabilities such as ground conditions, physical and geomorphological processes, and human activities cannot be determined on a continuous basis, making it challenging to predict the time of slope failure accurately [2]. Hence, instead of developing a

phenomenological model of slope failure, practitioners have relied on a detailed analysis of slope deformation [3].

Deformation data, the most relevant data for time series analysis of slope failures, is readily available from the monitoring equipment used for geotechnical risk management analysis [4, 5]. Some of the modern and traditional monitoring technologies include but are not limited to tension crack mapping, survey networks, wireline extensometers, synthetic aperture radar, satellite-based synthetic aperture radar, and ground-based real aperture radar [6]. The radar systems usually record the increase in deformation accurately until the slope movement becomes too fast for the radar to capture or until slope collapses. The time and deformation data acquired from these monitoring systems will provide the opportunity to observe the prefailure evolution of a moving slope till the time of collapse. The three prefailure stages include primary, secondary, and tertiary movement (Figure 1). The primary stage displays a decreasing strain rate, the secondary stage displays a constant strain rate, and the tertiary stage represents an accelerating strain rate leading to

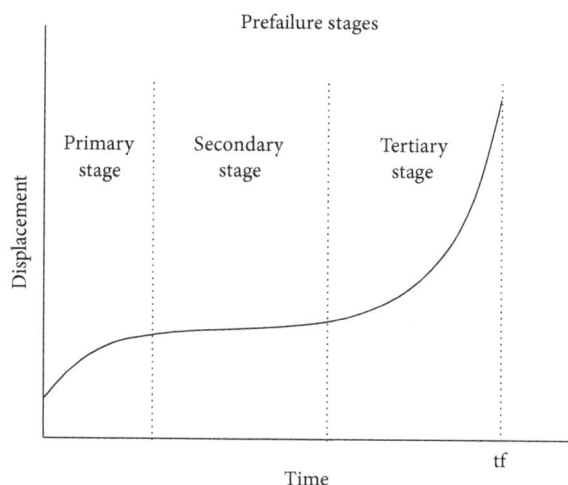

FIGURE 1: Primary stages of prefailure evolution.

FIGURE 2: Flow chart of supervised machine learning.

failure. The prefailure evolution of slope movement exhibits similar characteristics to the creep observed in the study of geomaterials [7–10].

Based on the core understanding of the prefailure evolution, many attempts have been made to develop suitable methods to predict an accurate time of slope failure. Many of the slope failure studies have used the inverse velocity (IV) method proposed by Fukuzono in 1985 [11]. The fuzzy neural network approach is another method that gained popularity in the civil engineering industry and was slowly adapted for slope stability analysis. Predicting slope failure is a common practice in active mines to prevent injuries and fatalities due to ground movement issues. With a view to make a prediction that allows for ample evacuation time, the forecast should provide a time before the actual slope failure. As all operations are different, the time required to evacuate the area will solely depend on the size of the mine and the resources available. A prediction that occurs before the actual time of failure is considered a safe prediction, whereas a forecast that occurs after the actual time of failure would be regarded as an unsafe prediction. It is therefore highly desirable to make a safe failure prediction to evacuate the area if required.

The aim of this paper is to investigate the use of machine learning (ML) to make safe predictions. Mitchell defined machine learning as the question of how to build computer programs that improve their performance at some task through experience in 1997 [12]. Machine learning enables the computer to recognize patterns and explore the data and uses algorithms to help make predictions based on the input data. There are many algorithms available today to sort through the data, learn visible and invisible patterns, and use the learnt pattern to make better decisions. There are three main types of machine learning, namely, supervised learning, unsupervised learning, and semisupervised learning. The three types of machine learning are briefly explained below:

(i) Supervised learning: a set of data with the known solution is used as the input data; the input data is called training data. The training data is used to train the computer to learn trends and build a model to make informed predictions. Corrections are made to the model for the training process till the desired results are achieved. Classification and regression are examples of supervised learning [13].

(ii) Unsupervised learning: a set of data with unknown solution is used as the input data. For unsupervised learning, a model is built by assuming the presence of structures in the input data by looking for redundancy or similarity in the data [13].

(iii) Semisupervised learning: dataset is a mixture of known and unknown solutions. For this learning, the model is built to understand structure and make predictions [13].

The application of the fuzzy neural network in slope stability studies is an example of supervised ML. The fuzzy set theory has been used in the past to analyze the potential of a slope failure. These studies successfully demonstrated how the fuzzy neural network could assist preparing for a potential slope failure; however, the fuzzy neural networks have not been used for predicting slope failure [14–18]. Fuzzy set theory is a machine learning system based on the real-life model of the neuron's work in a human brain. In 1965, Zadeh first introduced the fuzzy set theory, which was adopted for analysis that can be probabilistic or deterministic [19].

The primary goal of slope failure analysis is to predict the time of slope failure in the presence of evidence that demonstrates signs of a possible slope failure. Similarly, the aim of supervised machine learning is to build a model that can make predictions based on evidence in the data. Using adaptive algorithms, the prediction network learns from the training data and builds a model that is used to make predictions. A large set of training data provides more observations for the training set to learn from and improve its predictive performance. Figure 2 presents a flow chart of how supervised machine learning is used in this study. The idea to attempt a machine learning approach was inspired by a previous study conducted by the authors, where they proposed the use of minimum inverse velocity (MIV) method to improve the accuracy of slope failure predictions [20].

2. Methodology

The approach we propose to predict the time of failure is based on nonlinear regression using a two-layer feed-forward network. A feed-forward network is a unit in which the processes would not form a cycle; the processes would flow from start to finish like a chain reaction (Figure 2).

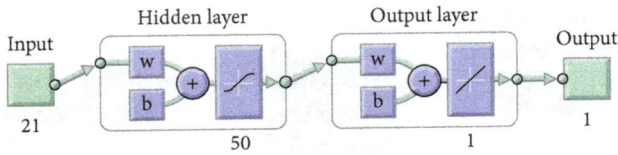

FIGURE 3: Flow chart of the prediction network.

After preprocessing the 22 datasets at our disposal, the time-deformation data is divided into training, validation, and test sets, a network architecture is selected, the network is trained, and multistep prediction is performed. The prediction network was designed in MATLAB using the Neural Network Toolbox. Figure 3 shows the structure of the network.

The datasets collected at open-pit mining operations have different sampling intervals and deformation rates. To obtain the best predictions, preprocessing must be performed to standardize the time series. This step consists of resampling and rescaling the data. Resampling involves changing the sampling interval in each time-deformation series to match or set to a value less than the smallest sampling interval among all the datasets. In this study, a linear interpolation method was used. The deformation data from different mine sites have a wide range of values. When the testing was initiated, the trial tests confirmed that the prediction network is sensitive to magnitude because features with larger values impose more influence on the training set. Transforming data, to have values between 0 and 1, considerably improved the accuracy of predictions.

As mentioned above, this research has been inspired by a recent study conducted by the authors [20], where they introduced the concept of the minimum inverse velocity. MIV was shown to improve the time of failure predictions compared to the traditional IV method proposed by Fukuzono in 1985 [11]. The predictions from the MIV were obtained using data from a two-hour time window before slope failure. In the machine learning study, we predicted deformation for the same two-hour window to compare the performances of the ML and MIV methods.

To predict the failure time in each deformation curve, we created an eight-hour training set consisting of the resampled and rescaled data from the other 21 datasets. Different algorithms that update the weight and bias values in the network training function were tested. We settled on the Levenberg-Marquardt algorithm, since it provides the fastest convergence and better overall prediction values. The property DIVIDEMODE was set to TIMESTEP forcing the targets to be divided into training, validation, and test sets according to timesteps. Various segmentations of training, validation, and test sets were tried, and no significant impact on the final results was observed. For the remainder of the study, 70% of training, 15% of validation, and 15% of testing data were specified from the total dataset of 22 available records.

A curvature index (the normalized area between the time-deformation curve and a straight line connecting the first point in the time series to the failure point) was calculated for each of the 22 datasets used in this research.

12 time series exhibited a linear type deformation, curvature index close to zero. Nine datasets had a curvature index less than -0.1, representing a regressive type deformation. One dataset had a curvature index greater than $+0.1$, indicative of a progressive deformation. To perform a multistep prediction for a given dataset, the values for the last two-hour window in the time-deformation series were set to NaN (not a number). After much experimentation, we settled on a value of 50 nodes in the hidden layer as it provided the smallest number of nodes and most consistent prediction values.

During training, the network performance over 250 to 400 timesteps produced a mean square error of less than 10^{-5} as shown in Figure 4. In Figure 4, the x-axis represents the error and the Zero Error line represents the mean square error, whereas the y-axis represents the number of instances the algorithm ran in order to produce the results. Next, the first peak in the output sequence of the network was used to determine the time of failure for the slope. Figure 5 provides a plot of the original dataset and the predicted values for mine site 20.

3. Results

The predictions obtained from the MIV resulted in a 75% improvement in comparison to the IV method [20]. Table 1 presents a summary of the results comparing the IV and MIV methods including the predicted time of failures. The predictions based on both methods have been compared to the real time of failure. The negative values in the column "IV-MIV" represent a success (prediction in the safe zone) for the IV method whereas the positive values represent a success for the MIV method. Based on the results below, we can see that MIV method results in a 75% improvement in slope failure predictions.

Some machine learning algorithms perform well when used for fitting data especially if the training set contains strong features. The curvature indices in the 22 datasets range from -0.324 to $+0.127$ with most of the deformation curves displaying a linear behavior (a curvature index close to zero). Because the time series in the training set have a relatively similar form, we surmised that ML would provide prediction values that are closer to the real time of failure. Twenty-two historical failure datasets with a time span of 8 hours were used to generate the training dataset. Table 2 summarizes the results obtained using ML to predict the time of failure. All the predictions were made based on two hours of missing data prior to the failure. The results of ML are compared to the traditional IV method. The column "IV-ML" provides the time difference between the two approaches. Positive values represent the number of hours ML prediction is closer to the real time of failure compared to IV. Each positive value in the "IV-ML" column represents a success for ML, whereas each negative value represents a success for IV. Based on the results, 19 cases demonstrate a better prediction. The results show an 86.4% success rate in the predictions obtained using the ML method.

The predictions obtained using ML method for the slope failure were compared to the results based on the MIV method. Table 3 shows a comparison between MIV and

TABLE 1: Results of the comparison between inverse velocity (IV) method and minimum inverse velocity (MIV) method from 22 different failure examples. For the analysis demonstrated below, all calculations are based on a 60-minute averaging window.

#	Actual time of failure	Prediction: IV	Delta time (hr)	Prediction: MIV	Delta time (hr)	IV – MIV (hr)
(1)	3/10/09 7:12	3/10/09 7:51	0.65	3/10/09 7:42	0.52	0.13
(2)	3/13/13 9:48	3/13/13 9:19	−0.48	3/13/13 8:45	−1.04	−1.51
(3)	3/5/12 4:09	3/5/12 4:34	0.43	3/5/12 4:31	0.38	0.05
(4)	8/3/13 22:06	8/3/13 23:53	1.79	8/3/13 22:15	0.16	1.63
(5)	7/27/12 18:50	7/27/12 19:38	0.81	7/27/12 19:00	0.18	0.64
(6)	10/24/13 22:39	10/24/13 22:40	0.03	10/24/13 20:58	−1.68	−1.65
(7)	5/5/14 5:04	5/6/14 2:41	21.62	5/4/14 20:23	−8.67	12.96
(8)	6/16/14 6:58	6/16/14 7:10	0.21	6/16/14 4:01	−2.95	−2.74
(9)	1/28/12 9:14	1/28/12 12:04	2.84	1/28/12 9:08	−0.10	2.74
(10)	10/29/12 12:15	10/31/12 18:58	54.73	10/29/12 11:58	−0.28	54.45
(11)	9/25/12 8:40	9/26/12 18:16	33.61	9/25/12 4:33	−4.12	29.49
(12)	3/26/13 21:03	4/10/13 23:08	362.10	3/26/13 23:58	2.92	359.18
(13)	2/24/12 11:27	2/24/12 17:03	5.61	2/24/12 13:42	2.26	3.36
(14)	3/2/14 4:23	3/2/14 5:55	1.54	3/2/14 5:48	1.42	0.12
(15)	10/25/12 16:17	10/25/12 16:47	0.51	10/25/12 15:35	−0.69	−0.18
(16)	4/22/13 13:27	4/22/13 13:58	0.52	4/22/13 10:02	−3.40	−2.88
(17)	3/14/10 4:01	3/14/10 4:16	0.26	3/14/10 4:02	0.03	0.24
(18)	7/11/13 0:20	7/11/13 0:25	0.10	7/11/13 0:13	−0.11	−0.01
(19)	7/21/11 15:01	7/21/11 15:17	0.28	7/21/11 14:35	−0.42	−0.14
(20)	6/20/09 22:24	6/21/09 1:09	2.76	6/20/09 23:58	1.57	1.20
(21)	2/9/10 12:38	2/9/10 14:05	1.47	2/9/10 13:20	0.71	0.75
(22)	1/30/15 18:25	1/30/15 21:55	3.51	1/30/15 21:11	2.78	0.74

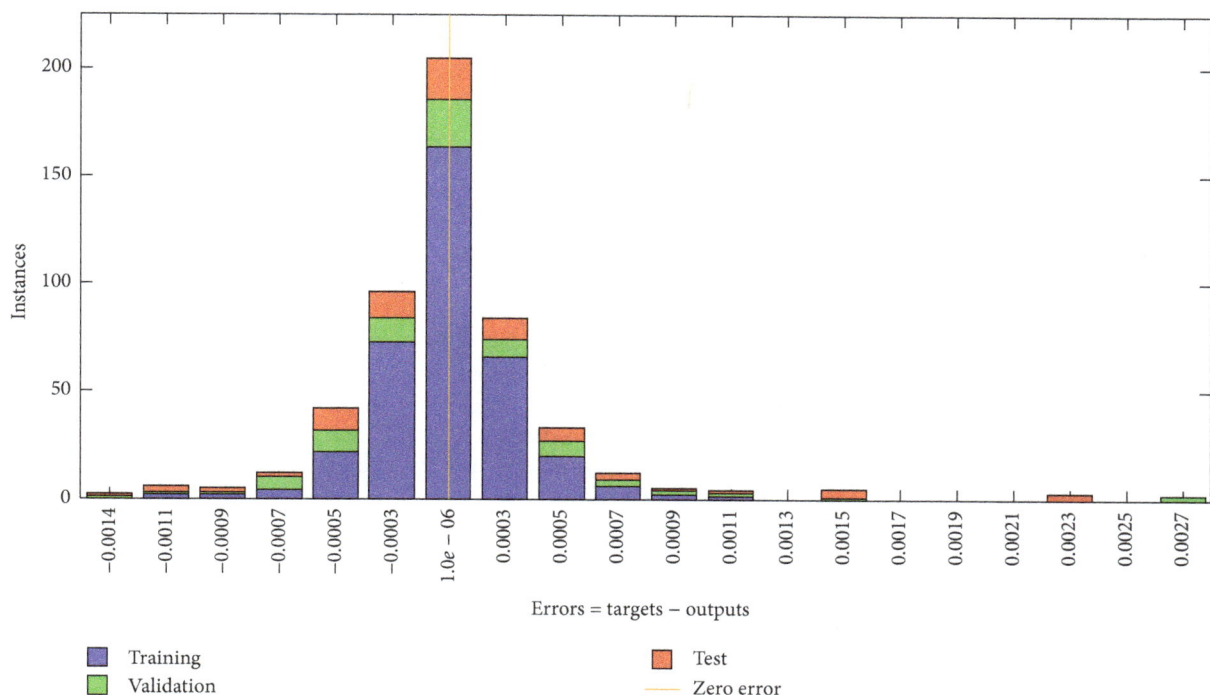

FIGURE 4: Error histogram from a training period.

TABLE 2: Results of the comparison between inverse velocity (IV) method and machine learning (ML) method from 22 different failure examples.

#	Actual time of failure	Prediction: IV	Delta time (hr)	Prediction: ML	Delta time (hr)	IV – ML (hr)
(1)	3/10/09 7:12	3/10/09 7:51	0.65	3/10/09 6:17	−0.92	−0.27
(2)	3/13/13 9:48	3/13/13 9:19	−0.48	3/13/13 9:52	0.07	0.41
(3)	3/5/12 4:09	3/5/12 4:34	0.43	3/5/12 3:24	−0.75	−0.32
(4)	8/3/13 22:06	8/3/13 23:53	1.79	8/3/13 21:56	−0.17	1.62
(5)	7/27/12 18:50	7/27/12 19:38	0.81	7/27/12 18:51	0.02	0.79
(6)	10/24/13 22:39	10/24/13 22:40	0.03	10/24/13 22:40	0.02	0.01
(7)	5/5/14 5:04	5/6/14 2:41	21.62	5/5/14 4:13	−0.85	20.77
(8)	6/16/14 6:58	6/16/14 7:10	0.21	6/16/14 6:57	−0.02	0.19
(9)	1/28/12 9:14	1/28/12 12:04	2.84	1/28/12 8:19	−0.92	1.92
(10)	10/29/12 12:15	10/31/12 18:58	54.73	10/29/12 11:59	−0.27	54.46
(11)	9/25/12 8:40	9/26/12 18:16	33.61	9/25/12 8:44	0.07	33.54
(12)	3/26/13 21:03	4/10/13 23:08	362.10	3/25/13 19:40	−1.38	360.72
(13)	2/24/12 11:27	2/24/12 17:03	5.61	2/24/12 11:25	−0.03	5.58
(14)	3/2/14 4:23	3/2/14 5:55	1.54	3/2/14 4:22	−0.02	1.52
(15)	10/25/12 16:17	10/25/12 16:47	0.51	10/25/12 15:24	−0.88	−0.37
(16)	4/22/13 13:27	4/22/13 13:58	0.52	4/22/13 13:31	0.07	0.45
(17)	3/14/10 4:01	3/14/10 4:16	0.26	3/14/10 3:50	−0.18	0.08
(18)	7/11/13 0:20	7/11/13 0:25	0.10	7/11/13 0:19	−0.02	0.08
(19)	7/21/11 15:01	7/21/11 15:17	0.28	7/21/11 15:00	−0.02	0.26
(20)	6/20/09 22:24	6/21/09 1:09	2.76	6/20/09 22:23	−0.02	2.74
(21)	2/9/10 12:38	2/9/10 14:05	1.47	2/9/10 12:30	−0.13	1.34
(22)	1/30/15 18:25	1/30/15 21:55	3.51	1/30/15 16:34	−1.85	1.66

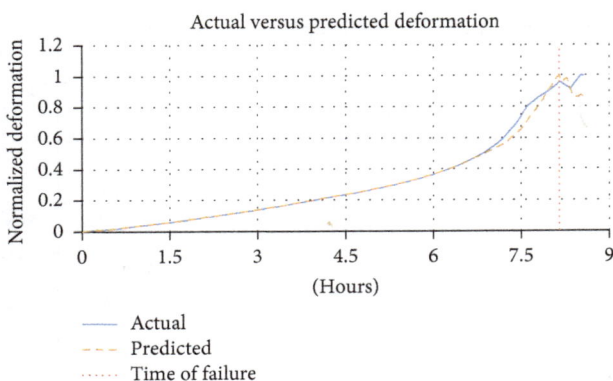

FIGURE 5: Actual and predicted slope failure data from mine site 20.

ML methods. The column "MIV-ML" provides the time difference between the two approaches. A negative value in the MIV-ML value represents a success (prediction in the safe zone) for the MIV method whereas all the positive values represent a success for the ML method. A comparison of the MIV and ML methods provides a 72% success rate for the machine learning technique. Out of the 22 cases studies, ML method gave better results in 16 cases. Based on the overall results, ML performed significantly better than IV and MIV methods.

After comparing the three methods, IV, MIV, and ML, it is concluded that ML gives the results that are the closest to the real time of failure. A 95% confidence interval was calculated for the three methods. The results are displayed in Table 4. Based on the confidence interval calculations, it is concluded that 95% of the slope failure predictions using the IV method will fall between −131 and 176 hours from the real time of failure. As the confidence level was applied to the datasets used for the analysis, 21 of the prediction fell in the 95% confidence interval using IV method. The confidence interval calculated for MIV indicates that 95% of the slope failure predictions calculated using MIV will fall between −6 and 5 hours away from the real time of failure. Twenty-one of the 22 datasets analyzed gave a failure prediction that fell into the 95% confidence interval with the MIV method. The confidence interval calculated for ML indicates that 95% of the slope failure predictions calculated using ML will fall between −1.45 and 0.72 hours away from the real time of failure. Twenty-one of the 22 datasets analyzed using ML method gave a failure prediction that fell in the 95% confidence interval. In addition to giving the best results, ML also has the smallest time window between the lower and upper limit of the 95% confidence interval.

In addition to getting a prediction time that is close to the real time of failure, another aim of this study was to provide a time of failure prediction that falls in the safe

TABLE 3: Results of the comparison between minimum inverse velocity (MIV) and machine learning (ML) method from 22 different failure examples.

#	Actual time of failure	Prediction: MIV	Delta time (hr)	Prediction: ML	Delta time (hr)	MIV − ML (hr)
(1)	3/10/09 7:12	3/10/09 7:42	0.52	3/10/09 6:17	−0.92	−0.40
(2)	3/13/13 9:48	3/13/13 8:45	−1.04	3/13/13 9:52	0.07	0.97
(3)	3/5/12 4:09	3/5/12 4:31	0.38	3/5/12 3:24	−0.75	−0.37
(4)	8/3/13 22:06	8/3/13 22:15	0.16	8/3/13 21:56	−0.17	−0.01
(5)	7/27/12 18:50	7/27/12 19:00	0.18	7/27/12 18:51	0.02	0.16
(6)	10/24/13 22:39	10/24/13 20:58	−1.68	10/24/13 22:40	0.02	1.66
(7)	5/5/14 5:04	5/4/14 20:23	−8.67	5/5/14 4:13	−0.85	7.82
(8)	6/16/14 6:58	6/16/14 4:01	−2.95	6/16/14 6:57	−0.02	2.93
(9)	1/28/12 9:14	1/28/12 9:08	−0.10	1/28/12 8:19	−0.92	−0.82
(10)	10/29/12 12:15	10/29/12 11:58	−0.28	10/29/12 11:59	−0.27	0.01
(11)	9/25/12 8:40	9/25/12 4:33	−4.12	9/25/12 8:44	0.07	4.05
(12)	3/26/13 21:03	3/26/13 23:58	2.92	3/25/13 19:40	−1.38	1.54
(13)	2/24/12 11:27	2/24/12 13:42	2.26	2/24/12 11:25	−0.03	2.23
(14)	3/2/14 4:23	3/2/14 5:48	1.42	3/2/14 4:22	−0.02	1.40
(15)	10/25/12 16:17	10/25/12 15:35	−0.69	10/25/12 15:24	−0.88	−0.19
(16)	4/22/13 13:27	4/22/13 10:02	−3.40	4/22/13 13:31	0.07	3.33
(17)	3/14/10 4:01	3/14/10 4:02	0.03	3/14/10 3:50	−0.18	−0.15
(18)	7/11/13 0:20	7/11/13 0:13	−0.11	7/11/13 0:19	−0.02	0.09
(19)	7/21/11 15:01	7/21/11 14:35	−0.42	7/21/11 15:00	−0.02	0.40
(20)	6/20/09 22:24	6/20/09 23:58	1.57	6/20/09 22:23	−0.02	1.55
(21)	2/9/10 12:38	2/9/10 13:20	0.71	2/9/10 12:30	−0.13	0.58
(22)	1/30/15 18:25	1/30/15 21:11	2.78	1/30/15 16:34	−1.85	0.93

TABLE 4: Confidence interval calculated for the IV and MIV methods. The calculations use $\mu \pm 2\sigma$ to get the upper and lower bounds of the 95.5% confidence interval.

	95.5% confidence interval		
	IV method	MIV method	ML method
Mean (μ)	22.5	−0.48	−0.37
Standard deviation (σ)	77.04	2.57	0.54
Upper limit	176.52	4.66	0.72
Lower limit	−131.39	−5.62	−1.46

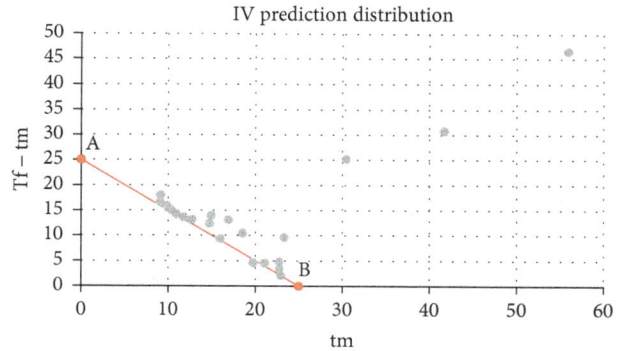

FIGURE 6: Distribution of failure predictions using IV method. Line AB represents failure time.

zone. Figures 6–8 demonstrate the distribution of the failure prediction using IV, MIV, and ML method. The distribution of the time of failure predictions is compared to the real time of failure, distinguishing the predictions as safe or unsafe. A failure prediction is considered a safe prediction when the failure occurs after the predicted time, whereas if the failure occurs before the predicted time, it is deemed to be an unsafe prediction. In the figures demonstrating the prediction distribution, line AB represents the life expectancy of the moving slope; any prediction below line AB is a safe prediction whereas any prediction above line AB is considered an unsafe prediction. The results were rotated 45 degrees and plotted on an x-y plot to distinguish between safe and unsafe predictions. In Figures 6–8, let us assume that the

x-axis represents the predicted time of failure whereas the y-axis represents the actual time of failure minus the predicted time.

Tables 1–3 provide all the results using IV, MIV, and ML method to predict the time of failure. The results are represented in a graphical format in Figures 6–8. Figure 6 represents the prediction distribution of the IV method; the outlier with a time difference of 362 hours was eliminated from the graph to demonstrate a better visualization of the rest of the data. Figure 6 shows that some of the predictions

TABLE 5: Comparison between failure predictions 1, 2, 3, 4, and 5 hours prior to failure for locations 8 and 20.

| Location | Failure time | Prediction: 1, 2, 3, 4, and 5 hours before failure | | | | |
		1	2	3	4	5
08	**6/16/14 6:58**	6/16/14 7:01	6/16/14 6:57	6/16/14 5:39	6/16/14 4:51	6/16/14 6:26
	Time difference	0.05	−0.02	−1.37	−2.20	−0.56
20	**6/20/09 22:24**	6/20/09 22:17	6/20/09 22:23	6/20/09 21:48	6/20/09 21:46	6/20/09 21:28
	Time difference	−0.12	−0.02	−0.60	−0.63	−0.93

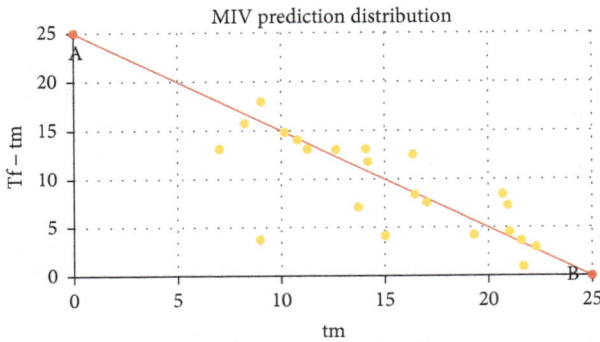

FIGURE 7: Distribution of failure predictions using MIV method. Line AB represents failure time.

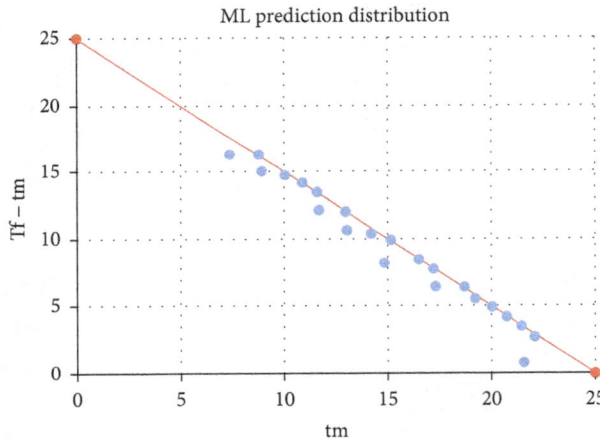

FIGURE 8: Distribution of failure predictions using ML method. Line AB represents failure time.

Machine learning analysis was also used to analyze the accuracy of the failure predictions with respect to the proximity of the real time of failure to current data. Two datasets were chosen from the twenty-two records utilized in this study, and predictions were made for time series 1, 2, 3, 4, and 5 hours before the failure. This analysis showed that predictions improve as the actual time of failure approaches. Theoretically, slope failure predictions should get closer to the real time of failure as the size of the collected dataset increases. When an accelerating movement is observed in the data, the trend is defined as progressive or regressive as time goes on. If predictions are made with a well-defined progressive curve, the chance of a better prediction improves as the time of actual failure approaches. The investigation related to making predictions 1, 2, 3, 4, and 5 hours prior to failure confirmed the above hypothesis. Results in Table 5 show a trend of decreasing time difference as the real time of failure approaches.

From the two datasets analyzed, it can be concluded that failure predictions made two hours before the failure are the closest to the time of failure, resulting in the best time of failure predictions. In general, as a slope approaches failure, the rate of deformation increases rapidly. If the rate of slope movement is faster than the scan rate, it is likely that the monitoring system does not capture the entire deformation. When the radar is not able to record the movement correctly, the time-deformation curve appears to drop or to slow down. Due to this limitation of the monitoring systems, it has been observed that the gap between the predicted and actual failure time increases.

4. Discussion

Risk identification, risk management, and risk mitigation processes benefit from reliable slope stability monitoring and forecasting. The desired outcome of slope stability monitoring is to be able to make safe predictions. Any prediction that occurs before the actual failure time is considered a safe prediction. ML approach resulted in 17 failure predictions that occurred in the safe zone and five predictions that were in the unsafe zone. The five unsafe predictions were within 5 minutes of the real time of failure. Therefore, for all practical purposes, they could be considered as safe predictions.

The selection of alarm thresholds at a mine site is often based on historical behavior of slopes and could present challenges as one or several factors controlling slope deformation change suddenly. Increasing the scan rate can help improve the reliability of capturing unexpected acceleration. However, in many situations, due to the large distance from the face

are close to the real time of failure, but there is only one safe prediction. Figure 7 represents the distribution of MIV method, demonstrating that 50% of the failure predictions are in the safe zone. Figure 8 represents the distribution of ML method; this graph shows that all the predictions are very close to the real time of failure. It is hard to see from the graph but only 4 of the predictions using ML method occur after the actual failure time and all the unsafe predictions are within a 5-minute interval from the real time of failure and can be considered as a safe prediction. Statistically, 82% of the predictions using ML fall in the safe zone. The comparisons between IV, MIV, and ML show a significant improvement in the time of failure predictions using the ML method.

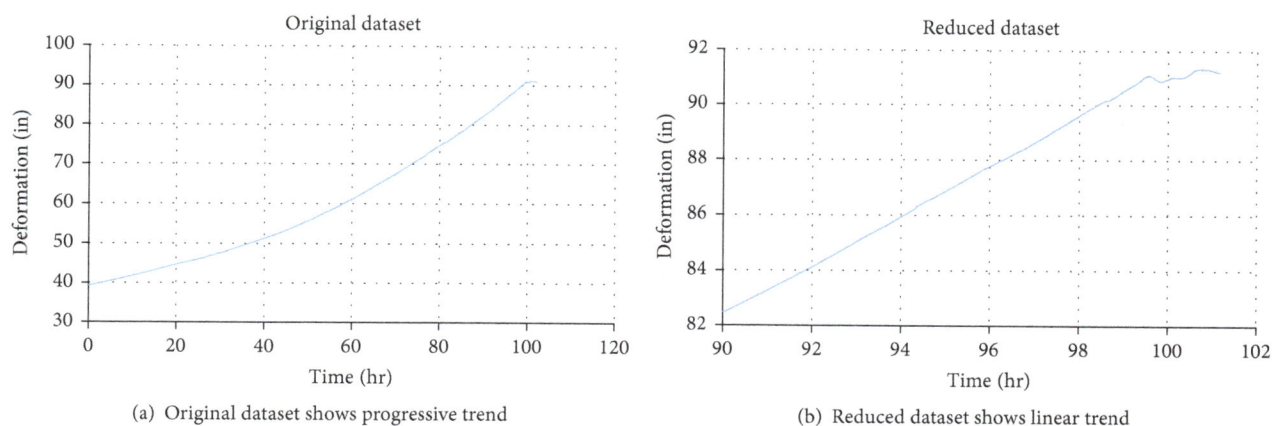

(a) Original dataset shows progressive trend

(b) Reduced dataset shows linear trend

FIGURE 9

and the size of the area to cover, operating the monitoring equipment at high scan rates may not be feasible.

Assessing the potential for failure based on the traditional inverse velocity and the minimum inverse velocity method proposed by the authors in a previous publication depends to a large extent on establishing an appropriate trend line. A velocity curve that is noisy or presents a strong bias effectively limits the potential for making a reliable prediction. The ML method, however, is less prone to noise and bias because it uses time-deformation data. Velocity is calculated by differentiating the deformation curve. Inherently, the process of differentiation amplifies higher frequencies in the time series, hence the requirement to smooth out the velocity data before using the inverse velocity or minimum inverse velocity techniques.

The analysis presented in Table 5 provides another opportunity to quantitatively assess the imminence of failure using the ML method. When creating the training set, one of the requirements is to align the datasets with respect to the actual time of failure in each set. In this study, the training set contained data from eight hours prior to failure. As expected, the prediction network performed best when a multistep prediction was carried out on a test set with a one- and two-hour window of missing data prior to failure. In this case, the peak in the prediction curve aligns closely with the peaks in the training set. However, when a test set with a time window far from the actual time of failure is used, a significant shift in the failure peak of the predicted time series is observed. This measure could be used to further verify the accuracy of predictions.

A progressive trend in the deformation data would be most concerning as it has a higher probability of failure. The seven datasets in Table 3 with a difference of more than 30 minutes from the actual time of failure did not show a smooth progressive trend. Datasets that contain a high amount of atmospheric noise (due to rain, wind, or dust) or large movement resulting from mining activity could adversely affect the predicted values. When a slope has been moving for an extended period of time, or if the movement is accelerating, the deformation data would show a linear trend

as it moves away from the inflection point that marks the beginning of a progressive curve.

For ML to perform well, all the datasets are required to be of the same length. One could reason that this is a drawback of the ML method. For all the datasets to be of the same length, it might be necessary to shorten the length of the data from an area that has been moving for an extended period of time. Shortening a dataset may remove the inflection point that marks the beginning of a progressive trend. If the data does not include the inflection point, the deformation curve will tend, in general, to show a linear trend line instead of a progressive trend. If the training set does not contain strong progressive features, predictions on the test set with a linear trend could lead to inaccurate failure predictions. The deformation curve from mine site 7 is an example of a dataset that has a progressive trend; however, the data in the eight-hour window before the slope failure displays a linear trend (Figures 9(a) and 9(b)).

To improve the performance of the prediction network, it is therefore recommended to include in the training set deformation data that show similar behavior in terms of the degree of curvature around the inflection point. If a large dataset is available, several training sets could be created based on a judicious grouping of curvature indexes calculated for each time-deformation curve.

5. Conclusion

Geotechnical risk management analysis is the key to successfully manage the risks posed to personnel, equipment, and production at an active mine [6]. Slope failures have been an issue in the past and continue to be a threat today. To effectively mitigate the risks of unstable slopes, it is important to make more reliable predictions. Slope failure predictions are only helpful when they allow sufficient time to remove people and equipment from the unsafe areas. As all mining operations are different, they may have different escape routes planned in case of emergencies. For this study, the term ample evacuation time is tied together with a safe prediction. It is assumed that each mine will have an estimated required evacuation time based on the size of the mine and the resources available. For the purpose of this study, it is believed that if the predicted time of failure is very close to the real failure or before the collapse, it will provide sufficient time for evacuation. In other words, we want to make a time of failure prediction that gives us the confidence to evacuate if required. For example, if a mine requires 2 hours of evacuation time and the failure was predicted to happen in the next 4 hours, it would be necessary to evacuate the mine 2 hours after the prediction was made or 2 hours before the expected time of failure. It is important to understand that no predetermined amount of time can be considered as ample evacuation time as it can vary from one mine to another and it can also vary between different sections of a single mine.

The current study proposed the use of machine learning (ML) to predict the time of failure. The results of the study show that ML provided prediction values that are 86% of the time closer to the actual time of failure when compared to the traditional IV method. When compared to MIV, ML had a 72% success rate. All the failure predictions using ML method were within 2 hours of the actual time of failure. Fifteen out of the 22 datasets analyzed gave a time of failure prediction that was within 30 minutes of the actual time of failure. Only 2 datasets gave a failure prediction that was over 60 minutes away from the real time of failure. The ML method resulted in 17 datasets with safe predictions and only 5 sets with an unsafe prediction. The five sets with an unsafe prediction were within 5 minutes of the actual failure time, making the unsafe predictions reliable. A larger training set containing carefully selected data based on the similarity of the deformation curves would further improve the reliability of slope failure predictions.

References

[1] K. S. Osasan and T. R. Stacey, "Automatic prediction of time to failure of open pit mine slopes based on radar monitoring and inverse velocity method," *International Journal of Mining Science and Technology*, vol. 24, no. 2, pp. 275–280, 2014.

[2] L. Nie, Z. Li, Y. Lv, and H. Wang, "A new prediction model for rock slope failure time: a case study in West Open-Pit mine, Fushun, China," *Bulletin of Engineering Geology and the Environment*, vol. 76, no. 3, pp. 975–988, 2016.

[3] H. Chen, Z. Zeng, and H. Tang, "Landslide deformation prediction based on recurrent neural network," *Neural Processing Letters*, vol. 41, no. 2, pp. 169–178, 2015.

[4] Z. Liu, J. Shao, W. Xu, H. Chen, and C. Shi, "Comparison on landslide nonlinear displacement analysis and prediction with computational intelligence approaches," *Landslides*, vol. 11, no. 5, pp. 889–896, 2014.

[5] P. Mazzanti, F. Bozzano, and I. Cipriani, "New insights into the temporal prediction of landslides by a terrestrial SAR interferometry monitoring case study," *Landslides*, vol. 12, no. 1, pp. 55–68, 2015.

[6] U. P. Chandarana, M. Momayez, and K. Taylor, "Monitoring and predicting slope instability: a review of current practices from a mining perspective," *International Journal of Research in Engineering and Technology*, vol. 5, no. 11, pp. 139–151, 2016.

[7] A. Federico, M. Popescu, G. Elia, C. Fidelibus, G. Internò, and A. Murianni, "Prediction of time to slope failure: a general framework," *Environmental Earth Sciences*, vol. 66, no. 1, pp. 245–256, 2012.

[8] G. B. Crosta and F. Agliardi, "Failure forecast for large rock slides by surface displacement measurements," *Canadian Geotechnical Journal*, vol. 40, no. 1, pp. 176–191, 2003.

[9] M. Saito, "Forecasting time of slope failure by tertiary creep," in *Proceedings of the Proceedings of the 7th International Conference on Soil Mechanics and Foundation Engineering*, pp. 677–683, 1996.

[10] Q. Xu, Y. Yuan, and Y. Zeng, "Some new pre-warning criteria for creep slope failure," *Science China Technological Sciences*, vol. 54, no. 1, pp. 210–220, 2011.

[11] T. Fukuzono, "A new method for predicting the failure time of a slope," in *Proceedings of the Fourth International Conference and Field Workshop on Landslides*, pp. 150-150, Japan Landslide Society, Tokyo, Japan, 1985.

[12] T. Mitchell, *Machine Learning*, Science/Engineering/Math, McGraw-Hill, New York, NY, USA, 1997.

[13] M. Paluszek and S. Thomas, *MATLAB Machine Learning*, vol. 1, 2017.

[14] S. H. Ni, P. C. Lu, and C. H. Juang, "Fuzzy neural network approach to evaluation of slope failure potential," *Microcomputers in Civil Engineering*, vol. 11, pp. 56–66, 1996.

[15] M. G. Sakellariou and M. D. Ferentinou, "A study of slope stability prediction using neural networks," *Geotechnical and Geological Engineering*, vol. 23, no. 4, pp. 419–445, 2005.

[16] H. B. Wang, W. Y. Xu, and R. C. Xu, "Slope stability evaluation using Back Propagation Neural Networks," *Engineering Geology*, vol. 80, no. 3-4, pp. 302–315, 2005.

[17] S. Hwang, I. F. Guevarra, and B. Yu, "Slope failure prediction using a decision tree: a case of engineered slopes in South Korea," *Engineering Geology*, vol. 104, no. 1-2, pp. 126–134, 2009.

[18] H.-M. Lin, S.-K. Chang, J.-H. Wu, and C. H. Juang, "Neural network-based model for assessing failure potential of highway slopes in the Alishan, Taiwan Area: pre- and post-earthquake investigation," *Engineering Geology*, vol. 104, no. 3-4, pp. 280–289, 2009.

[19] L. A. Zadeh, "Fuzzy sets," *Information and Control*, vol. 8, no. 3, pp. 338–353, 1965.

[20] U. C. Kothari and M. Momayez, "New approaches to monitoring, analyzing and predicting slope instabilities," *Journal of Geology and Mining Research*, vol. 1, pp. 1–14, 10.

5

Integrated Resistivity and Ground Penetrating Radar Observations of Underground Seepage of Hot Water at Blawan-Ijen Geothermal Field

Sukir Maryanto,[1] Ika Karlina Laila Nur Suciningtyas,[2] Cinantya Nirmala Dewi,[2] and Arief Rachmansyah[3]

[1]Department of Physics, Faculty of Mathematics and Science, University of Brawijaya, Malang 65145, Indonesia
[2]Postgraduate Program of Physics, Faculty of Mathematics and Science, University of Brawijaya, Malang 65145, Indonesia
[3]Department of Civil Engineering, Faculty of Engineering, University of Brawijaya, Malang 65145, Indonesia

Correspondence should be addressed to Sukir Maryanto; sukir@ub.ac.id

Academic Editor: Yun-tai Chen

Geothermal resource investigation was accomplished for Blawan-Ijen geothermal system. Blawan geothermal field which located in the northern part of Ijen caldera presents hydrothermal activity related with Pedati fault and local graben. There were about 21 hot springs manifestations in Blawan-Ijen area with calculated temperature about 50°C. We have performed several geophysical studies of underground seepage of hot water characterization. The geoelectric resistivity and GPR methods are used in this research because both of them are very sensitive to detect the presence of hot water. These preliminary studies have established reliable methods for hydrothermal survey that can accurately investigate the underground seepage of hot water with shallow depth resolution. We have successfully identified that the underground seepage of hot water in Blawan geothermal field is following the fault direction and river flow which is evidenced by some hot spring along the Banyu Pahit river with resistivity value less than 40 Ωm and medium conductivity.

1. Introduction

Indonesia has high potential of geothermal energy spread over 265 locations [1]. Since the geothermal resource exploration in Indonesia, the Blawan-Ijen geothermal field has been identified as a region with high geothermal potential approximately 110 MW. Unfortunately the Blawan-Ijen geothermal resource has not been developed for power generation until now. However, a significant rise in population and a great increase in energy and water demand lead to identifying and assessing Blawan-Ijen geothermal field for future commercial development. To determine the geothermal potential in Blawan-Ijen geothermal field, a study of subsurface conditions is required. Some geothermal researches in Blawan-Ijen have been done [2–4] but have never been done using geoelectric resistivity and ground scan penetrating radar (GPR) method.

The presence of water will reduce resistivity; meanwhile the presence of air in voids will increase subsurface resistivity [5, 6]. If the temperature rises, the resistivity value of rocks will be smaller [7]. However, the Ground Penetrating Radar (GPR) is a noninvasive electromagnetic geophysical technique with high resolution for identifying and mapping in subsurface exploration [8]. Detectability of objects in the ground depends upon several geological factors such as hot water, salt water, mineralogical clay, soils, and other electrical properties through conduction losses.

Delineation of underground seepage using geophysical methods has gained wide interests in the past few decades. Geoelectrical resistivity and GPR are the most frequently used geophysical techniques in exploring the groundwater. It offers quick and cost-effective imaging of the shallow subsurface with acceptable resolution [9]. Some uses of this method in groundwater are determination of the thickness, boundary, and depth of different layers of an aquifer [10–13]. Due to the successful application of geoelectrical resistivity and GPR over the years in groundwater exploration, this propels me to adopt the method to investigate the underground seepage of hot water in the study area. Therefore, the main aim of this study is to investigate the subsurface structure and to delineate the underground seepage of hot water in Blawan geothermal field.

2. Geological Setting

Ijen caldera complex shaped elliptical caldera wall with diameter about 15-16 km which resulted from Kendeng volcano eruption. The eruption left only the northern caldera arch, while the southern caldera arch was covered by volcanic sediment [14].

In the past, there is Blawan Lake that existed in the caldera Ijen after caldera formed. The geological conditions in this region allow rain water trapped inside the caldera basin to accumulate and eventually formed a lake that is quite spacious with a diameter greater than 5 km. The lake is in a fairly long period of time characterized by the thick presence of sedimentary clastic rocks near Banyu Pahit river, Sat Kali river, and Kali Sengon river [15].

The research area located in Blawan geothermal field, Bondowoso district, East Java, as shown in Figure 1. The hot springs distribution in Blawan geothermal field is shown in Figure 1. According to Sidarto, the hot springs flow in Blawan geothermal field in two different rock types, namely, Old Ijen volcanic rocks (Qpvi) and Young Ijen volcanic rock (Qhvi) [16]. The hot springs manifestations are clustered based on the discharge location of hot water [17].

3. Materials and Methods

The data acquisition strategy was to select acquisition site very close to hot water distribution (Figure 1), obtain typical resistivity evidence, and situate additional possible sites with similar morphological expression but unknown features. The entire measurement was carried out along the possible sites in the study area. To get good coupling between electrodes and ground, we scatter some water to ensure a wet surface condition. Based on hot spring distributions and possible location, we have determined 9 locations for geoelectrical resistivity and GPR field data acquisition (Figure 1).

3.1. Geoelectric Resistivity. Geoelectrical resistivity field data are acquired by dipole-dipole configuration using Mac-Ohm, OYO Japan Resistivity meters. The data was collected at various selected sites, namely, PB1, PB2, PB3, PB4, PB5, PB6, PB7, PB8, and PB9. Geoelectrical resistivity methods consist in a ground injection of an electrical current between two electrodes then measuring the induced potential difference between two potential electrodes [18, 19]. The dipole-dipole is the most convenient configuration on the field especially for large spacing [16]. The dipole-dipole resistivity technique consists of a collinear array with current dipole separation of length a, potential dipole separation of length a, and a total distance between the dipoles of length na (Figure 2). The

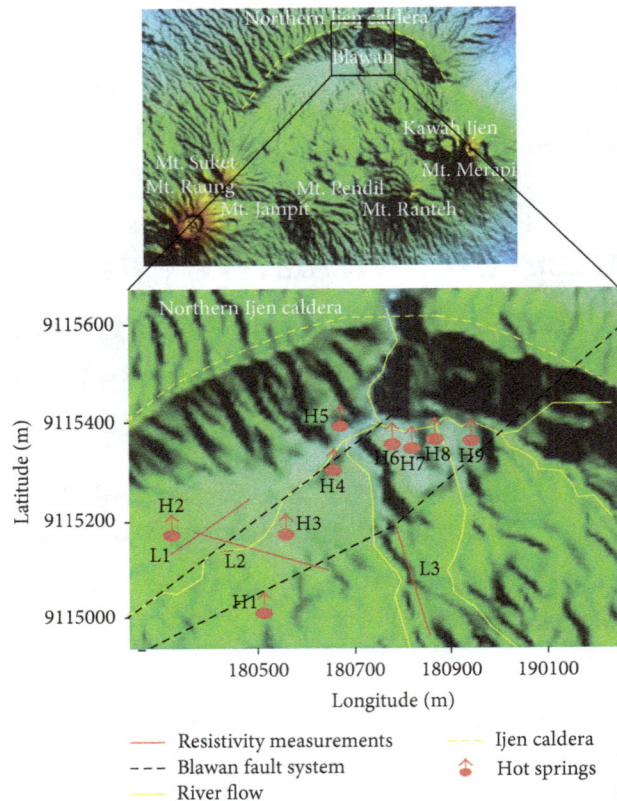

FIGURE 1: Research area.

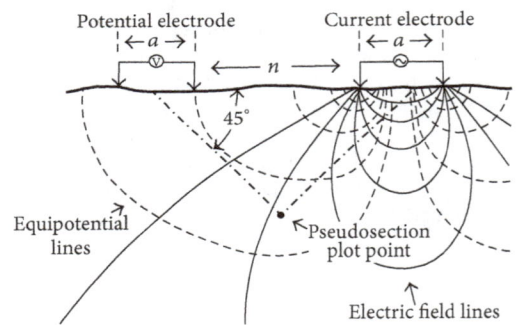

FIGURE 2: Dipole-dipole configuration [20].

current and potential electrodes are moved along a profile with constant spacing between electrodes. The apparent resistivity value is plotted along intersecting 45-degree lines centered on the dipoles [20].

Generally the data obtained during geoelectric field measurements classically presented as apparent resistivity pseudosections, which give an approximate picture of the subsurface resistivity [9]. The apparent resistivity equation applied in data processing is

$$\rho_a = \frac{V}{I}\pi a n(n+1)(n+2) = \frac{V}{I}K, \quad (1)$$

where is ρ_a apparent resistivity (Ωm); V is voltage; I is current; a is electrode spacing; n is datum point; K is geometry factor.

Geometry factor is very important to estimate the quality and quantity of resistivity data. The lower of a value will give decrease in K value and resulted in high quality data. The longer line measurement makes n value increase and then the K value will increase and resulted in increasing the depth measurement.

The apparent resistivity value is the resistivity of a homogeneous ground which will give the same resistance value for the same electrode arrangement. The relationship between the apparent resistivity and the true resistivity is a complex relationship. To determine the true subsurface resistivity, an inversion of the measured apparent resistivity values using a computer program software package will be used [21].

The processing and modelling of geoelectrical data was done with RES2DINV which provides inverse pseudosections of resistivity and reflects the true resistivity of the study area [18, 22]. It is a Windows based computer program that automatically determines a two-dimensional (2D) subsurface resistivity model for data obtained from electrical imaging surveys [23, 24]. The inversion procedure using RES2DINV software is based on the regularized least-squares optimization method [9, 25]. The conventional smoothness-constrained least-squares method attempts to minimize the square of the changes in the model resistivity values and to smooth the boundaries. As a result, a model with a smooth variation in the resistivity values is obtained [26]. 2D resistivity images prospecting field information about both lateral and vertical distributions of the study area's resistivity can be used in both qualitative and quantitative ways for the identification of underground seepage of hot water at shallow depths.

3.2. Ground Penetrating Radar (GPR). The GPR method is based on transmitting a high frequency electromagnetic pulse (radio waves) from a transmitting antenna to the subsurface to probe lossy dielectric material and recording of the pulse responses reflected from the interfaces and objects below the subsurface [27, 28]. GPR data are acquired using GPR Future 2005. Our GPR units consist of some components that are capable of being operated by a single user: the antenna, control unit, horizontal probe, USB Bluetooth dongle, PC to display, and external power supply (Figure 3). The Future 2005 is a multihead air coupled GPR instrument that comes packaged with a geophysical electromagnetic sensor (GEM). It has eight simultaneous sampling heads that cover an overlapping one-meter swath at the surface that provides multiple testing of anomalies. The radio wave produced is 450 MHz [29]. Future 2005 and its accessories serve for analysis of detecting objects deposited and changes performed in the ground.

The data was taken at the same locations as the geoelectric resistivity sites, that is, at PB1, PB2, PB3, PB4, PB5, PB6, PB7, and PB8 location. Ground scan operating mode is used in this survey. This operating mode records a graphical measurement of the measured data. The device will send out the impulses and we have to walk continuously in the measure line. If all impulses of the first measured line were sent out, the device will stop automatically [30].

The registered data of the ground structure in real time will be transmitted and stored digitally to a PC for visual

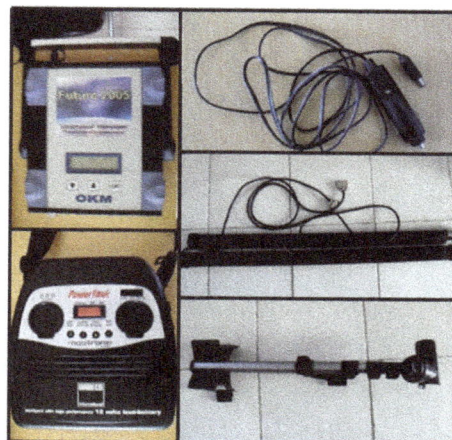

FIGURE 3: GPR Future 2005 set.

representation in a special software program. An anomaly is an area that has a sufficiently different wave signature to be identified as being separate from the background. Each anomaly detected below the surface reflects and inhibits the radio wave in unique ways [29]. The graphical representation should mainly include green, red, or blue color. The green color represents normal ground. Normally metallic objects are represented in red color and the blue color represents the cavities, water deposits, and diggings [30]. The focus of this study is on blue color which is considered as a hot water layer.

4. Results and Discussion

The data collected in the survey was interpreted by using RES2DINV 2D inversion software in which this software uses the rapid least-squares inversion method to model the final resistivity section [31]. The depth of the resistivity image depends on the distance between the electrodes, the used array, and the used equipment [32].

The 2D geoelectrical resistivity data have been processing and provide a description of rock layers with different resistivity values at each location. The 2D resistivity models at each location are grouped and shown in Figure 4.

Figure 4 shows the result resistivity data processing at 9 measurement locations. Sequentially, Figures 4(a), 4(b), 4(c), 4(d), 4(e), 4(f), 4(g), 4(h), and 4(i) are the result of resistivity data processing at locations PB1, PB2, PB3, PB4, PB5, PB6, PB7, PB8, and PB9. In Figure 3, low resistivity values ($<40\,\Omega$m) are dominant at locations PB2, PB5, PB6, and PB9. At locations PB1, PB3, and PB4, low resistivity values are also shown ($<40\,\Omega$m), but not as much as the previous location. At locations PB7 and PB8, despite low resistivity values, in those locations there is no manifestation of hot springs.

At location PB1, there are 3 resistivity and GPR lines in the same location, namely, PB1A (Figure 5(a)), PB1B (Figure 5(b)), and PB1C (Figure 5(c)). Figure 5 shows the data of geoelectric resistivity and GPR on PB1; GPR results are shown in the horizontal direction, while the results of geoelectric

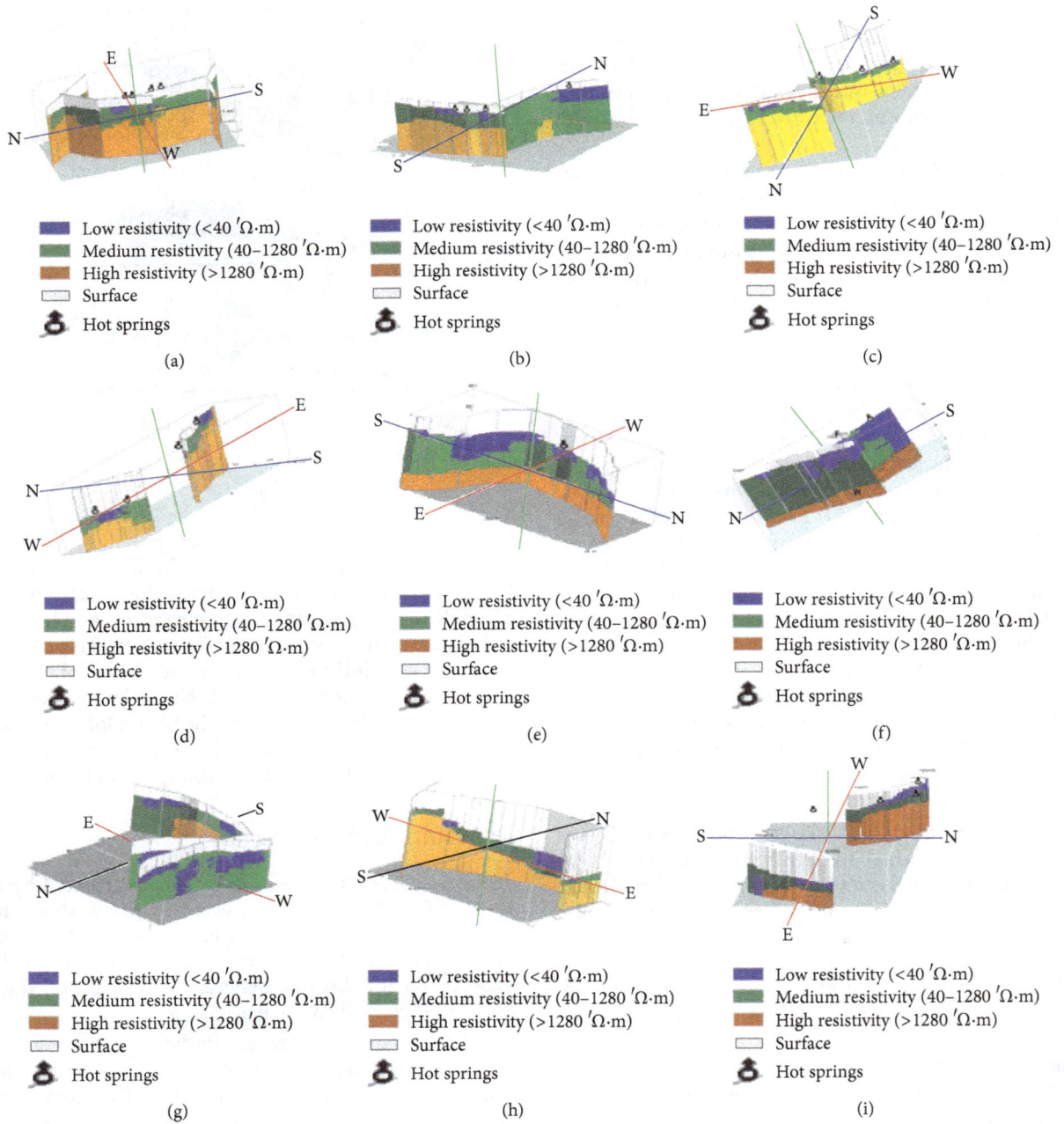

FIGURE 4: Result of resistivity data processing at 9 measurement locations in Blawan geothermal field.

resistivity are shown in the vertical direction. The total anomaly length of geoelectric resistivity is approximately 100 m and the depth range of the anomalies seems to be more than 25 m. Based on the results of geoelectrical data, processing can be interpreted that the subsurface conditions at the location of PB1 are composed of a layer of silty sandstone that contains a lot of water indicated by the green layer (40–1280 Ωm) and lava rock represented by orange layer (>1280 Ωm), and the blue layer (<40 Ωm) indicates a layer that contains hot water. Figure 5 shows that the rock layers

that contain hot water only present in PB1B (Figure 5(b)) at the horizontal profile of 55 m–60 m and extend down up to 5 m of depth under the surface, also at the horizontal profile of 75 m–80 m, and extend down up to 10 m of depth under the surface. In PB1C line (Figure 5(c)), there is also the rock layer containing hot water which is present at the horizontal profile of 20 m–25 m and reaches the depth of 5 m under the surface.

Meanwhile, GPR processing results show the ground scan of PB1 location with an estimated maximum depth until 5 m under the surface. The PB1 is dominated by green and blue

Model resistivity
with topography

PB1A (dipole-dipole configuration,
Blawan-Ijen Bondowoso)

Resistivity ($'\Omega \cdot m$)

(a)

Model resistivity
with topography

PB1B (dipole-dipole configuration,
Blawan-Ijen Bondowoso)

Resistivity ($'\Omega \cdot m$)

(b)

Model resistivity
with topography

PB1C (dipole-dipole configuration,
Blawan-Ijen Bondowoso)

Resistivity ($'\Omega \cdot m$)

(c)

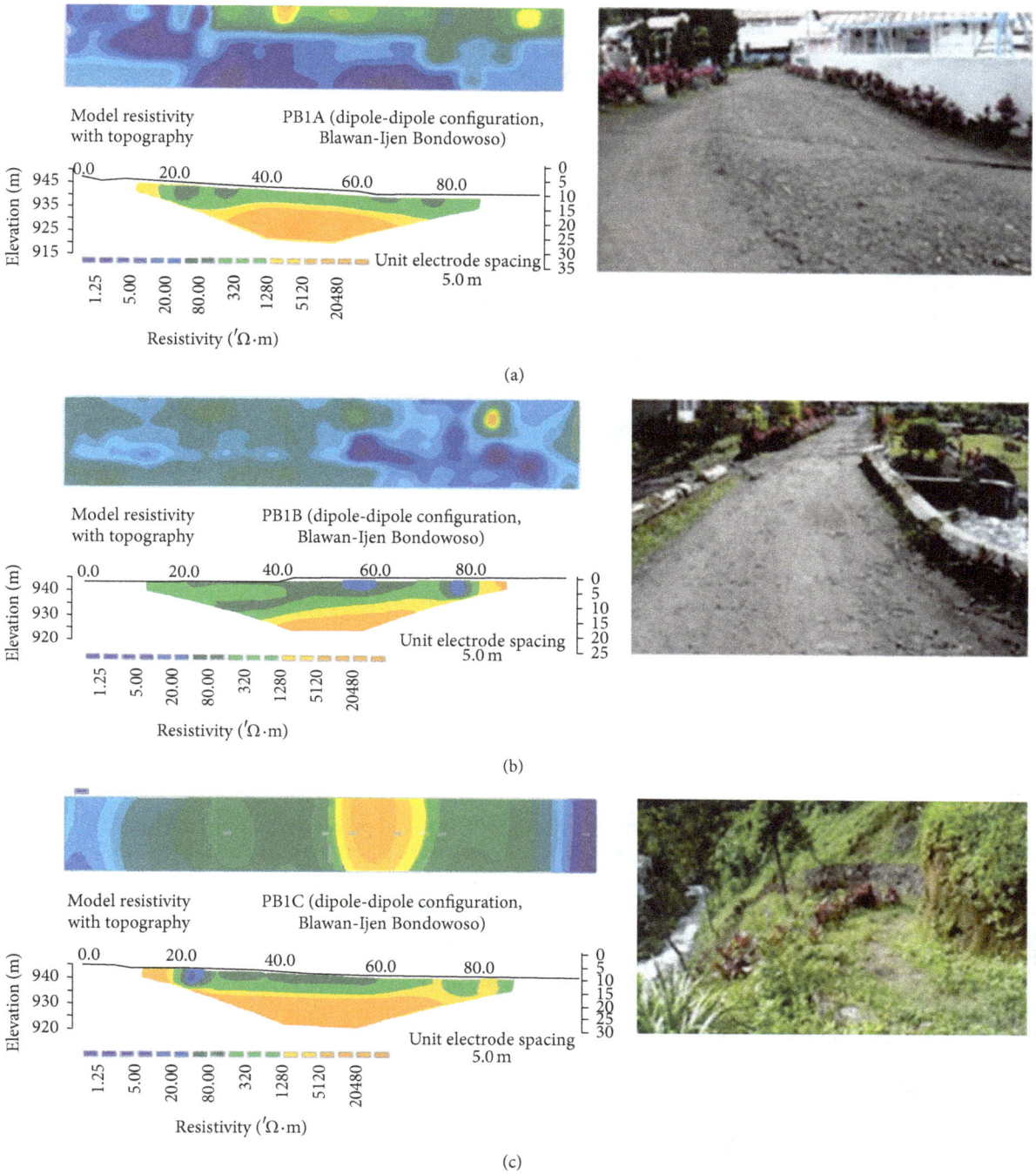

FIGURE 5: 2D Resistivity Mapping and GPR amplitude slice in locations (a) PB1A, (b) PB1B, and (c) PB1C.

layers that represent the layers of rock containing water. Conditions at the site showed that at PB1 there is a river flow in accordance with the GPR results. Yellow to orange layer from GPR processing results is a layer of dry solid and indicated by the presence of hills.

Location PB2 consists of 2 resistivity lines, namely, PB2A (Figure 6(a)) and PB2B (Figure 6(b)), and a line of GPR in the same direction with PB2B (Figure 6(b)). The total anomaly length of geoelectric resistivity results is approximately 100 m and the depth range of the anomalies reaches up to 25 m. Based on the results of data processing geoelectric resistivity in Figure 6, it can be interpreted that the blue layer (<40 Ωm) is a type of rock layers that contain hot water. The results of resistivity data processing are supported by the presence of hot springs on the location. In fact, PB2A line contained hot water layer parallel to 3 hot springs; they are present

(a)

Horizontal scale is 64.25 pixels per unit spacing
Vertical exaggeration in model section = 0.52
First electrode is located at 0.0 m
Last electrode is located at 100.0 m

(b)

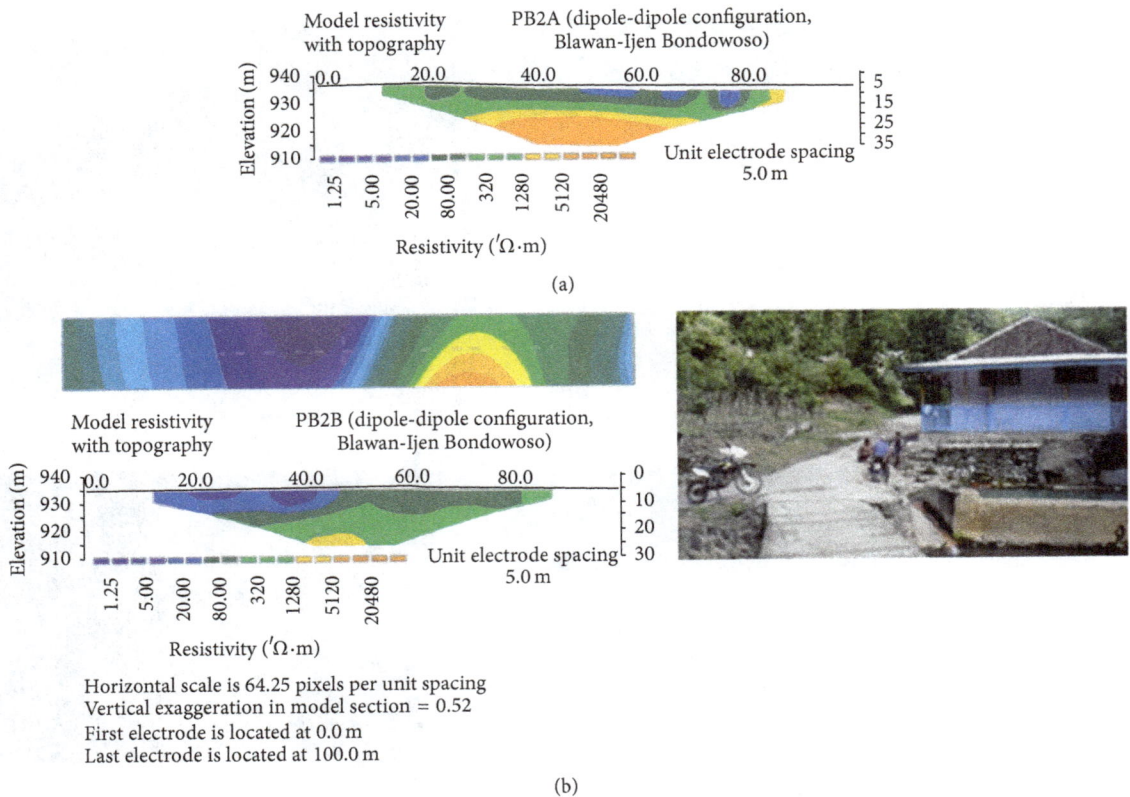

FIGURE 6: 2D Resistivity Mapping and GPR amplitude slice in locations (a) PB2A and (b) PB2B.

along 50 m–60 m and 65 m–70 m of horizontal profile with 5 m of depth under the surface and also along 50 m–60 m and 75 m–80 m of horizontal profile with 10 m of depth under the surface. However, the hot water in PB2B line is parallel to 1 hot spring, along 12 m–48 m horizontal profile, and reaches up to 15 m of depth under the surface. There is also a silty sandstone layer that contains a lot of water which is indicated by the green layer (40–1280 Ωm). In addition, the deeper subsurface is estimated consisting of lava rocks indicated by the orange layer (>1280 Ωm).

As supporting data, we used the GPR method taken in the same location with PB2B line. GPR results show the ground scan of PB2 location with an estimated maximum depth until 5 m under the surface. Green and blue layers in GPR results indicate a layer containing water at PB2 location. These results are in accordance with the conditions on the survey area which indicates the flow of the river. The presence of hot springs, indicated by the blue layer, is found in the GPR result. Meanwhile, orange layer on the results of the GPR is a representation of nonaqueous layer.

PB3 site consists of 4 resistivity lines, namely, PB3A (Figure 7(a)), PB3B (Figure 7(b)), PB3C (Figure 7(c)), and PB3D (Figure 7(d)). The result of resistivity data processing in Figure 7 shows that the subsurface at location PB3 is composed of hot water layer which is indicated by the blue layer (<40 Ωm). In PB3A line, the hot water layer is present along 11 m–12.5 m, 27 m–30 m, 32 m–35 m, and 38 m-39 m

of horizontal profile with 3 m of thickness. In PB3B line, the hot water layer is present along 22 m–23 m, 32 m–35 m, and 76 m–79 m of horizontal profile with 5 m of thickness. The presence of hot water layer in Figure 7(b) is supported by hot springs located in PB3A. The green layer (40–1280 Ωm) is a representation of sand that fills with water. In PB3D line, the hot water layer is present along 40 m–43 m of horizontal profile with 2.5 m of thickness and also along 55 m–65 m and 71 m–83 m of horizontal profile with 12.5 m of thickness. Overall, silty sandstone layers (40–1280 Ωm) contained water presence in all geoelectric resistivity line. Orange layer (>1280 Ωm) is a representation of lava rock present in all lines with a varied depth.

As supporting data, GPR data are taken exactly at the PB3A location (Figure 7(a)). GPR results show the ground scan of PB3 location with an estimated maximum depth until 5 m under the surface. Green and blue layers in GPR results indicate a layer containing water at PB3 location. GPR processing results as shown in Figure 7(a) show the predominance of blue layer which means that there is a layer which contains a lot of water above the hot springs and supports the presence of hot water.

Location PB4 consists of 2 resistivity lines, namely, PB4A (Figure 8(a)) and PB4C (Figure 8(b)). PB4C line is located in a higher elevation than the PB4A line. Each line has the manifestation of hot springs around. The total anomaly length of geoelectric resistivity results is approximately 100 m and

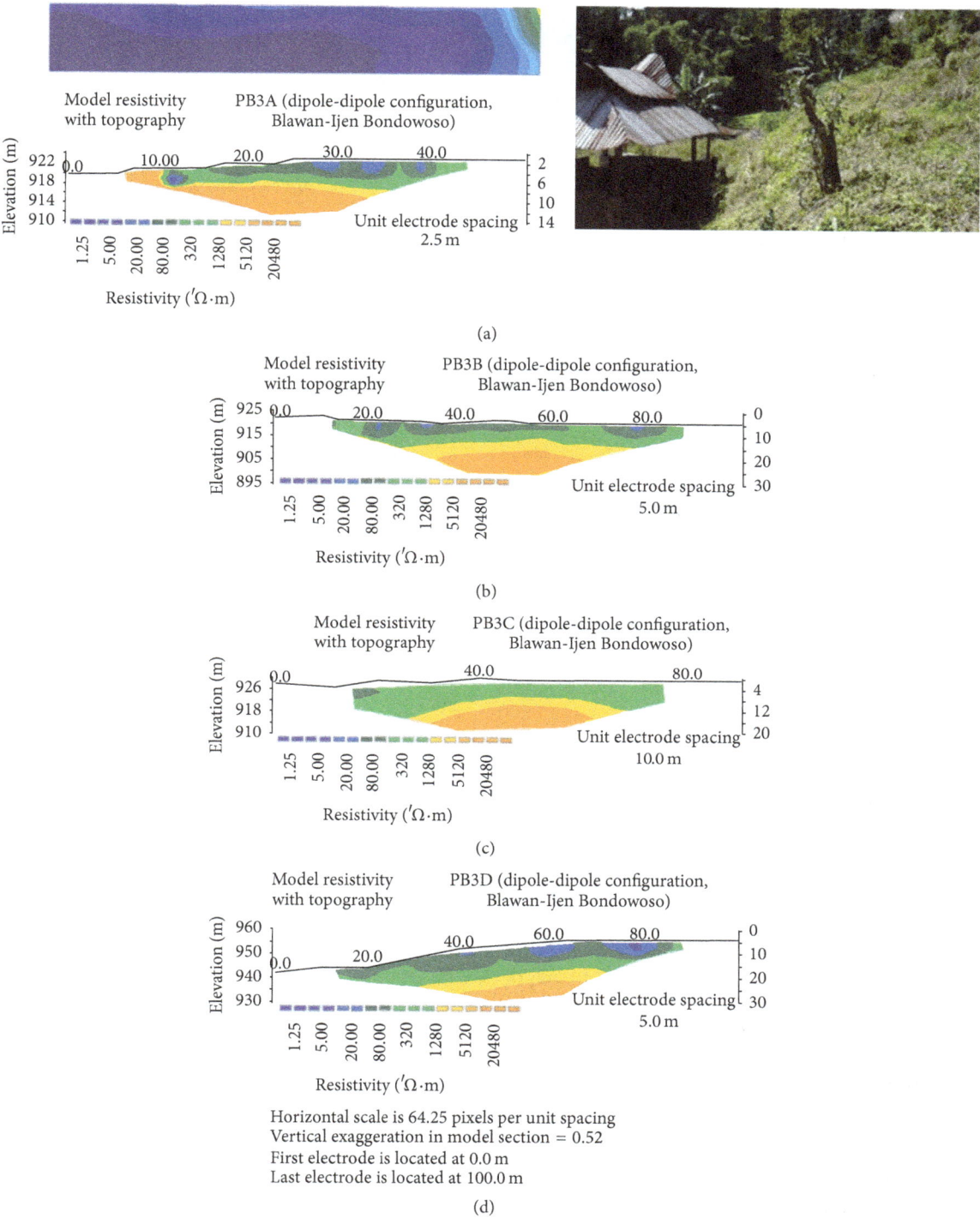

Model resistivity
with topography

PB3A (dipole-dipole configuration,
Blawan-Ijen Bondowoso)

Unit electrode spacing
2.5 m

Resistivity ($'\Omega\cdot$m)

(a)

Model resistivity
with topography

PB3B (dipole-dipole configuration,
Blawan-Ijen Bondowoso)

Unit electrode spacing
5.0 m

Resistivity ($'\Omega\cdot$m)

(b)

Model resistivity
with topography

PB3C (dipole-dipole configuration,
Blawan-Ijen Bondowoso)

Unit electrode spacing
10.0 m

Resistivity ($'\Omega\cdot$m)

(c)

Model resistivity
with topography

PB3D (dipole-dipole configuration,
Blawan-Ijen Bondowoso)

Unit electrode spacing
5.0 m

Resistivity ($'\Omega\cdot$m)

Horizontal scale is 64.25 pixels per unit spacing
Vertical exaggeration in model section = 0.52
First electrode is located at 0.0 m
Last electrode is located at 100.0 m

(d)

FIGURE 7: 2D Resistivity Mapping and GPR amplitude slice in locations (a) PB3A, (b) PB3B, (c) PB3C, and (d) PB3D.

150 m and the depth range of the anomalies reaches up to 25 m. The hot water flow shown in blue layer (<40 Ωm) in PB4A line (Figure 8(a)) is present along 21 m-22 m of horizontal profile and extends down up to 5 m of depth under the surface. While, in PB4C line (Figure 8(b)), the hot water is present along 22 m–24 m, 32 m–34 m, 115 m–120 m, and 126 m-127 m of horizontal profile with 7.5 m of thickness, in PB4C line, its subsurface is formed by a layer which is interpreted as the presence of limestone and allows the flow of hot water in the rocks crack. PB4C line taken in Damarwulan and adjacent to the limestone cave was once at the deepest part of Blawan Lake.

(a)

(b)

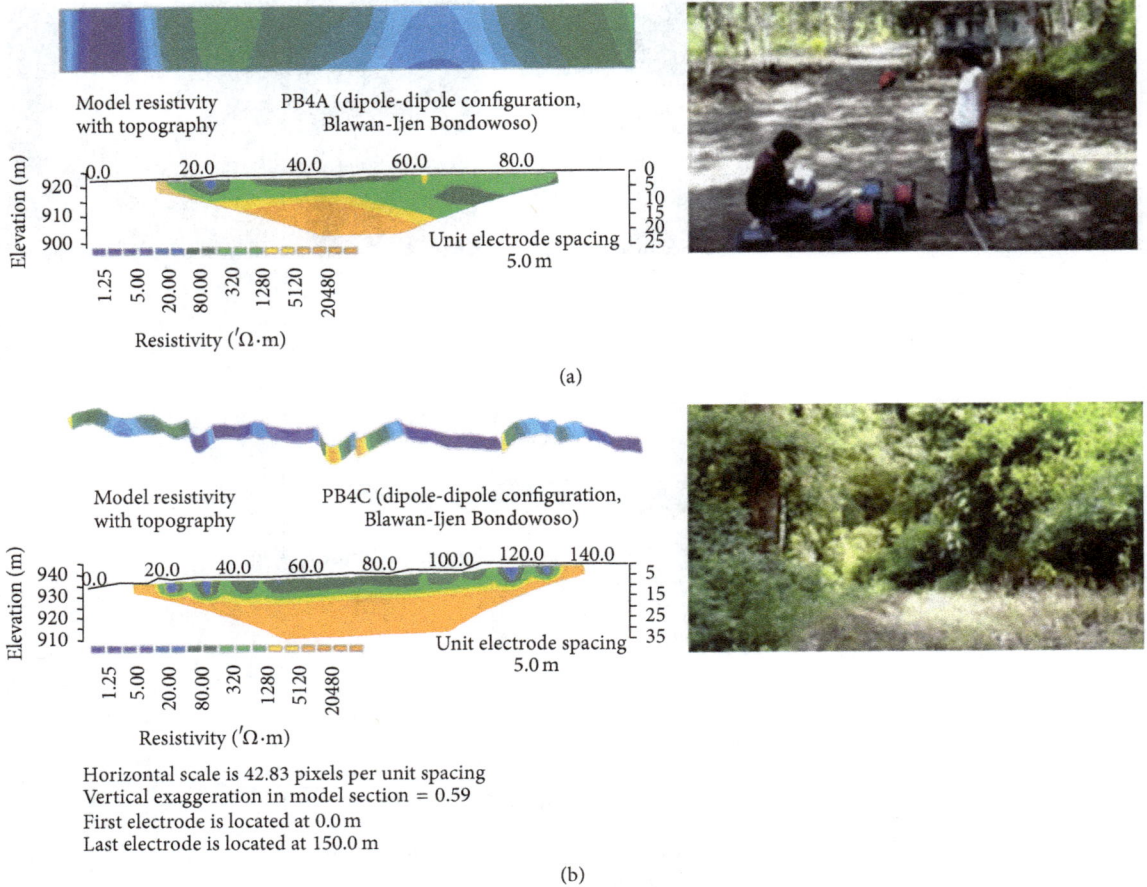

FIGURE 8: 2D Resistivity Mapping and GPR amplitude slice in locations (a) PB4A and (b) PB4C.

As geoelectric resistivity supporting data, GPR method is used in the same location with resistivity. GPR results show the ground scan of PB4 location with an estimated maximum depth until 5 m under the surface. Green and blue layers in GPR results indicate a layer containing water at PB4 location. Interpretation of the dominance blue and green layers in Figures 8(a) and 8(b) shows a layer that contains a lot of water. Blue colors in GPR result match with those in resistivity result, which means there is a flow of hot water.

There are 2 resistivity lines in PB5, namely, PB5A and PB5B. The total anomaly length of geoelectric resistivity is approximately 50 m and the depth range of the anomalies is up to 10 m. Figure 9(a) is the result of PB5A data processing, and the blue layer (<40 Ωm) indicates the hot water presence along 9 m–15 m, 22 m–26 m, 29 m–34 m, and 46 m–39 m of horizontal profile, close to the surface until 3 m of depth. The presence of hot water layer at PB5A is supported by the presence of the hot springs. PB5A line also consists of the aquifer layer (40–1280 Ωm) located above the lava rock layers (>1280 Ωm). Figure 9(b) is the result of PB5B data processing showing the same result as the PB5A line. The hot water in PB5B line (Figure 9(b)) is present along 19 m–43 m of horizontal profile and reaches up to 6 m of depth from the surface. The elevation of PB5B line is higher than PB5A, so hot water is

gathering towards PB5A as a lower location. Figure 9(b) also shows the result of GPR including the condition of the survey location. GPR results show the ground scan of PB5 location with an estimated maximum depth until 5 m under the surface. Green and blue layers in GPR results indicate a layer containing water at PB5 location. The results of GPR dominated with blue layer indicate a layer which contains hot water.

PB6 site consists of 2 resistivity lines, namely, PB6A and PB6B. Total depth generated by each line is different, 25 m for PB6A line and 14 m for PB6B line depending on the separation of the current and potential electrodes used in PB6A line (100 m) and PB6B line (50 m). The resistivity result of PB6A (Figure 10(a)) shows the hot water layer (<40 Ωm) presence along 13 m–41 m, 50 m–55 m, and 75 m–79 m of horizontal profile and up to 15 m of depth under the surface. However, in PB6B (Figure 10(b)) the hot water layer almost covers all of horizontal profile and up to 9 m of depth under the surface. Besides that, the subsurface in PB6 is formed by silty sandstone layer (40–1280 Ωm) and lava rock layers (>1280 Ωm).

At PB6 there are 3 GPR lines, 2 of them taken in the same position with PB6A (Figure 10(a)) and PB6B (Figure 10(b)) and the third one located in the way towards hot springs (Figure 10(c)). The GPR results indicate the presence of the hot water flow support PB6A resistivity result. GPR results

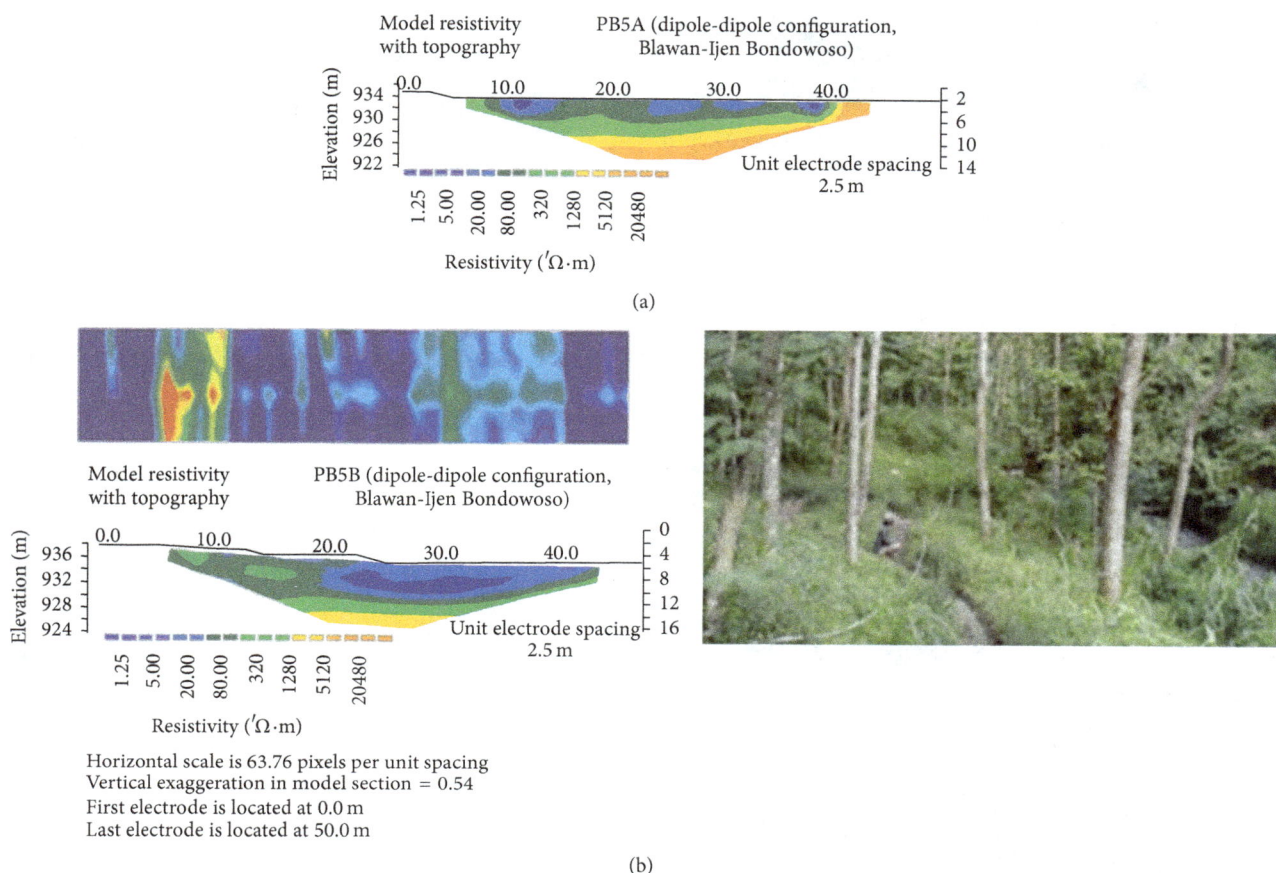

FIGURE 9: 2D Resistivity Mapping and GPR amplitude slice in locations (a) PB5A and (b) PB5B.

show the ground scan of PB6 location with an estimated maximum depth until 5 m under the surface. Green and blue layers in GPR results indicate a layer containing water at PB6 location. The blue color in Figure 10(c) shows the presence of a hot spring.

At the location of PB7 site, there are 3 resistivity lines, namely, PB7A (Figure 11(a)), PB7B (Figure 11(b)), and PB7C (Figure 11(c)). The total anomaly length of geoelectric resistivity is approximately 100 m and the depth range of the anomalies is up to 25 m. This region is dominant with groundwater recharge demonstrated by the construction citizen's water reservoir. In all of resistivity lines, it seems that there is hot water layer which is indicated by the blue layer (<40 Ωm). The resistivity result of PB7A (Figure 11(a)) shows the hot water layer (<40 Ωm) presence along 12 m–59 m of horizontal profile and spreads over from the surface up to 25 m of depth under the surface, also along 77 m–87 m of horizontal profile with 15 m of thickness. However, in PB7B (Figure 11(b)), the hot water layer almost covers all of horizontal profile and spreads over up to 20 m of depth under the surface. The hot water layer in PB7C (Figure 11(c)) is present along 20 m–26 m and 62 m–85 m of horizontal profile and extends down up to 7 m of depth under the surface. Overall, the subsurface in PB7 is formed by aquifer layer (40–1280 Ωm) as shown in

Figures 11(a) and 11(b) by the green layer. But Figure 11(c) has different characteristics; there is an intrusion lava rocks layer (>1280 Ωm) in PB7C line. The hot water coming from the hot water layer in PB7 will flow toward a lower place. This possibility is evidenced by the presence of hot springs in PB6 which has lower elevation than PB7.

As geoelectric resistivity supporting data, GPR method is used in the same location with resistivity line. GPR results show the ground scan of PB7 location with an estimated maximum depth until 5 m under the surface. Green and blue layers in GPR results indicate a layer containing water at PB7 location. GPR data interpretation on PB7 (Figure 11) shows the presence of potential hot water on this line. In accordance with the results of geoelectric resistivity, PB7A, PB7B, and PB7C are formed by aquifer layers.

At location PB8, there are 3 resistivity lines, namely, PB8A (Figure 12(a)), PB8B (Figure 12(b)), and PB8C (Figure 12(c)). The subsurface is dominated by a high resistivity rock (>1280 Ωm) shown in Figure 12. The presence of hot water layer (<40 Ωm) is less than the igneous layers (>1280 Ωm). In PB8B (Figure 12(b)), the hot water layer is present along 20 m–24 m and 77 m–84 m of horizontal profile with 5 m of thickness, while the hot water layer in PB8C (Figure 12(c)) is present along 10.5 m–12.5 m and 15 m–23 m of horizontal

(a)

(b)

(c)

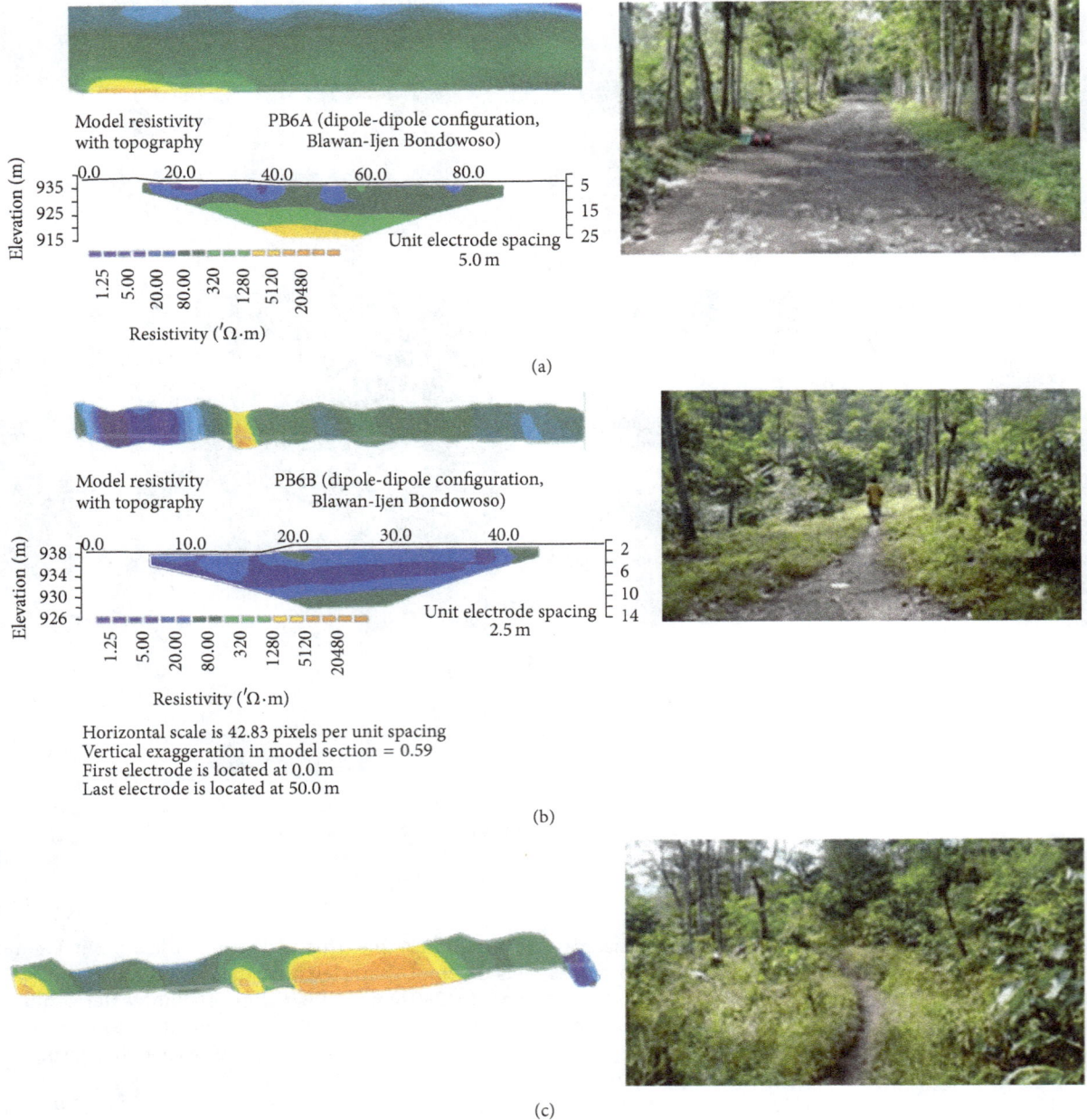

FIGURE 10: 2D Resistivity Mapping and GPR amplitude slice in locations (a) PB6A, (b) PB6B, and (c) PB6C.

profile with 3 m of thickness. It is estimated that the hot water is towards a lower place in PB4. This assumption is strengthened by the presence of hot spring in PB4.

As resistivity supporting data, GPR data are taken in all of resistivity lines. GPR results show the ground scan of PB8 location with an estimated maximum depth until 5 m under the surface. Green and blue layers in GPR results indicate a layer containing water at PB7 location. The GPR results in Figure 12 show the dominance of the aquifer layer there in all lines.

PB9 site consists of 2 resistivity lines, namely, PB9A and PB9B. The total anomaly length of geoelectric resistivity is approximately 50 m and the depth range of the anomalies is up to 12 m. The subsurface in PB9A line (Figure 13(a)) shows the presence of low resistivity values (<40 Ωm) which is a representation of the hot water. There is a hot spring along PB9A line along 11 m–13 m of horizontal profile and it reaches up to 3 m of depth under the surface. The result of PB9B resistivity line (Figure 13(b)) shows that aquifer layer (40–1280 Ωm) and hot water layer (<40 Ωm) form near the surface with 3 m of

Model resistivity
with topography

PB7A (dipole-dipole configuration,
Blawan-Ijen Bondowoso)

Resistivity ($'\Omega \cdot m$)

(a)

Model resistivity
with topography

PB7B (dipole-dipole configuration,
Blawan-Ijen Bondowoso)

Resistivity ($'\Omega \cdot m$)

(b)

Model resistivity
with topography

PB7C (dipole-dipole configuration,
Blawan-Ijen Bondowoso)

Resistivity ($'\Omega \cdot m$)

Horizontal scale is 64.25 pixels per unit spacing
Vertical exaggeration in model section = 0.52
First electrode is located at 0.0 m
Last electrode is located at 100.0 m

(c)

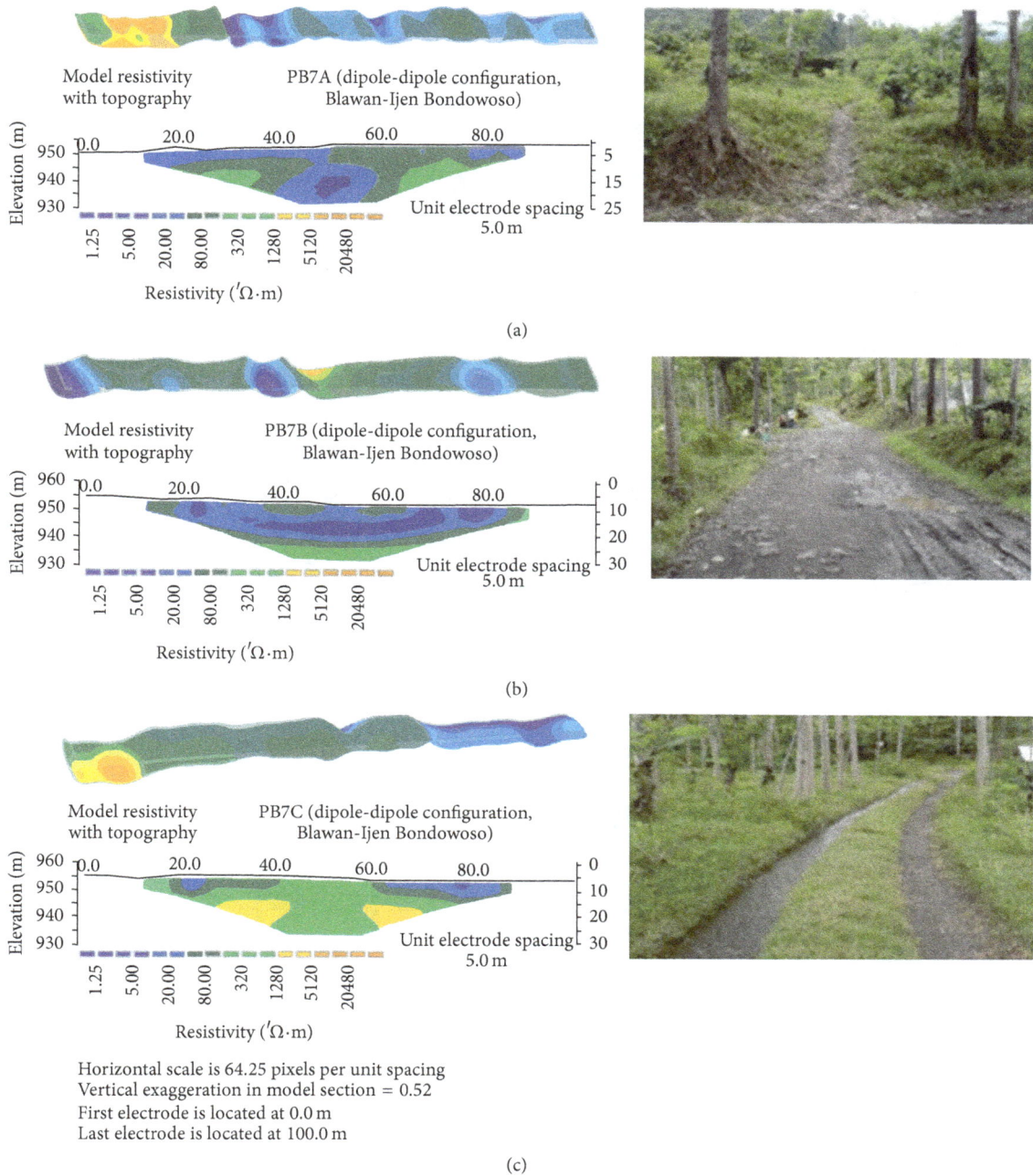

FIGURE 11: 2D Resistivity Mapping and GPR amplitude slice in locations (a) PB7A, (b) PB7B, and (c) PB7C.

thickness. Meanwhile, the bottom layer of the subsurface is lava rock with high resistivity values (>1280 Ωm).

In Figure 14, the low resistivity values appear clumped and spread. Based on visual observations, the hot springs are distributed to the right and left of the hill and follow the river flow. In accordance with the characteristic of water that flows to a lower level, the underground seepage of hot water is towards the Northeast (Kendeng Caldera Arc) that has a lower topography with fault system and the presence of a waterfall.

5. Conclusions

In Blawan area, low resistivity values (<40 Ωm) are interpreted as a hot water layer. Furthermore, the resistivity values between 40 and 1280 Ωm are interpreted as aquifer layers and the high resistivity values (>1280 Ωm) are interpreted as lava rock. The underground seepage of hot water in Blawan geothermal field follows existing fault. Distribution of hot springs mostly to the Northeast follows the river flow pattern. Some hot springs located along the river showed that

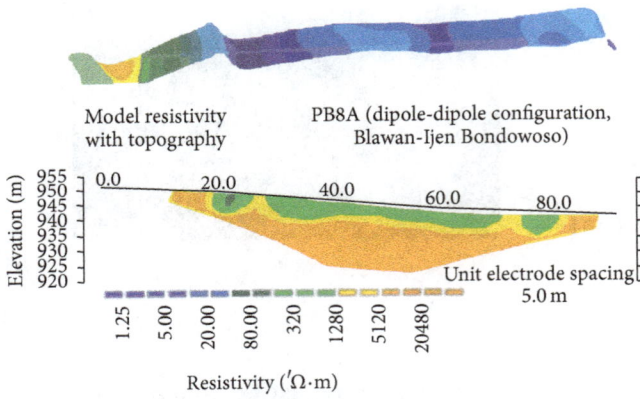

(a)

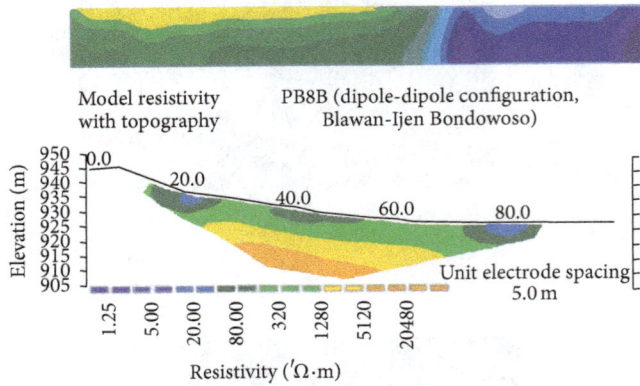

(b)

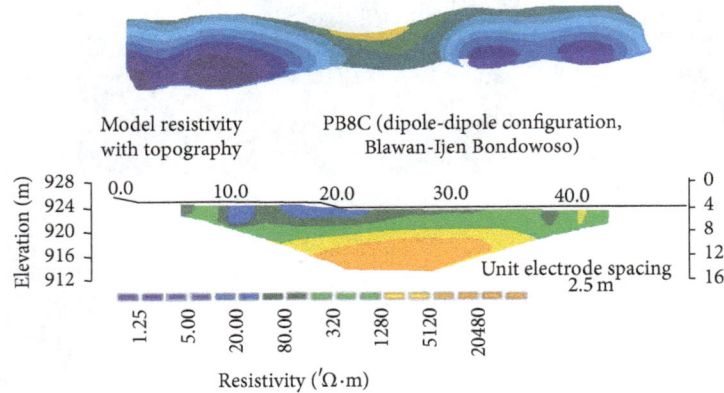

Horizontal scale is 63.70 pixels per unit spacing
Vertical exaggeration in model section = 0.54
First electrode is located at 0.0 m
Last electrode is located at 50.0 m

(c)

FIGURE 12: 2D Resistivity Mapping and GPR amplitude slice in locations (a) PB8A, (b) PB8B, and (c) PB8C.

underground seepage of hot water is impermeable layer with a resistivity value less than 40 Ωm.

Competing Interests

The authors declare that there are no competing interests regarding the publication of the paper.

Acknowledgments

Thanks are due to the Geophysical Laboratory of Physics, University of Brawijaya, and staff who helped in the study. Thanks are due to Blawan-Ijen Team from Indonesia Multimedia and all of friends for their help and cooperation during data acquisition. This study was partially funded by PUPT, PHK, and USAID.

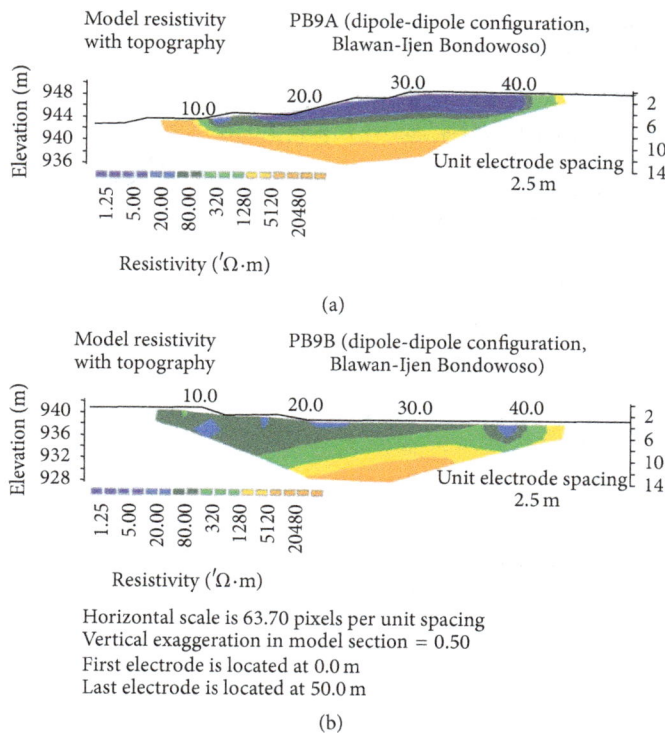

FIGURE 13: 2D Resistivity Mapping and GPR amplitude slice in locations (a) PB9A and (b) PB9B.

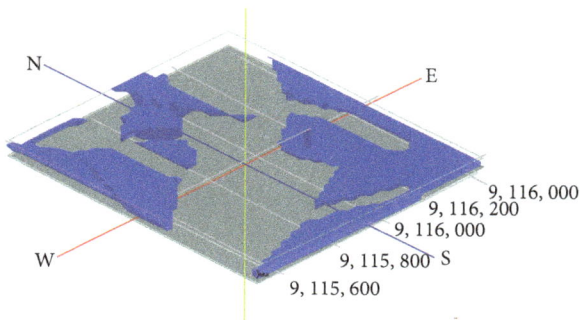

FIGURE 14: The underground seepage of hot water in Blawan geothermal field.

References

[1] Department of Energy and Mineral Resources, *Geothermal Potential in East Java*, 2012, http://esdm.jatimprov.go.id/esdm/attachments/article/37/Data%20Eka-potensi%20panasbumi%20jatim%202012.pdf.

[2] A. Afandi, S. Maryanto, and A. Rachmansyah, "Identification reservoir geothermal based on geomagnetic method in area blawan sempol district bondowoso," *Journal of Neutrino*, vol. 6, no. 1, pp. 1–10, 2013.

[3] R. A. Raehanayati and S. Maryanto, "Study of Blawan-Ijen geothermal energy potential, East Java based on gravity method," *J. Neutrino*, vol. 6, no. 1, pp. 31–39, 2013.

[4] C. N. Dewi, S. Maryanto, and A. Rachmansyah, "Blawan geothermal system, East Java based on Magnetotelluric survey,"

[5] D. Santoso, *Introduction to Geophysical Engineering*, ITB Press, Bandung, Indonesia, 2002.

[6] P. Vingoe, *Electrical Resistivity Surveying*, Atlas Copco ABEM, Stockholm, Sweden, 1972.

[7] A. S. Ogungbe, J. A. Olowofela, O. O. Oresanya, and A. A. Alabi, "Mapping of unconfined aquifer using vertical electrical sounding (VES) at Lagos State University (Lasu), Ojo," *Applied Science Research*, vol. 2, no. 2, pp. 24–34, 2010.

[8] A. A. R. Zohdy, "A new method for differential resistivity sounding," *Geophysics*, vol. 34, no. 6, pp. 924–943, 1969.

[9] M. Metwaly and F. Alfouzan, "Application of 2-D geoelectrical resistivity tomography for subsurface cavity detection in the eastern part of Saudi Arabia," *Geoscience Frontiers*, vol. 4, no. 4, pp. 469–476, 2013.

[10] G. D. Khan, Waheedullah, and A. S. Bhatti, "Groundwater investigation by using resistivity survey in Peshawar, Pakistan," *Journal of Resource Development and Management*, vol. 2, pp. 9–20, 2013.

[11] A. A. R. Zohdy and D. B. Jackson, "Application of deep electrical soundings for groundwater exploration in Hawaii," *Geophysics*, vol. 34, no. 4, pp. 584–600, 1969.

[12] M. E. Young, R. G. M. De Bruijn, and A. Bin Salim Al-Ismaily, "Exploration of an alluvial aquifer in Oman by time-domain electromagnetic sounding," *Hydrogeology Journal*, vol. 6, no. 3, pp. 383–393, 1998.

[13] P. M. Soupios, M. Kouli, F. Vallianatos, A. Vafidis, and G. Stavroulakis, "Estimation of aquifer hydraulic parameters from surficial geophysical methods: a case study of Keritis Basin in Chania (Crete-Greece)," *Journal of Hydrology*, vol. 338, no. 1-2, pp. 122–131, 2007.

Jurnal RISET Geologi dan Pertambangan, vol. 25, no. 2, p. 111, 2015.

[14] A. Neal, "Ground-penetrating radar and its use in sedimentol-ogy: principles, problems and progress," *Earth-Science Reviews*, vol. 66, no. 3-4, pp. 261–330, 2004.

[15] A. Zaennudin, D. Wahyudin, M. Surmayadi, and E. Kusdinar, "Forecasts of volcanic eruption danger Ijen in East Java," *Jurnal Lingkungan dan Bencana Geologi*, vol. 3, no. 2, pp. 109–132, 2012.

[16] S. T. Sidarto and D. Sudana, *Geological Map of the Quadrangle Banyuwangi Java 1707-4, Scale 1:100.000*, Geological Research and Development Center, Bandung, Indonesia, 1993.

[17] K. Sitorus, *Volcanic stratigraphy and geochemistry of the Ijen Caldera Complex, East Java, Indonesia [M.S. thesis]*, Victoria University of Wellington, Wellington, New Zealand, 1990.

[18] N. P. Claude, N.-M. Théophile, A. S. Patrick, and K. T. Crepin, "Evidence of iron mineralization channels in the Messondo area (Centre-Cameroon) using geoelectrical (DC & IP) methods: a case study," *International Journal of Geosciences*, vol. 5, no. 3, pp. 346–361, 2014.

[19] M. N. Tijani, O. O. Osinowo, and O. Ogedengbe, "Mapping of sub-surface fracture systems using integrated electrical resistiv-ity profiling and VLF-EM methods: a case study of suspected gold mineralization," *RMZ—Materials and Geoenvironment*, vol. 56, pp. 415–436, 2009.

[20] A. S. Ogungbe, J. A. Olowofela, O. J. Da-Silva, A. A. Alabi, and E. O. Onori, "Subsurface characterization using electrical resistivity (Dipole-Dipole) method at Lagos State University (LASU) Foundation School, Badagry," *Applied Science Research*, vol. 1, no. 1, pp. 131–145, 2010.

[21] J. E. Sunday, *Appication of geolectrical resisitivity imaging to investigate groundwater potential in Atan, Ogun State Southwest-ern Nigeria [M.S. thesis]*, College of Science and Technology, Covenant University, Ota, Nigeria, 2012.

[22] M. H. Loke and R. D. Barker, "Rapid least-squares inversion of apparent resistivity pseudosections by a Quasi-Newton method," *Geophysical Prospecting*, vol. 44, no. 1, pp. 131–152, 1996.

[23] M. S. Barseem, T. A. A. El Lateef, H. M. E. El Deen, and A. A. A. A. Abdel Rahman, "Geoelectrical exploration in South Qantara Shark area for supplementary irrigation purpose-Sinai-Egypt," *Hydrology: Current Research*, vol. 6, no. 2, pp. 1–10, 2015.

[24] D. H. Griffiths and R. D. Barker, "Two-dimensional resistivity imaging and modelling in areas of complex geology," *Journal of Applied Geophysics*, vol. 29, no. 3-4, pp. 211–226, 1993.

[25] M. Loke, I. Acworth, and T. Dahlin, "A comparison of smooth and blocky inversion methods in 2D electrical imaging surveys," *Exploration Geophysics*, vol. 34, no. 3, pp. 182–187, 2003.

[26] M. Ezersky, "Geoelectric structure of the Ein Gedi sinkhole occurrence site at the Dead Sea shore in Israel," *Journal of Applied Geophysics*, vol. 64, no. 3-4, pp. 56–69, 2008.

[27] E. Novakova, M. Karous, A. Zajícek, and M. Karousova, "Evaluation of ground penetrating radar and vertical electrical sounding methods to determine soil horizons and bedrock at the locality dehtaře," *Soil and Water Research*, vol. 8, no. 3, pp. 105–112, 2013.

[28] A. P. Annan, "GPR—history, trends, and future developments," *Subsurface Sensing Technologies and Applications*, vol. 3, pp. 253–270, 2002.

[29] S. D. Smith, J. B. Legg, T. S. Wilson, and J. Leader, *Obstinate and Strong : The History and Archaeology of the Siege of Fort Motte*, University of South Carolina, Columbia, SC, USA, 2007.

[30] OKM Ortungstechnik GmbH, *User's Manual: FUTURE 2005*, 2007, http://www.okm-gmbh.de.

[31] L. B. Conyers and J. Leckebusch, "Geophysical archaeology research agendas for the future: some ground-penetrating radar examples," *Archaeological Prospection*, vol. 17, no. 2, pp. 117–123, 2010.

[32] M. H. Loke and R. D. Barker, "Practical techniques for 3D resis-tivity surveys and data inversion," *Geophysical Prospecting*, vol. 44, no. 3, pp. 499–523, 1996.

6

Discovery of Naturally Etched Fission Tracks and Alpha-Recoil Tracks in Submarine Glasses: Reevaluation of a Putative Biosignature for Earth and Mars

Jason E. French[1] and David F. Blake[2]

[1]Department of Earth and Atmospheric Sciences, University of Alberta, 1-26 Earth Science Building,
Edmonton, Alberta, Canada T6G 2E3
[2]NASA Ames Research Center, Exobiology Branch, MS 239-4, Moffett Field, CA 94035-1000, USA

Correspondence should be addressed to Jason E. French; jef@ualberta.ca

Academic Editor: Ghaleb H. Jarrar

Over the last two decades, conspicuously "biogenic-looking" corrosion microtextures have been found to occur globally within volcanic glass of the *in situ* oceanic crust, ophiolites, and greenstone belts dating back to ~3.5 Ga. These so-called "tubular" and "granular" microtextures are widely interpreted to represent *bona fide* microbial trace fossils; however, possible nonbiological origins for these complex alteration microtextures have yet to be explored. Here, we reevaluate the origin of these enigmatic microtextures from a strictly nonbiological standpoint, using a case study on submarine glasses from the western North Atlantic Ocean (DSDP 418A). By combining petrographic and SEM observations of corrosion microtextures at the glass-palagonite interface, considerations of the tectonic setting, measurement of U and Th concentrations of fresh basaltic glass by ICP-MS, and theoretical modelling of the present-day distribution of radiation damage in basaltic glass caused by radioactive decay of U and Th, we reinterpret these enigmatic microtextures as the end product of the preferential corrosion/dissolution of radiation damage (alpha-recoil tracks and fission tracks) in the glass by seawater, possibly combined with pressure solution etch-tunnelling. Our findings have important implications for geomicrobiology, astrobiological exploration of Mars, and understanding of the long-term breakdown of nuclear waste glass.

1. Introduction

Understanding and successfully identifying examples of preserved microbial life from extreme environments on planet Earth are pertinent to the astrobiological exploration of Mars, and this was highlighted during recent debates over Martian meteorite ALH84001 (e.g., [1–6]). Accordingly, a flurry of recent studies have focussed on understanding terrestrial examples of life-harbouring extreme environments/paleoenvironments that could have analogs on Mars, including evaporite deposits [7], thermal spring deposits [8], Antarctic paleolake deposits [9], deep sea hydrothermal vent systems and deep subsurface aquifers [5, 10], and the glassy margins of submarine pillow basalts [11, 12]. Knowledge about the geomorphology and geological setting of these environments at the macroscopic scale on Earth can help with landing site selection for Mars astrobiology missions [13]; however, even more imperative to the successful astrobiological exploration of Mars is the ability of scientists to distinguish with absolute certainty whether or not relict signs of life are present in a returned rock sample (e.g., in samples acquired and returned to Earth during future robotic rover or manned missions to Mars). Numerous lines of evidence will probably be necessary to indicate that a true biosignature is present in such a sample and, among others, may include geochemical and stable isotopic constraints [14, 15], the identification of biologically produced minerals [16, 17], detection of biomolecules [18], and paleontological arguments such as recognition of microscopic morphological biomarkers [3, 19].

Among these possible microbial biosignatures in rocks, probably the most contentious of all are the recognition of nano- to microscopic morphological biomarkers, especially because there is typically a great deal of subjectivity involved in deciding which of these tiny shapes and forms appear to look like microbial remains/traces based only on visual interpretations and comparisons to known terrestrial biotic microstructures, and, moreover, many such micro-/nanofeatures also have straightforward and readily deduced nonbiological explanations. In fact, it is quite common at this scale of observation (e.g., under petrographic microscope or in high resolution scanning electron microscope (SEM) images) that there may be multiple explanations for such tiny physical structures in rock samples, including both biogenic and nonbiogenic (e.g., mineralogical) explanations, and three well-known examples of this include (1) abiotically produced nanoscopic mineral grains (i.e., calcite) in carbonate rocks that exhibit spherical, rod, and ovoid shapes resembling bacterial remains [20]; (2) filamentous and segmented carbonaceous microstructures in the ~3.5 Ga Apex cherts of Western Australia resembling bacterial and cyanobacterial remains [21] that have also been reinterpreted as abiogenic amorphous graphite [22]; and (3) concentrically zoned carbonate globules in Martian meteorite ALH84001 originally interpreted as bacterially induced carbonate precipitates [3] that were later reinterpreted as abiotic, high temperature, hydrothermally deposited minerals associated with volcanic activity—based on the discovery of similar carbonate globules in Spitsbergen, Norway [1, 6]. These three examples clearly demonstrate the value in seeking both biological and nonbiological explanations for the origin of putative microscopic morphological biomarkers in rocks, especially when found in extreme environments on Earth that may have similar counterparts at or below the surface of Mars.

Over the last two decades, conspicuously "biogenic-looking" corrosion microtextures have been found to occur globally at the glass-alteration interface within submarine volcanic glass of the *in situ* oceanic crust, ophiolites, and greenstone belts dating back to ~3.5 Ga [11, 23–29], and more recently in terrestrial impact glasses as well [30]. These micron-scale petrographic textures, that is, the so-called *tubular texture* (Figures 1(a), 1(c), and 1(e)) and *granular* [palagonite] *texture* (Figures 2(a) and 2(c)) (see summaries in [11, 26–28, 31]), have emerged as the "prime evidence" (p. 4 in [26]) or "strongest evidence" (p. 157 in [27]) for bioalteration of basaltic glass and have been used to define what is arguably the most geographically vast and long-lived lithoautotrophic microbial ecosystem on Earth [24, 26, 27, 32], representing an important part of the deep biosphere [11, 33, 34], and they currently represent the oldest unrefuted microscopic morphological biomarkers in the geological record [35–38], as well as a key biomorphic (trace fossil) target to look for in the astrobiological search for ancient microbial life on Mars [12]. All of these claims, however, need to be scrutinized, questioned, and tested by the scientific community, but, remarkably, this is something that has not yet taken place for this vast *putative* microbial ecosystem in volcanic glass on planet Earth.

Throughout this 23-year period (i.e., since the publication of Thorseth et al. [23]—the first study to invoke microbial bioalteration of basaltic glass in the development of complex corrosion microtextures at the glass-palagonite interface), it is surprising that none of these studies characterizing putative biogenic microtextures in basaltic glass have ever seriously considered possible nonbiological explanations for these tiny etch features. Certainly, when investigating the origin of conspicuous microtextures in petrographic thin sections of volcanic rocks, a petrological (i.e., abiotic-geological or petrogenetic) origin should first be considered, especially given that these putative biogenic microtextures are typically associated with subaqueously altered (i.e., partially palagonitized) regions of *igneous* rocks (pillow rim, hyaloclastite, or tuffaceous basaltic glasses). Historically, this was actually the case for several earlier studies on partially palagonitized basaltic glasses that identified the presence of microchannels or etch-pits in fresh basaltic glass immediately adjacent to the glass-palagonite interface (e.g., [39], "mist zone" of Morgenstein and Riley [40], and [41–44]). And although their significance was not initially addressed to much extent in the literature (as highlighted by Zhou and Fyfe [44]), some of these authors suggested (in passing) that such microscopic cavities are simply the result of dissolution processes associated with the incipient stage of palagonitization during aqueous alteration of the glass [40], such as corrosion [41] or the formation of etch-pits [44], and saw that there is no need to invoke microbial activity in the formation of these tiny etch features. In contrast, however, during the more recent time period (~1992–2014), this has not been the case virtually in all studies evaluating the origin of complex microscopic rock textures at the glass-palagonite interface in volcanic glass, where only a *biogenic* origin has been sought for these micron-sized (i.e., microbe-sized) etch features, during the accumulation of at least 77 scientific papers documenting such grooved, tubular, and granular "bioalteration" microtextures in submarine glasses from geological sites spanning a large part of the globe and dating back to ~3.5 Ga [11, 12, 23–28, 31–38, 45–105]. It is quite amazing that in the face of this daunting body of scientific research purporting to have documented *bona fide* microscopic morphological biomarkers in volcanic glasses worldwide—complex microtextures supposedly resulting from microbial bioalteration, biocorrosion, or biogenic microboring of volcanic glass (i.e., *tubular, granular,* and *microgroove* textures)—to date, there have only been a handful of scientific studies that have actually proposed nonbiological origins for such complex microtextures in volcanic glass and this includes (1) linear to curvilinear *microgrooves* on glass shards attributed to preferential etching of thermal cracks (i.e., as an alternative explanation for biogenic grooving [106]); (2) complex patterns of *dendritic nanogrooves* on vesicle walls in submarine basaltic glass reminiscent of microbial trace fossils, which are attributed instead to the abiotic, fluid mechanical process of viscous fingering between magmatic vapour and hot glass surrounding vesicles upon cooling through the glass transition [107]; and (3) titanite-mineralized *tubular* textures (similar to previously reported microbial trace fossils)

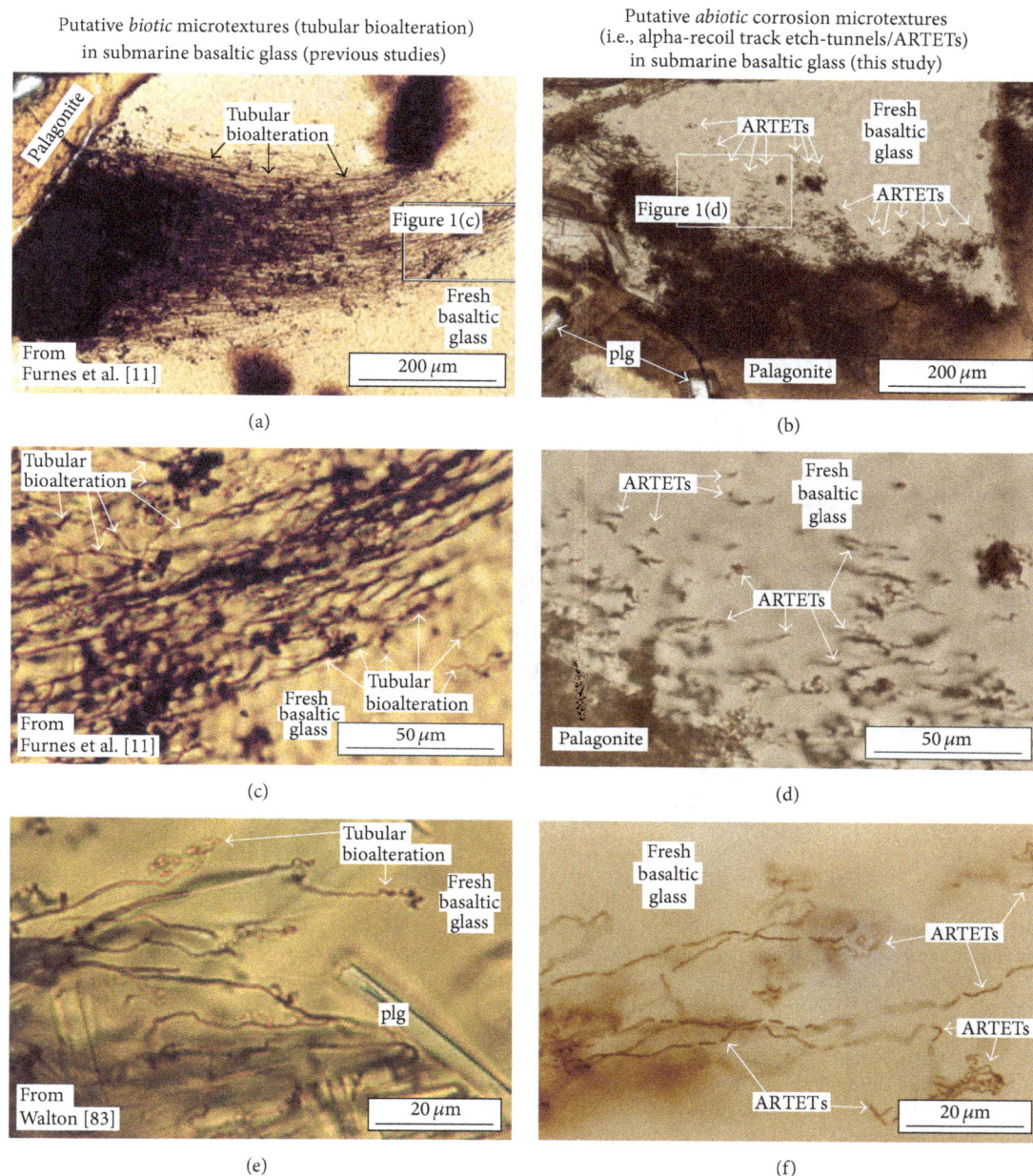

FIGURE 1: Scale comparisons (1 : 1) of previously reported "tubular" bioalteration microtextures in submarine basaltic glass (a, c, and e) with abiotic corrosion microtextures (alpha-recoil track etch-tunnels) in DSDP 418A basaltic glass (this study) (b, d, and f). All images (a–f) are photomicrographs of polished petrographic sections of submarine basaltic glasses taken in plane polarized light (uncrossed nicols). (a) and (c) are from [11] (sample *DSDP-418A-62-4-[64–70]*), and (e) is from [83] (Hawaii Scientific Drilling Project (HSDP) sample 4656.7). (b) and (d) are from sample *DSDP-418A-68-3[40–43]*, and (f) is from sample *DSDP-418A-75-3[120–123]*. Note the similarities when comparing (a), (c), and (e) to (b), (d), and (f), respectively, despite the differences in their inferred origin (biotic versus abiotic). ARTETs: alpha-recoil track etch-tunnels; plg: plagioclase phenocryst.

found in metaglassy Archean rocks attributed to nonbiological metamorphic processes [108, 109]. A few studies have touched on the topic of examining possible nonbiological origins for microscopic etch-tunnels in submarine glasses (e.g., [83, 93]), but in both cases biotic explanations were favoured in the end. One recent study [110] also noted (in

passing) the occurrence of empty "tubules" at the glass-alteration interface in Hawaiian glasses, but it offered no explanation for them. Therefore, for most of these putative microbially produced "complex" etch features documented in basaltic glasses around the globe (especially tubular and granular textures), possible nonbiological explanations have

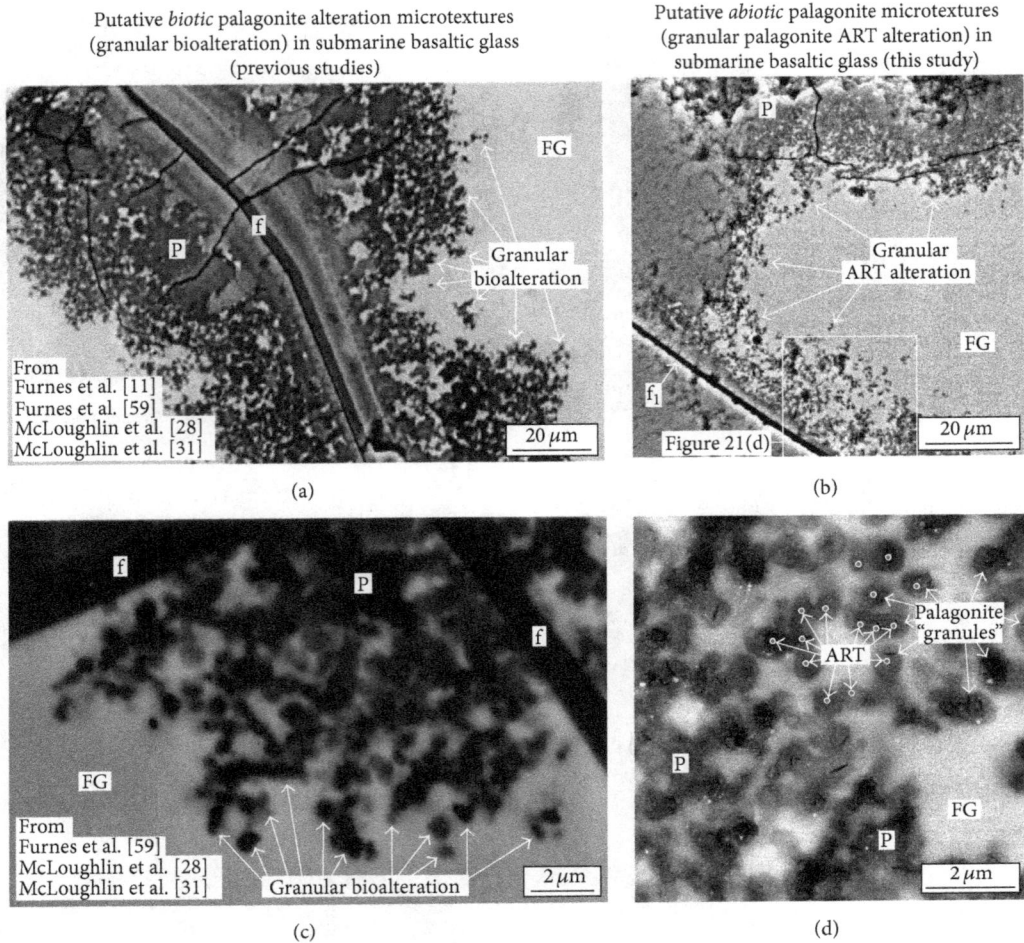

FIGURE 2: Approximate scale comparisons (near ~1:1) of previously reported "granular" bioalteration microtextures in submarine basaltic glass (a, c) with *abiotic* corrosion microtextures (granular palagonite ART alteration) in DSDP 418A basaltic glass (b, d) (this study; sample *DSDP-418A-68-3[40–43]*). All images are BSE images (acquired by SEM) of polished petrographic sections of submarine basaltic glasses. (a) is either from sample *DSDP-417D-30-6-[20–24]* (as reported in Figure 1(e) of [11]) or sample *DSDP-418A-62-4-[64–70]* (as reported in Figure 5(e) of [31] and Figure 2(c) of [28]), and (c) is from sample *DSDP-504b-4-2-[0–20]* (as reported in Figure 3(b) of [59], Figure 5(f) of [31], and Figure 2(d) of [28]). Note the similarities when comparing (a) and (c) with (b) and (d), respectively, despite the differences in their inferred origin (biotic versus abiotic). The 120 nm diameter pink circles in (d) correspond to "hypothetical" previously existing alpha-recoil tracks that have undergone selective palagonitization resulting in the formation of palagonite "granules" (or "granular" palagonite texture). ART: alpha-recoil track; f: fracture; f_1: early fracture; P: palagonite.

yet to be explored and this task is especially important in light of humanity's impending astrobiological exploration of Mars. Accordingly, after initial suggestions (by [107, 111–113]) that "abiotic" explanations can *also* be sought for the origin of corrosion microtextures in submarine glasses (especially "tubular" and "granular" microtextures), some more recent studies have begun to follow suit (e.g., [114]).

Searching for possible nonbiological origins for microscopic tunnels in volcanic glass is quite logical, especially because *abiotic* natural and experimental chemical etching of minerals has long been known to produce elongate microscopic etch-tunnels that exhibit a wide diversity of morphological forms, ranging from straight to curvilinear tubes, branched tubes, and those exhibiting spiral-/helical-, ribbon-, zigzag-, and worm-like shapes [115–123] that collectively are

somewhat similar to the morphological diversity of naturally formed microscopic channels found in basaltic glass (i.e., straight to curvilinear microscopic tubular channels exhibiting spiral/helical, vermicular (worm-like), branched, and annulated tubular textures attributed to microbial activity [11, 26–28, 31]). In addition, morphologically similar (i.e., having straight to curvilinear, spiral/helical, and branched forms), elongate microscopic etch-tunnels/tubes known as "ambient inclusion trails" are known to develop within microcrystalline silica (e.g., agates and cherts) by the migration of pyrite grains and/or organic materials through the process of abiotic, chemical etching, possibly caused by the corrosive products resulting from the diagenetic breakdown of organic matter [93, 124]. Moreover, experimental microscopic etch-tunnelling of volcanic glass (and tektite glasses) is routinely

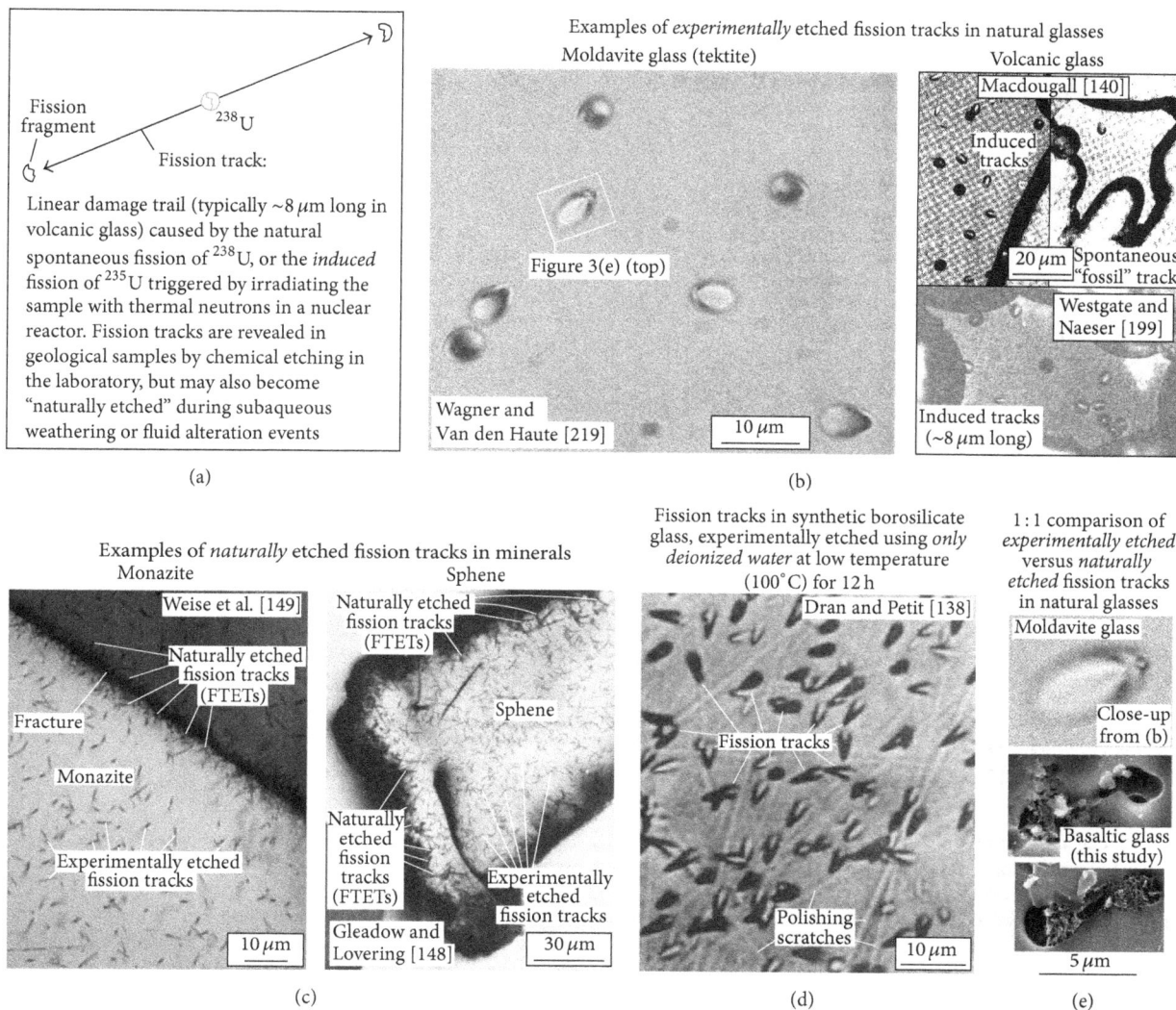

FIGURE 3: Background on fission tracks and examples of fission track etching. (a) Schematic illustration of a fission track. (b–d) Transmitted light photomicrographs of experimentally etched fission tracks in natural glasses (b), naturally etched fission tracks in minerals (c) (i.e., revealed by fluid alteration along a fracture in monazite (left) or by weathering and corrosion around the outer surface of a sphene grain (right)), and fission tracks in synthetic borosilicate glass (simulated nuclear waste glass) etched only with deionized water (d). (e) Scale comparison (1:1) of an experimentally etched fission track in natural moldavite glass (top), with two "naturally etched" fission tracks in submarine volcanic glass (sample *DSDP-418A-75-3[120–123]*; see close-ups and petrographic context of these two fission track etch-tunnels in Figures 11 and 13).

carried out in the laboratory during fission track dating studies (Figure 3), in which the polished surfaces of volcanic glass samples etch preferentially along elongate (typically ~6–8 μm long) latent damage trails in the glass caused by the spontaneous fission of ^{238}U during aging of the sample (or the induced fission of ^{235}U) [125–128]. Similarly, experimental chemical etching of alpha-recoil tracks (Figure 4) caused by the radioactive decay of U and Th (e.g., in alpha-recoil track dating studies of micas [129–134]) represents another style of etch-tunnelling of geological samples that could potentially be relevant to understanding the origin of naturally formed etch-tunnels in minerals and glasses. Furthermore, radiation damage in synthetic borosilicate nuclear waste glasses (for

which basaltic glass is an ideal natural analog: [135, 136]), including fission tracks and simulated alpha-recoil damage, also leads to preferential etching of glass in a matter of hours, even at low temperatures (20–100°C) and using very weak etchants including deionized water or seawater (Figure 3(d)) [137, 138]. In fact, preliminary research indicates that preferential etching and corrosion of randomly distributed fission tracks and alpha-recoil tracks in submarine volcanic glasses—that is, during the *abiotic* encroachment of seawater during palagonitization along fractures—is a feasible abiotic geochemical process to form such "biogenic-looking" tubular and granular microtextures (e.g., p. 10 in French and Muehlenbachs [107] [111–113, 139]).

(a)

Examples of *experimentally* etched alpha-recoil tracks in micas

(b)

Examples of *naturally* etched alpha-recoil tracks in basaltic glass (this study)

(c)

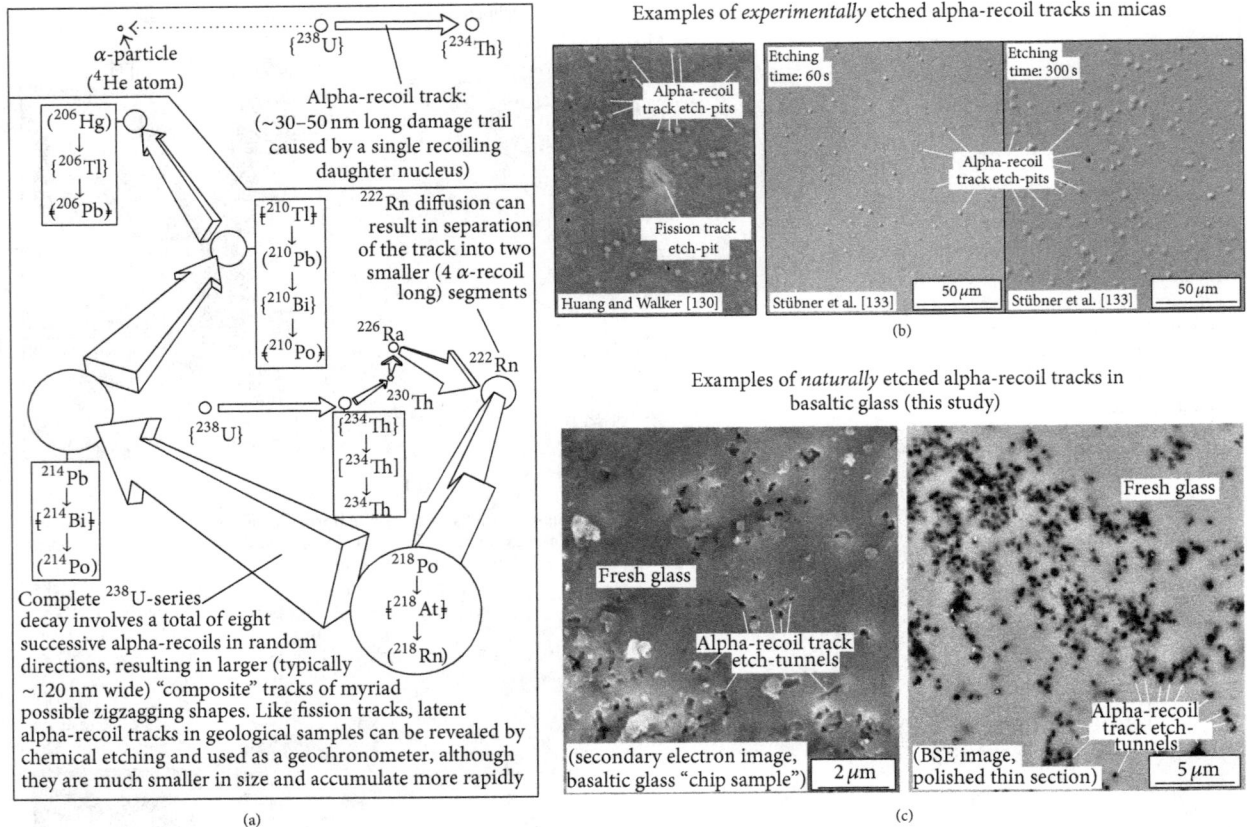

FIGURE 4: Background on alpha-recoil tracks and examples of alpha-recoil track etching. (a) Schematic illustration of alpha-recoil track development during complete ^{238}U-series radioactive decay (note: alpha-recoil tracks also originate from ^{235}U- and ^{232}Th-series decay—not shown). Beta decays are indicated by small arrows pointing down (enclosed in boxes/circles), while alpha decay parent-daughter pairs are denoted by identical bracket styles. (b) Etched alpha-recoil tracks in micas, revealed by experimental chemical etching of fresh cleavage surfaces. The image at left is of muscovite (etched for 2 hours at 20°C using 48% HF) and was obtained using phase contrast microscopy [130] (no scale bar is available for this image). The other two images (center and right) are of phlogopite (etched using 40% HF and then coated with a thin layer of carbon and imaged via an optical microscope equipped with an incident light device combined with Nomarski-Differential Interference Contrast [133]). Note the increase in average size of alpha-recoil track etch-pits with time, which is evident when comparing the 60 s and 300 s images. (c) Naturally occurring alpha-recoil track etch-tunnels (ARTETs)—identified at the glass-palagonite interface in the glassy margins of midocean ridge pillow basalts (this study). Left image: sample *DSDP-418A-75-3[120–123]* (see Figure 11 for petrographic context). Right image: sample *DSDP-418A-68-3[40–43]*.

Consequently, in the present study, we evaluate the origin of complex microscopic alteration textures commonly observed at the glass-palagonite interface in submarine basaltic glasses worldwide (i.e., tubular and granular textures: Figures 1 and 2), from a strictly nonbiological standpoint. In particular, we consider the likely effects of the accumulation of radiation damage—that is, in the form of randomly distributed spontaneous fission tracks and alpha-recoil tracks caused by radioactive decay of U and Th—on the process of natural abiotic corrosion (i.e., low temperature alteration) and dissolution (i.e., etch-tunnelling) of basaltic glass by seawater. Rationale for this study comes from the well-known fact that spontaneous fission tracks and alpha-recoil tracks in silicate glasses are known to etch preferentially over undamaged regions of glass [137, 138, 140–145], commonly resulting in microscopic etch-pits (e.g., Figures 3(b), 3(d), and 3(e)). Coupled with the observations that midocean

ridge basaltic glasses worldwide are known to contain trace amounts of U and Th [146] and are routinely dated by the fission track method [125, 127, 147], this provides direct evidence that radiation damage could potentially play a very important role in the natural corrosion and dissolution of basaltic glass by seawater (i.e., as opposed to microbial activity). Moreover, naturally etched fission tracks (i.e., fission track etch-tunnels formed during low temperature aqueous weathering or fluid alteration) have also been documented previously in a variety of minerals including the outer surfaces of sphene grains (Figure 3(c)) [148], along fractures within monazite (Figure 3(c)) [149], and also within apatite grains [150].

Numerous previous studies of submarine volcanic glasses have claimed that "tubular" and "granular" corrosion microtextures are simply "too complex and too reminiscent of biological processes to be explicable by an abiotic process" [59], that "abiotic" palagonitization produces only very simple (i.e., straight and sharp) glass-palagonite interfaces without such tubular or granular textures (e.g., Section 4.1 in Furnes et al. [59]; Figure 1 in Staudigel et al. [75]; Figure 13(a) in Furnes et al. [11]; Figure 1 in Furnes et al. [26]; Figure 2 in McLoughlin et al. [31]; Figure 1 in Staudigel et al. [27]), and that (other than microbial etching) no alternative abiotic process has yet been proposed to explain such globally distributed complex microtextures found in volcanic glasses worldwide [25]. In the present study, we challenge these three points by demonstrating unequivocally that such complex corrosion microtextures at the glass-palagonite interface in submarine volcanic glasses (e.g., see Figures 1 and 2) can form as a direct consequence of the preferential dissolution (etch-tunnelling) and corrosion of randomly distributed spontaneous fission tracks and alpha-recoil tracks (i.e., radiation damage) in the glass by seawater and are therefore *not* necessarily a consequence of biological processes such as microbial boring. Accordingly, in this study we question the widely accepted "biocorrosion" model for the development of complex alteration microtextures in submarine volcanic glass and, therefore, the biogenicity of such "tubular" and "granular" corrosion microtextures found in basaltic glasses worldwide.

To accomplish these objectives, we have focussed this multidisciplinary petrographic (rock-textural) and theoretical modelling study on partially palagonitized, 120.6 Ma basaltic glass "pillow margin" samples recovered from drill core of Deep Sea Drilling Project (DSDP) Hole 418A, in the southwestern North Atlantic Ocean (Figure 5)—a drill hole which in several previous studies has yielded many such basaltic glass samples exhibiting classic "tubular" and "granular" microtextures attributed to microbial activity (e.g., Figures 1(a), 1(c), 2(a), and 5(b)) [11, 26–28, 31, 33, 59, 61, 93, 94, 100]. Our study combines petrographic observations, high resolution SEM imaging, considerations of the geological setting, determination of trace element concentrations (U and Th) in fresh basaltic glass by Inductively Coupled Plasma-Mass Spectrometry (ICP-MS), and theoretical modelling of radiation damage in basaltic glass caused by radioactive decay of U and Th (i.e., the distribution of spontaneous fission tracks and alpha-recoil tracks in the glass). Ultimately, we discuss the implications that this new abiotic "U–Th–Pb radiogenic" paradigm for basaltic glass corrosion and dissolution has on the understanding of microbial ecology and microbial trace fossil identification on Earth (particularly within volcanic basement rocks of *in situ* Layer 2 oceanic crust, ophiolites, Archean greenstone belts, or impact glasses), as well as the vital implications that this new paradigm has for the future astrobiological exploration of Mars—a planet that importantly, appears to be dominated at the surface (geologically) by the widespread occurrence of basalt [151–155], palagonite [151, 152, 156], weathered volcanic/basaltic glass [151, 153, 157], and evidence for past action of liquid water [154, 155, 158, 159]—and which therefore may contain volcanic or impact glasses bearing analogous *microgrooves*, *granular* palagonite textures, or *tubular* etch-tunnels.

The natural breakdown and corrosion of basaltic glass have also long been considered to represent an important

natural analog process pertinent to understanding the long-term breakdown of high level nuclear waste glasses stored in geological repositories [135, 136]. Therefore, we also highlight that our discovery/identification of naturally etched fission tracks and alpha-recoil tracks in 120.6 million-year-old submarine basaltic glass at DSDP 418A—formed by incremental encroachment of seawater through radiation damaged sites—also has profound implications for predicting the long-term behaviour of borosilicate nuclear waste glasses stored in deep geological aquifers (i.e., repositories), providing an ideal "natural laboratory" for carrying out such long-term studies.

2. Geological Setting

2.1. Sample Locations and Previous Work on DSDP-418A Basaltic Glasses. For this case study on the origin of complex microtextures at the glass-palagonite interface in submarine volcanic glasses, partially palagonitized basaltic glass pillow margin samples showing evidence of "tubular" and "granular" microtextures (cf. [11, 25–28, 31, 33]) were selected from rocks recovered from DSDP Hole 418A. The drill hole is situated under 5511 m of water in the western North Atlantic Ocean (Figure 5; latitude: 25°02.10'N; longitude: 68°03.44'W) and was drilled to a depth of 868 m below the seafloor [160]. At this site, sediments of Layer 1 oceanic crust are 324 m thick, below which occurs 120.6 Ma (for age constraints, see Figures 5(c) and 5(d) and Section 2.2), Layer 2 volcanic basement [160] that is considered to represent typical eruptive oceanic crust, comprising a succession of basaltic pillow lavas and massive flows, with lesser intercalations of breccias and sediments, as well as cross-cutting mafic dykes lower down in the succession (Figure 5(b)) [161]. Basaltic glass pillow margin samples from this study originate from core samples *DSDP-418A-68-3[40–43]*, *DSDP-418A-72-4[13–15]*, and *DSDP-418A-75-3[120–123]* collected from depths of 732, 760, and 785 m below the seafloor, respectively (Figure 5(b)). Detailed geochemical analyses (i.e., determination of major element oxide compositions) of pillow rim basaltic glasses throughout much of the entire sequence of lava flows at DSDP 418A—along with a comprehensive evaluation of the observed down-hole variations in volcanic chemical stratigraphy and delineation of volcanic eruptive units for these ancient midocean ridge pillow basalts, flows, and breccias—was carried out previously [161, 162]. According to their down-hole depths and relative positions in the volcanostratigraphic sequence at DSDP 418A (Figure 5(b)), all three of the basaltic glass pillow margin samples in the present study occur within chemical (i.e., glass) type "J" [162] of lithologic unit "13" of volcanic eruptive unit "Vb" [161]. Type "J" basaltic glasses are characterized by an average glass composition of 51.03 ± 0.35 wt.% SiO_2, 14.19 ± 0.14 wt.% Al_2O_3, 11.31 ± 0.24 wt.% $FeO(T)$, 7.13 ± 0.09 wt.% MgO, 11.86 ± 0.11 wt.% CaO, 2.34 ± 0.04 wt.% Na_2O, 0.11 ± 0.01 wt.% K_2O, 1.54 ± 0.03 wt.% TiO_2, and 0.14 ± 0.01 wt.% P_2O_5 ($n = 18$, errors given in standard deviation [162]). Of particular interest to the present study (aimed at understanding the likely role of self-incurred radiation damage on the development of complex corrosion microtextures in submarine basaltic

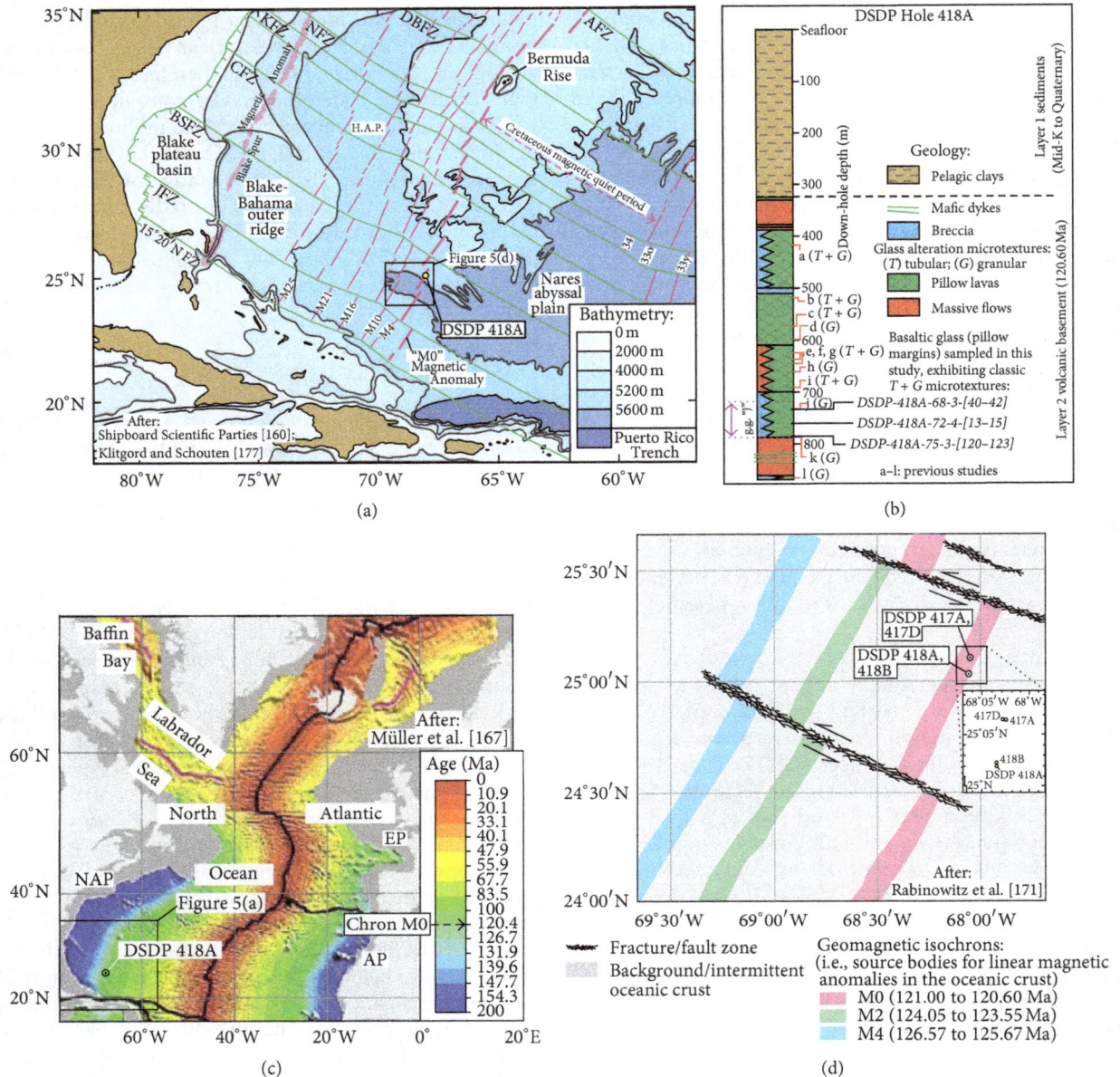

FIGURE 5: Geological maps (and stratigraphic column) of the North Atlantic region highlighting the position of DSDP Hole 418A within the context of regional patterns of lithospheric age and seafloor spreading magnetic anomalies. (a) Bathymetry, seafloor spreading magnetic anomalies (pink lines), and fault/fracture zones (green lines) of the southwestern North Atlantic Ocean (after [160, 177]). HAP denotes Hatteras Abyssal Plain. Fracture zones (FZ) include the Atlantis (AFZ), Delaware Bay (DBFZ), Norfolk (NFZ), Kane (KFZ), Carolinas (CFZ), Blake Spur (BSFZ), Jacksonville (JFZ), and 15°20′N FZ. (b) Stratigraphic column for DSDP 418A (after [160–162]) showing the locations of basaltic glass pillow margin samples in which classic "tubular" (T) and "granular" (G) alteration microtextures have been documented at the glass-palagonite interface in this (black) and previous studies (red). The black dashed line demarcates the top of volcanic basement, and zigzags represent interbedding of lithologic units. The volcanostratigraphic zone in which glass group "J" occurs [161, 162] is denoted in purple (g.g. "J") and corresponds to a geochemically coherent volcanic (glass) interval from which all three glass samples were collected in this study. Samples a–l from previous studies (shown in red) correspond to a—*418A-30-3-[4–6]* in Furnes and Staudigel [33] and Furnes et al. [59]; b—*418A-43-1-[80–82]* in Fliegel et al. [100]; c—*418A-49-2-[41–45]* in Staudigel et al. [27] and Furnes et al. [59]; d—*418A-52-5-[75–80]* in Furnes et al. [11] and Furnes et al. [59]; e—*418A-55-4-[112–114]* in Furnes et al. [11], Staudigel et al. [27], Furnes et al. [59], and Fliegel et al. [100]; f—*418A-56-5-[129–132]* in McLoughlin et al. [28] and Furnes et al. [59]; g—*418A-57-5-[12-13]* in Furnes et al. [59]; h—*418A-59-3* in Furnes et al. [59]; i—*418A-62-4-[64–70]* in Furnes et al. [11], Staudigel et al. [27], McLoughlin et al. [28], McLoughlin et al. [31], Furnes et al. [59], and Furnes et al. [61]; j—*418A-68-3-[32–38]* in Furnes et al. [59]; k—*418A-76-1-[4–7]* in Furnes et al. [59]; l—*418A-86-5-[24–32]* in Furnes et al. [59]. (c) Map showing lithospheric age in the North Atlantic Ocean, highlighting the location of DSDP 418A and Chron "M0" on which it lies (after Müller et al. [167]). Tectonic plates: AP: African Plate; EP: Eurasian Plate; NAP: North American Plate. (d) Close-up geological/geophysical map showing the location of DSDP 418A (and other nearby holes) in the context of local patterns of seafloor spreading magnetic anomalies and fracture/fault zones [160, 171]. The age intervals for geomagnetic isochrons M0, M2, and M4 are from Channell et al. [176].

glass), fission track dating of basaltic glasses from DSDP 418A was actually carried out early on as well [125]. In that study, it was found that spontaneous fission tracks (revealed by experimental chemical etching) are abundant enough in DSDP 418A basaltic glass to successfully carry out fission track dating, which (along with glasses from nearby holes 417A and D) yielded a thermally corrected, combined fission track age of 108.3 ± 1.3 Ma [125]. Furthermore, U concentrations in basaltic glass samples *DSDP-418A-30-2[71-72]* and *DSDP-418A-45-2[34–37]* (stratigraphically higher up in the volcanic succession than the rocks from this study) were determined to be 21.4 and 18.4 ppb, respectively [125]. The success of this early fission track dating study, coupled with the measurement of trace U in basaltic glasses at DSDP Hole 418A, provides important rationale for the present study on the corrosion of radiation damage in DSDP 418A basaltic glasses—because it already proves for us that such radiation damage (and U) is actually there, implying that fission tracks might indeed play an important role in controlling the microscopic patterns of corrosion during preferential dissolution and palagonitization of basaltic glass by seawater.

The present study on corrosion of DSDP 418A basaltic glasses also compliments a previous companion study [107] of branching, nanoscopic grooves on vesicle walls in basaltic glass from sample *DSDP-418A-75-3[120–123]* that we believe represent another variety of "abiotic" microtextural features in basaltic glass that could potentially be misidentified as microbial etch features (see Section 6.5).

Several previous studies on basaltic glass pillow margin samples from DSDP 418A have documented widespread evidence of granular and/or tubular alteration microtextures at the glass-palagonite interface throughout much of the entire succession of volcanic basement rocks encountered by this drill hole (Figure 5(b)) and universally attributed the origin of these tubular and granular microtextures to microbial activity/biocorrosion (Figures 1(a), 1(c), 2(a), and 5(b)) [11, 26–28, 31, 33, 59, 61, 93, 94, 100]. The downhole depths at which such putative tubular and granular bioalteration microtextures have been documented in these previous studies are indicated in Figure 5(b) (in red), along with the positions of basaltic glass pillow margins sampled in this study (in black) in which we document the occurrence of identical but clearly *abiotic* tubular and granular microtextures arising from preferential corrosion of randomly distributed fission tracks and alpha-recoil tracks in basaltic glass.

2.2. Constraints on the Age of DSDP-418A Pillow Lavas and the M0 Magnetic Anomaly.
In order to carry out accurate theoretical modelling (below in Section 5) of the accumulation of randomly distributed radiation damage (i.e., fission track and alpha-recoil track areal densities) in DSDP 418A basaltic glass in this study, it is imperative to know the exact age (t) of quenching of basaltic glasses (i.e., pillow eruption) at DSDP 418A. Because the amount of material we had for each pillow margin sample in this study is quite small, and no suitable U-bearing minerals are present in these glasses (such as zircon or baddeleyite), we were not able to carry out precise and accurate U–Pb isotopic dating of these

rock samples (e.g., by Isotope Dilution Thermal Ionization Mass Spectrometry or Laser Ablation-Inductively Coupled Plasma-Mass Spectrometry). Nevertheless, there are several previous studies that do provide a number of different age estimates for the timing of formation of Layer 2 volcanic basement at DSDP 418A—employing various geochronological techniques including fission track, ^{40}Ar–^{39}Ar, and Rb/Sr radiometric dating, microfaunal biostratigraphy, and global correlation of linear magnetic anomalies in the oceanic crust (and their associated isotopic ages)—from which we can place somewhat reliable constraints on the timing of quenching of these DSDP 418A basaltic glass samples.

Direct dating of pillow rim basaltic glasses from DSDP 418A was actually carried out in a previous study by Storzer and Sélo [125] using the fission track method, which yielded a relatively precise—albeit thermally corrected, mean—age of 108.3 ± 1.3 Ma determined on a combination of basaltic glass samples originating from nearby DSDP Holes 417A, 417D, and 418A (Figure 5(d)). Accordingly, this fission track age was interpreted in that study as a reasonable estimate for the timing of pillow eruption, glass quenching, and formation of the M0 linear magnetic anomaly (intersected by these drill holes: Figures 5(a) and 5(d)), all of which formed during ancient (ca. Mid-Cretaceous) seafloor spreading during early ocean opening of the central Atlantic (Figure 5(c)). Presumably, the entire vertical sequence of pillow eruption recorded in DSDP Hole 418A (some 40 eruptive units [161]) represents a relatively contemporaneous lava succession that was emplaced during a relatively short (<100,000 year) geological timespan (i.e., by comparison with the fission track ages reported for samples from the FAMOUS area—see Storzer and Sélo [125] and references therein). Therefore, taken at face value, the fission track age of 108.3 ± 1.3 Ma reported by Storzer and Sélo [125] *could* effectively be used to estimate the age of eruption and glass quenching of all pillow lavas sampled from DSDP Hole 418A. Another radiometric dating study focussed on DSDP 418A volcanic basement rocks [163], employed Rb–Sr isotopic dating of secondary smectites, which yielded a similar but less precise Rb–Sr isochron age of 108 ± 17 Ma—adding support to the fission track age of Storzer and Sélo [125]. However, additional radiometric dating of secondary analcites, celadonites, and smectites showing greater overall Rb–Sr enrichments that originate from volcanic basement rocks of nearby DSDP Hole 417A (situated on the same M0 linear magnetic anomaly: Figure 5(d)) yielded a much more well-constrained Rb–Sr isochron age of 108 ± 3 Ma [163]—again, adding further support for the idea that the age of volcanic basement at DSDP Hole 418A is about 108 Ma. Although these smectites dated by Rb–Sr constitute "secondary" minerals originating from the extensive alteration of igneous basement rocks, they were considered in that study [163] to have formed more or less contemporaneously with volcanic basement (given the agreement between various age determinations at ~108 Ma). Consequently, according to this interpretation, the three basaltic glass pillow margin samples investigated in the present study from this drill hole (*DSDP-418A-68-3[40–43], DSDP-418A-72-4[13–15]*, and *DSDP-418A-75-3[120–123]*)

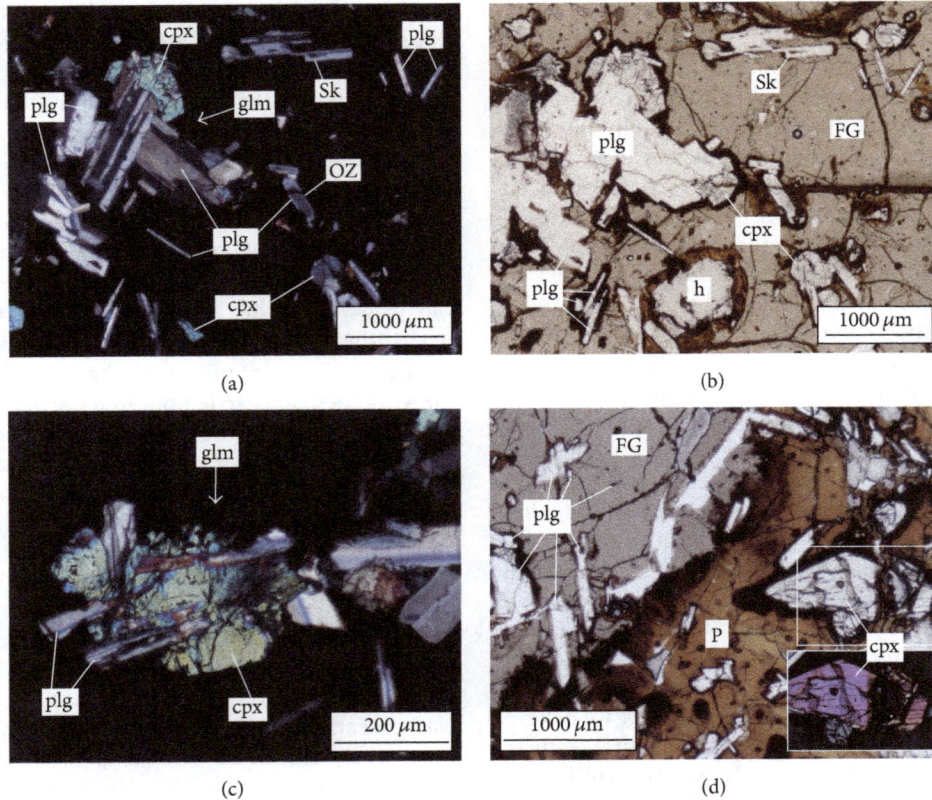

FIGURE 6: Photomicrographs of polished petrographic thin sections of DSDP 418A basaltic glass pillow margins, highlighting various igneous petrographic features. The photomicrographs shown in (a, c) and (d—"inset") are taken in cross polarized light (crossed nicols), whereas (b) and (d) are in plane polarized light (uncrossed nicols). (a-b): sample *DSDP-418A-75-3[120–123]*. (c-d): sample *DSDP-418A-68-3[40–43]*. Note that (a) and (b) depict the same petrographic area. cpx: clinopyroxene; FG: fresh basaltic glass; glm: glomerocryst; h: artificial hole in the petrographic section; OZ: oscillatory zoned plagioclase; plg: plagioclase phenocryst; Sk: skeletal plagioclase crystal.

would appear to have been emplaced at 108 ± 3 Ma (i.e., in agreement with the fission track age of 108.3±1.3 Ma, which is the radiometric age used in our original interpretation during previous preliminary work on these glasses [111, 112])—and indeed most previous geomicrobiological studies of DSDP 418A basaltic glasses have also considered the age of these glasses to be about 110 Ma [11, 26–28, 31, 33, 59, 94, 100]. However, given the occurrence of fresh basaltic glass in some pillow margin samples (e.g., the partially palagonitized samples in the present study, such as in Figures 6(b) and 6(d)), the timing of formation of some secondary smectites in volcanic basement rocks at DSDP 418A may not necessarily have been soon after pillow eruption, and so, in detail, the Rb–Sr age of 108 ± 3 Ma can only really be considered to represent a "minimum estimate" for the age of pillow eruption (e.g., could conceivably be millions or even tens of millions of years too young). Further evidence to support this type of scenario comes from the observation of a second generation of late-stage "off-axis" celadonite formation associated with renewed/continued alkali fixation documented locally in the oceanic crust [164]. It is also notable that due to problems associated with variable amounts of fission track fading, the initial (individual) fission track ages reported for various

samples of basaltic glass in the study by Storzer and Sélo [125] ranged widely between 46 and 76 Ma—collectively requiring the derivation of a series of quite significant "age corrections" that varied systematically according to depth, which could potentially affect the accuracy of the final "thermally corrected" fission track age of 108.3 ± 1.3 Ma because of the additional assumptions inherent in the derivation of these age corrections. Therefore, the widely accepted age of ~108 Ma (or ca. 110 Ma) for the timing of formation of volcanic basement at DSDP 418A (e.g., [11, 26–28, 31, 33, 59, 94, 100, 111, 112, 125, 163]) may not necessarily be that accurate based on these original age constraints—and this idea is furthered if we consider another ^{40}Ar–^{39}Ar radiometric dating study carried out on drilled basalts from DSDP Holes 417D and 418A [165], additional biostratigraphic controls from overlying Layer 1 sediments [166], and, most importantly, the regional tectonic setting during the formation of oceanic crust (and associated linear magnetic anomalies) in the western North Atlantic region, in the context of global models of seafloor spreading in the Mesozoic Era [167].

Firstly, another radiometric dating study of DSDP 418A (and 417D) basalts was carried out early on [165], which employed an ^{40}Ar–^{39}Ar stepwise degassing method of dating that was performed on seven whole rock samples of crystalline rocks (basalts) recovered from these drill holes. However, useful age information was only obtained for one of these samples (417D-22-3-[134–139])—a plagioclase phyric basalt—and therefore age constraints on the timing of pillow lava eruption at DSDP 418A can only be made by assuming that the timing of emplacement of Layer 2 basalts at these two nearby (within 8 km of one another) drill holes was about the same—which is a reasonable assumption given that both drill sites are situated on the same regional linear magnetic anomaly "M0" (Figure 5(d)). Stepwise ^{40}Ar–^{39}Ar dating of this basalt sample from DSDP Hole 417D yielded a relatively complex spectrum of apparent ages ranging from 98.6 to 185.3 Ma, although five out of seven fractions were found to plot along a ~120 Ma reference isochron [165], possibly indicating that the age of the rock (and the M0 anomaly) is about 12 million years older than was determined by the aforementioned fission track dating study [125] of basaltic glasses from DSDP Holes 417A, 417D, and 418A.

Additional support for a slightly older age of eruption of pillow basalts at DSDP 418A comes from paleontological (biostratigraphic) arguments derived from observations of calcareous nanofossils present within the 324 m succession of Layer 1 pelagic sediments overlying the volcanic basement containing these pillow lavas. Gartner [166] documented the presence of a distinctive nanofossil species (*Lithastrinus floralis*) in sediments, a short distance above the volcanic basement/sediment contact at DSDP 418A, 418B, and 417D, that are known to have first appeared in Middle Aptian times and to be common in the Upper Aptian. Accordingly, based on these biostratigraphic constraints, the age of volcanic basement intersected by these drill holes was estimated to be not younger than ~112 Ma (i.e., at least as old as about Late Aptian [166])—again, in contrast with the slightly younger fission track age of 108.3 ± 1.3 Ma determined by Storzer and Sélo [125]. Furthermore, the occurrence of another important nanofossil datum (*Corrolithion acutum*) even lower down in the sedimentary succession intersected by two of these three drill holes (417D and 418A) was interpreted to indicate that earliest sedimentation above volcanic basement began in lower Aptian times (i.e., closer to ~125 Ma; Figures 1 and 2 in Gartner [166]).

Probably one of the best methods for estimating the age of eruption of pillow lavas at DSDP 418A that is currently available right now (until precise and accurate U–Pb isotopic ages become available for these rocks) is to consider the geological context of these lavas within the broader scale regional age patterns of development of the oceanic crust (and associated linear magnetic anomalies) in the North Atlantic Ocean. Global correlation of geomagnetic isochrons (linear magnetic anomalies in the oceanic crust linked with seafloor spreading and the reversal history of Earth's magnetic field that often show twin "tape-recorder" like symmetries about the central axis of the spreading ridge in ocean basins [168, 169]; Figure 5(c)), coupled with plate tectonic theory, has become a very useful tool in determining the age of formation of vast regions of the oceanic crust and lithosphere [170], and in fact the general age patterns of the oceanic

crust and lithosphere present beneath all the world's oceans are already quite well-established [167]. Significant here is that one of the primary reasons for situating the DSDP 417 and 418 drill holes where they are (Figure 5(d)) was to intersect oceanic crust situated precisely on the Mid-Cretaceous linear magnetic anomaly/isochron "M0"—at a latitude of about 25°N [171, 172]. Therefore, before drilling was carried by the DV *Glomar Challenger*, early reconnaissance geophysical survey work was carried out by the USNS *Lynch*, in order to provide the necessary scientific means for drill site targeting [171]. From this reconnaissance work, a small geological/geophysical map was produced, highlighting local magnetic anomalies and fracture/fault zones, which situated the DSDP 417A, 417D, 418A, and 418B drill sites precisely on linear magnetic anomaly M0—with DSDP 418A situated right at the young (eastern) edge of the anomaly (Figure 5(d)). Consequently, the age of eruption of DSDP 418A pillow lavas and quenching of associated pillow rim basaltic glasses in this study coincides with the age of oceanic crust formation associated with the final stages of development of the M0 magnetic anomaly.

Linear magnetic anomalies recorded in the oceanic crust associated with the M0 Chron have now been correlated globally across the oceanic portions of several major tectonic plates—and beneath several different oceans [167, 173], notably including the North American, Eurasian, and African plates beneath the North Atlantic Ocean (note: in magnetostratigraphy, "chrons" are short intervals of geologic time, typically <1 million years in duration, often associated with a specific time period between reversals in polarity of the Earth's magnetic field [174]). On the basis of these global correlations of the M0 anomaly worldwide [167, 173], these narrow belts of oceanic crust/lithosphere are at present considered to have formed in the Mid-Cretaceous Period precisely at 120.4 Ma [167]—although some authors indicate a slightly older and more prolonged time interval of 121.00–120.60 Ma for the M0 Chron [175]. During the 1970's, 80's, and 90's, as more and more geochronological and geophysical data became available and as global models of seafloor spreading became continually more refined, age estimates for the time interval during which the globally correlated M0 magnetic anomaly was formed changed progressively between successive models from 109.01 to 108.19 Ma (LH75), from 118.7 to 118.0 Ma (KG85), from 124.88 to 124.32 Ma (GTS89), from 120.10 to 119.15 Ma (GRAD93), from 120.98 to 120.38 Ma (GRAD94), and from 121.00 to 120.60 Ma (CENT94) (see a summary of these data in Channell et al. [176]). The 121.00–120.60 Ma interval (i.e., from the base to the top of the M0 anomaly) proposed by Channell et al. [176] (based on global correlations of oceanic anomaly block models and magnetostratigraphy) is still currently considered to represent a robust estimate for the age of formation of the M0 linear magnetic anomaly worldwide [175], and it also coincides closely with the age of 120.4 Ma suggested by Müller et al. [167] (Figure 5(c)) for the age of this global magnetic anomaly. Therefore, in the present study on DSDP 418A basaltic glasses, we also consider this time interval of 121.00–120.60 Ma proposed by Channell et al. [176] to represent the current best estimate for the age of formation of the

M0 magnetic anomaly in the western North Atlantic Ocean. Accordingly, because DSDP Hole 418A was drilled directly into the "top" (or young edge) of magnetic anomaly M0 (Figure 5(d)), the timing of formation of volcanic basement at DSDP 418A is interpreted to coincide precisely with an age of 120.60 Ma. Consequently, we consider the timing of eruption of pillow lavas at DSDP 418A to be 120.60 Ma— which therefore coincides with the time of glass quenching and the age of basaltic glass "pillow margin" samples in this study (namely, *DSDP-418A-68-3[40–43], DSDP-418A-72-4[13–15],* and *DSDP-418A-75-3[120–123]*).

Here, it is important to note that the M0 Chron and associated seafloor spreading linear magnetic anomalies are quite crucial to understand many aspects of Earth evolution and plate tectonic theory, especially regarding the origin and opening of the Atlantic Ocean. In the North Atlantic Ocean, the M0 linear magnetic anomaly forms a prominent feature in the oceanic portions of both the North American (Figures 5(a) and 5(c) [167, 177]) and African (Figure 5(c)) plates and shows pronounced symmetry about the Mid-Atlantic ridge (Figure 5(c) [167]; also see Figure 1 in Bird et al. [175]) and in the southwestern North Atlantic Ocean (i.e., where DSDP Hole 418A is situated) it forms a well-defined linear magnetic anomaly that is traceable for a few thousand kilometers (Figures 5(a) and 5(c) [167, 177]). In terms of the geomagnetic and tectonic histories of the ocean basins (especially the North Atlantic), the linear magnetic anomaly M0 is quite significant, because it marks the commencement of the Cretaceous Magnetic Quiet Period (CMQP) [178, 179]—a relatively long time interval in Earth history between Chrons M0 and C34 (~120–84 Ma) during which no magnetic reversals took place (and consequently no linear magnetic anomalies are present in the oceanic crust—this CMQP is labelled in Figure 5(a))—which is also referred to as the "Cretaceous Normal Superchron" [180]. Furthermore, the M0 anomaly also coincides with an important time interval (Chron) in the geological history of the South Atlantic Ocean, because it corresponds to the time of initial ocean opening and seafloor spreading at ~120 Ma, which took place after the rifting of Africa from South America during the breakup of Gondwana, and therefore the M0 magnetic anomaly now fringes certain parts of the Atlantic-facing edges of these two continents [181]. In addition, the base of the globally correlated M0 magnetic anomaly, which is currently estimated at 121.00 Ma [175], is also an important stratigraphic time marker in that it is interpreted to coincide precisely with the Barremian–Aptian boundary of the Early Cretaceous Period [176].

3. Petrographic and SEM Studies of Dissolution/Alteration Microtextures in Basaltic Glass

To study and characterize alteration microtextures preserved in basaltic glass pillow margin samples in this study, we prepared polished petrographic thin sections and studied them (Sections 3.1–3.3) using both transmitted light microscopy (Figures 6–10; also see Figures 1(b), 1(d), and 1(f)) and

SEM analysis (Figures 8–10; also see Figures 2(b), 2(d), and 4(c)—right, and 7(c)). For the SEM studies (including backscattered electron (BSE) imaging and secondary electron imaging), polished sections were first coated with ~40 Å of iridium using a Xenosput XE200 and then analysed using a JEOL 6301F field emission scanning electron microscope equipped with a PGT IMIX model X-ray analysis system— the instrument used to obtain the energy dispersive X-ray spectroscopy (EDS) spectra shown in Figure 8(j). We also carried out high resolution SEM studies (secondary electron imaging) of corrosion microtextures exposed on freshly fractured surfaces of basaltic glass "chip samples" that we prepared—which are explained later (in Section 3.4).

In polished petrographic thin section, all three glassy pillow margin samples are very similar in appearance and comprise an original igneous assemblage of ~70% sideromelane (pale brown basaltic glass) with a phenocryst modal mineralogy of ~25% plagioclase and ~5% clinopyroxene (Figure 6)—similar to other previous petrographic descriptions of pillow basalts from lithologic Unit 13 [161]. Plagioclase crystals are typically euhedral (although rare skeletal crystals are also present, "Sk" in Figures 6(a) and 6(b)), show polysynthetic albite twinning (some grains show weak oscillatory zoning, "OZ" in Figure 6(a)), and exhibit a somewhat bimodal distribution with regards to crystal size, shape, and abundance. This bimodal distribution is defined by a few large (~0.5–1 mm across) subequant/stubby plagioclase crystals (i.e., phenocrysts) dispersed amongst a larger population of small and elongate plagioclase laths (~10–100 μm wide by a few hundred μm long), thus defining an overall porphyritic texture (Figures 6(a) and 6(b)). Clinopyroxene crystals are equant, range from subhedral to euhedral in form, vary in size from <100 μm up to ~1 mm across (Figure 6(d)), and are commonly observed to partially overgrow/envelope previously formed plagioclase grains (i.e., form "glomerocrysts" with plagioclase: "glm" in Figures 6(a) and 6(c); as noted by Flower et al. [161] as well), and this is true for both the large equant plagioclase crystals (Figure 6(a)) and also the smaller clusters of plagioclase laths (Figure 6(c)), thereby defining a sort of protoophitic petrographic texture. For the most part, all plagioclase and clinopyroxene crystals are fresh and unaltered in these pillow lavas; however, basaltic glass is variably altered to both orange-brown palagonite ("P" in Figures 6(d) and 7) and white (K-Al-Si)-rich devitrified zones ("D" in Figures 7(a), 7(b), and 8(a)–8(f)).

The most striking petrographic feature exhibited by these glassy pillow margin samples is pronounced irregular palagonitization (corrosion/alteration of basaltic glass) along fractures (e.g., Figures 1(b), 6(d), 7(a), 9(m)–9(o), and 10(a)). Accordingly, the basaltic glass in these samples ranges locally from fresh unaltered glass (e.g., "FG" in Figures 6–10) to highly altered zones along fractures (e.g., "P" in Figures 6–10), and at least two distinct episodes of fracturing and alteration/devitrification have affected these glasses: (1) early fracturing (labelled "f_1" in Figures 7–9) associated with incipient and ongoing palagonitization and (2) late fracturing (labelled "f_2" in Figures 7 and 8) associated with white (K-Al-Si)-rich devitrified zones (Figures 7(a), 7(b), and 8). Palagonitization refers to the formation of

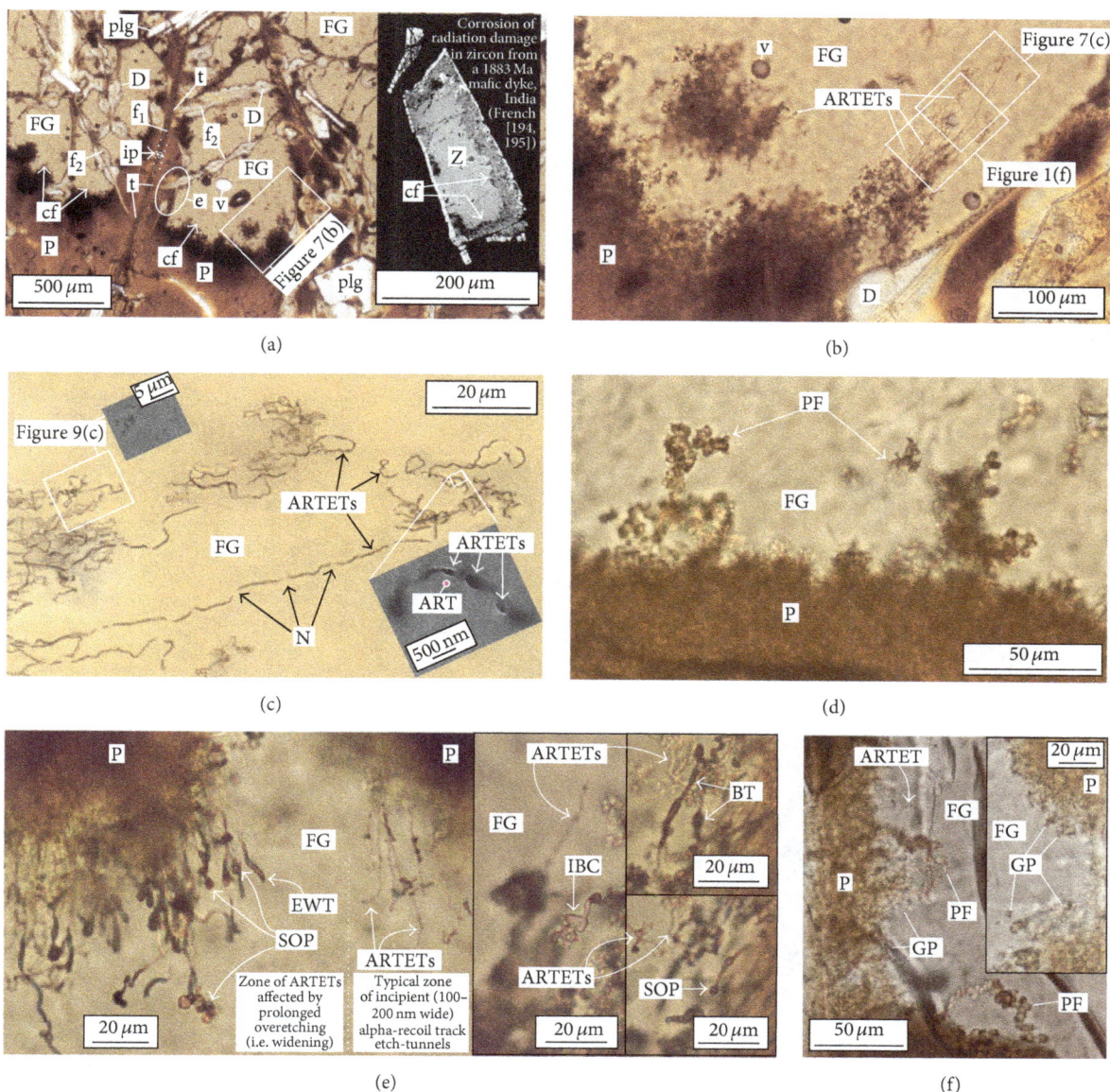

FIGURE 7: Diversity of *abiotic* corrosion microtextures in DSDP 418A basaltic glass linked with palagonitization (all thin section photomicrographs are taken in plane polarized light (uncrossed nicols); (a–e): sample *DSDP-418A-75-3[120–123]*; (f): sample *DSDP-418A-72-4[13–15]*). (a) Overview, highlighting fresh basaltic glass (FG), vesicles (v), plagioclase phenocrysts (plg), early fracturing (f_1) associated with incipient (ip) and ongoing palagonite (P), late fracturing (f_2) associated with white (K-Al-Si)-rich devitrified zones (D), regions where white devitrified zones have been enveloped (labelled "e") by ongoing palagonitization, and the corrosion front (cf) associated with etched radiation damage. Inset BSE image: Paleoproterozoic (1883.0 ± 1.4 Ma) zircon (Z) from the BD2 mafic dyke swarm, India, highlighting a similar corrosion front developed in the relatively U- and Th-rich (radiation damaged) zircon core during recent tropical weathering (Appendix A in French [194]; Figures 1 and 2 in French [195]). (b) Close-up from (a) highlighting fresh basaltic glass, vesicles, white (K-Al-Si)-rich devitrified zones exhibiting weak evidence of axiolitic internal microtextures, palagonite, and alpha-recoil track etch-tunnels (ARTETs). (c) Close-up from (b) highlighting the characteristic tortuosity of ARTETs in fresh basaltic glass and places where necking (N) has pinched off certain portions of these nanotunnels. This image is a photomosaic of numerous image fragments taken at 15 different focal depths throughout the entire depth of the ~30 μm thick petrographic thin section (a single image from this same area—see (b)—is shown in Figure 1(f), in which the alpha-recoil tracks appear to dive in and out of focus). Two inset SEM (secondary electron) images highlight the tiny size of these ARTETs, where they intersect the surface of the polished thin section (~100–200 nm wide), that is, about the same size as an alpha-recoil track (or ART; see ~120 nm pink dots in Figures 7(c) and 9(d)). (d) Region where palagonite fingers (PF) have now "overprinted" previously existing ARTETs. (e) Four regions where incipient ARTETs have been affected by prolonged overetching (i.e., etch-tunnel widening), possibly in relation to pressure solution, resulting in a diversity of etch-tunnel sizes and shapes, including elongate wide tunnels (EWT), string-of-pearls texture (SOP), irregular bulbous cavities (IBC), and boudinaged tunnels (BT). Note: the zone of incipient ARTETs (at right in the left image) is a photomosaic of five focal depths, and the two small images at far right are of the same region but different focal depths. (f) Two photomicrographs highlighting development of "granular palagonite ART alteration" (GP), as well as additional examples of palagonite fingers (PF) and a single (ARTET).

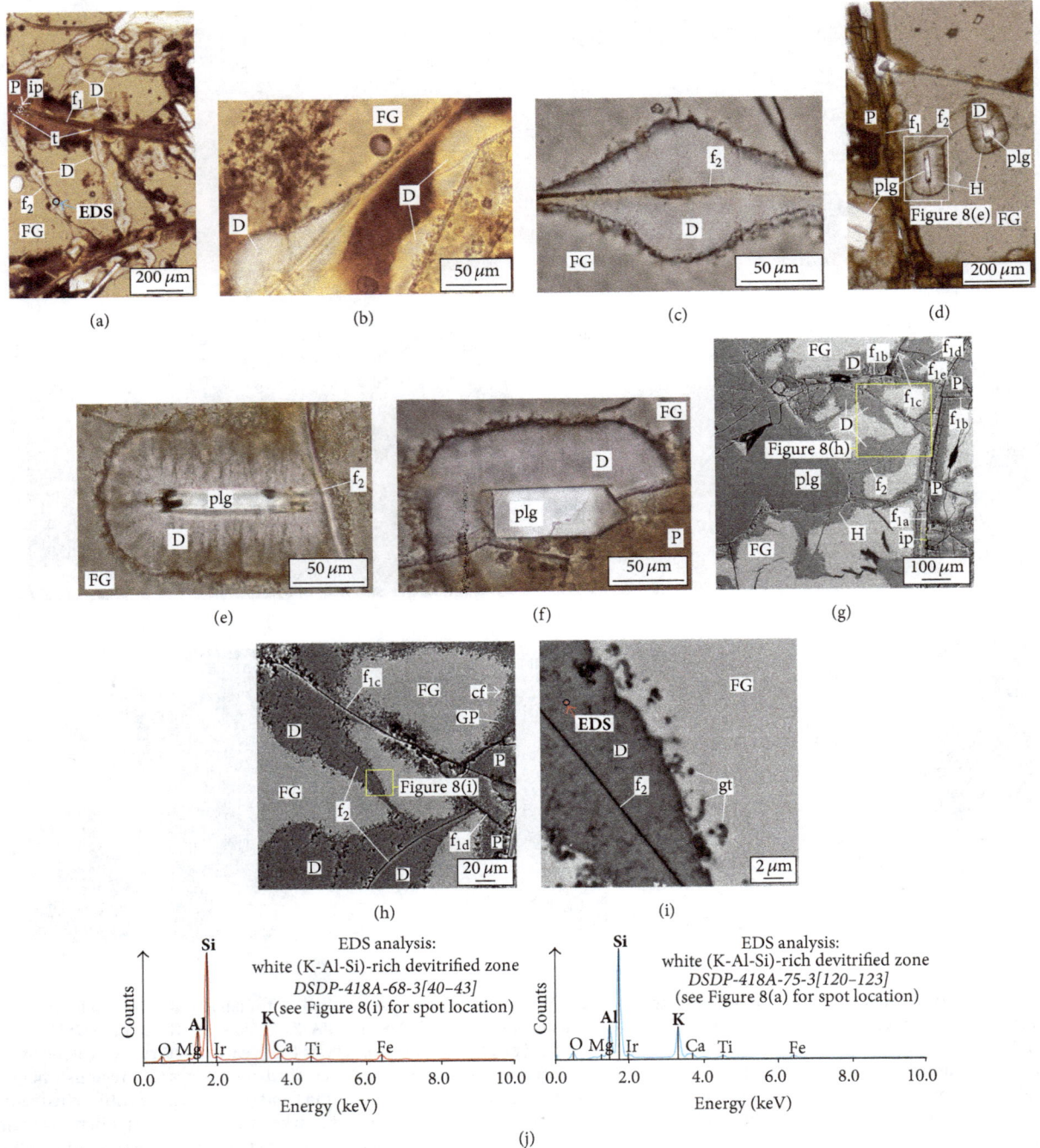

FIGURE 8: Petrographic and SEM study of white (K-Al-Si)-rich devitrified zones (in thin section). (a–f) Photomicrographs in plane polarized light, highlighting development of "lip-shaped" (K-Al-Si)-rich devitrified zones along late (f_2) fractures (a–c), as well as halos around some plagioclase crystals (d–f). (a, b): sample *DSDP-418A-75-3[120–123]*; (d, e): sample *DSDP-418A-68-3[40–43]*; (c, f): sample *DSDP-418A-72-4[13–15]*. (g–i) SEM (BSE) images of a polished thin section of sample *DSDP-418A-68-3[40–43]*, highlighting white (K-Al-Si)-rich devitrified zones. Note: palagonitization of basaltic glass along early (f_1) fractures (a, d, g, and h) and the stepwise/punctuated "cracking" of basaltic glass (from f_{1a} through f_{1e} in (g, h) and from f_1 through f_2 in (a, d, g, and h)), inferred based on the pattern of termination of successively younger fractures against older ones. (j) SEM EDS spectra for white (K-Al-Si)-rich devitrified zones (see petrographic context in (a, i)), showing abundant Si, Al, and K—consistent with microcrystalline K-feldspar ± quartz or cristobalite. cf: corrosion front; D: white (K-Al-Si)-rich devitrified zones; f_1 (including f_{1a}–f_{1e}): early fractures along which palagonite is rooted; f_2: late fractures along which white (K-Al-Si)-rich devitrified zones are rooted; FG: fresh basaltic glass; GP: granular palagonite ART alteration; gt: minor granular corrosion texture associated with devitrified zones; H: halos of white (K-Al-Si)-rich devitrified zones around plagioclase; ip: initial palagonite; P: palagonite; plg: plagioclase.

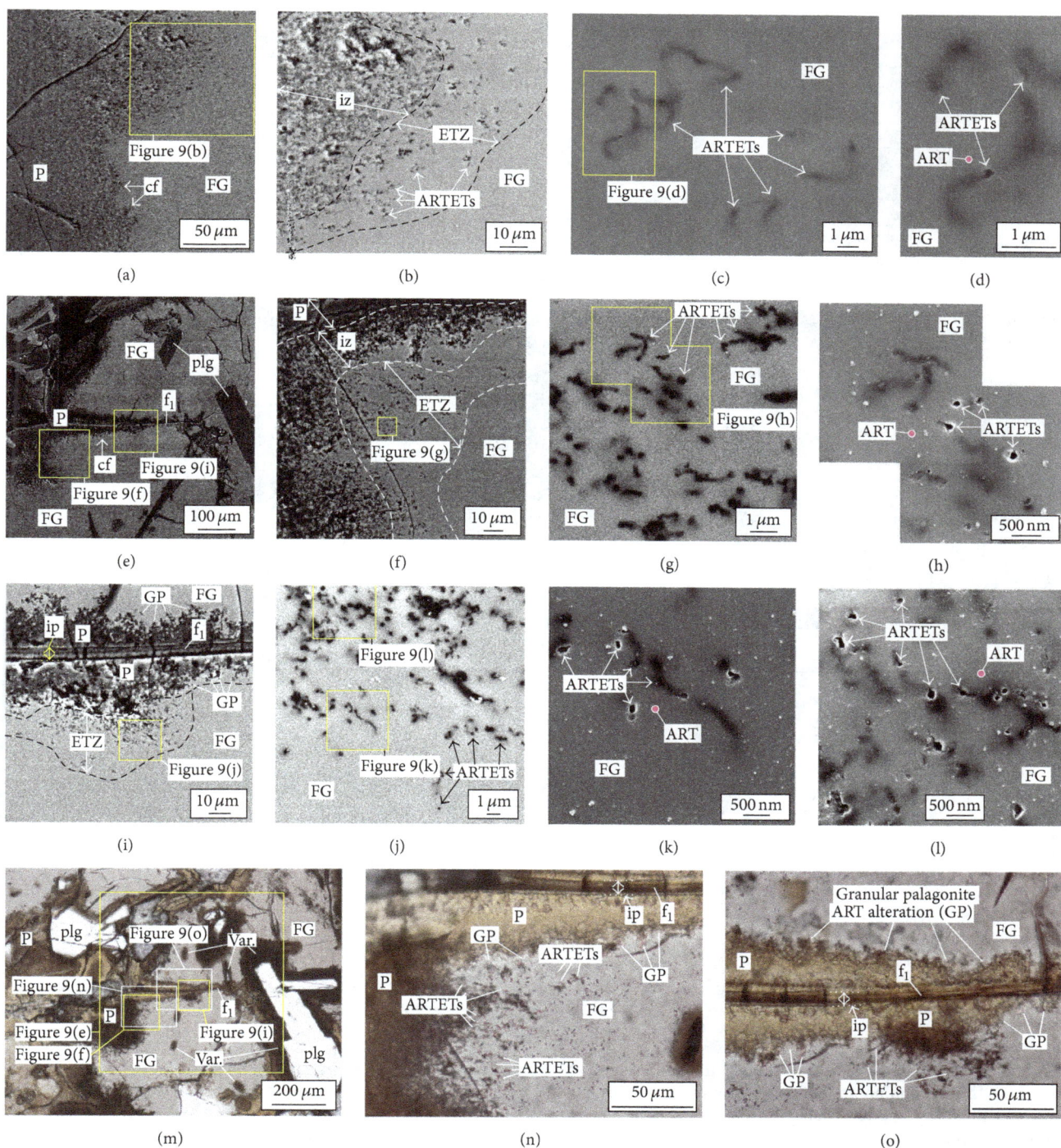

FIGURE 9: Petrographic and SEM study of the alpha-recoil track etch-tunnel (ARTET) zone at the glass-palagonite interface (in polished petrographic thin section). (a–d): sample *DSDP-418A-75-3[120–123]*; (e–o): sample *DSDP-418A-68-3[40–43]*. The images shown in (a, b, e–g, i, and j) are BSE images, while (c, d, h, k, and l) are secondary electron images. Note: (c) and (d) are close-ups from Figure 7(c). Pink dots in (d, h, k, and l) represent hypothetical ARTs—note the striking similarity in size between the real etch-tunnels and the ARTs. Photomicrographs (taken in plane polarized light) in (m–o) show the same petrographic area shown in (e–l). ART: alpha-recoil track; ARTETs: alpha-recoil track etch-tunnels; cf: corrosion front; ETZ: etch-tunnel zone; f_1: early fractures associated with palagonite; FG: fresh basaltic glass; GP: granular palagonite ART alteration; ip: initial palagonite; iz: intermediate zone (between the etch-tunnel and palagonite zones); P: palagonite zone; plg: plagioclase; Var.: varioles.

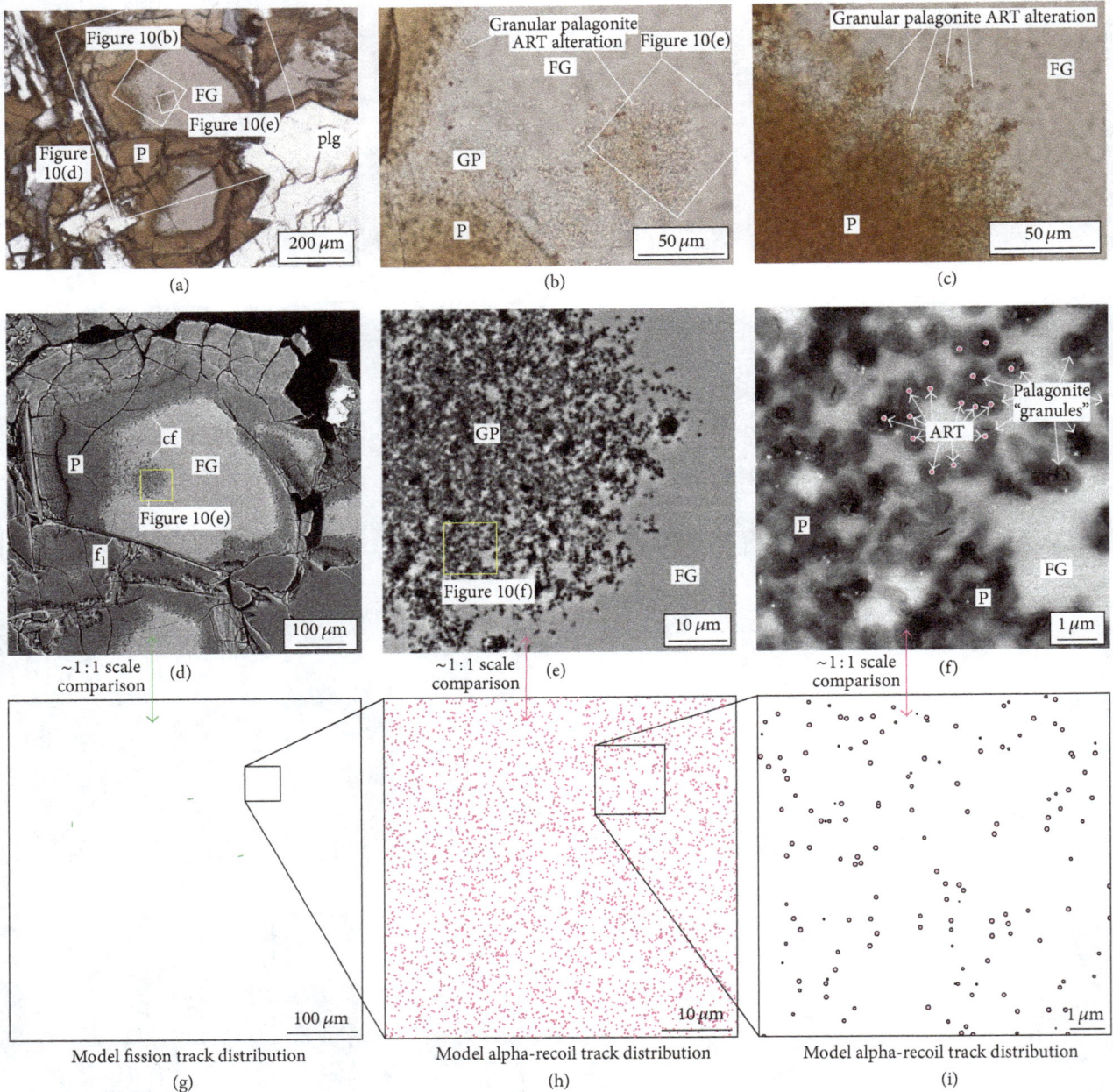

FIGURE 10: Petrographic, SEM, and theoretical modelling study of "granular palagonite ART alteration" microtexture in DSDP 418A basaltic glass. (a–c) Thin section photomicrographs (in plane polarized light) of samples *DSDP-418A-68-3[40–43]* (a, b) and *DSDP-418A-75-3[120–123]* (c), highlighting "granular palagonite ART alteration." (d–f) Close-up SEM (BSE) images from (a, b). (g–i) Theoretical plots of model fission track (g) and alpha-recoil track (h, i) areal distributions in DSDP 418A basaltic glass (calculated using (1) and (2); fission tracks are shown in green (g) and alpha-recoil tracks in pink (h, i)). The model track distributions in (g–i) are shown at approximately the same scale as the SEM images in (d–f), respectively—see green and pink arrows. Note that fission tracks are quite sparse (g), but alpha-recoil tracks are quite abundant and correlate very well with the observed pattern of development and areal distribution of ~0.3–1.0 μm palagonite "granules" (i.e., compare (h) and (e), and (i) and (f)), thus indicating that, during corrosion of basaltic glass by seawater, granular palagonite microtextures develop through selective palagonitization of alpha-recoil tracks (and not microbial activity). To emphasize this idea, several hypothetical, previously existing alpha-recoil tracks (120 nm pink dots) are plotted in (f), which would have acted as ideal "point sources" of radiation damage amenable to preferential corrosion/palagonitization. ART: alpha-recoil track; cf: corrosion front; f_1: early fractures associated with palagonite; FG: fresh basaltic glass; GP: granular palagonite ART alteration; P: palagonite; plg: plagioclase.

secondary orange-brown palagonite along fractures during low temperature aqueous alteration/corrosion, hydration, and dissolution of basaltic glass by infiltrating seawater (Figure 7; see further explanation in Sections 3.2 and 3.3—also see reviews on palagonite/palagonitization in Crovisier et al. [136] and Stroncik and Schmincke [182]), and the formation of white (K-Al-Si)-rich devitrified zones is explained below in Section 3.1. The development of such orange-brown palagonite during aqueous alteration of basaltic glass is a common feature in both submarine [44] and terrestrial (e.g., Icelandic [136, 183] and Hawaiian [110]) basaltic glasses, and data on the mineralogical and chemical characteristics of typical palagonites formed from weathered/altered basaltic glass can be found elsewhere [44, 110, 136, 182, 183].

3.1. White (K-Al-Si)-Rich Devitrified Zones.

The white devitrified zones ("D" in Figures 7(a), 7(b), and 8) are typically lip-shaped (i.e., pinch and swell shaped and symmetrical about a central fracture/axis: Figures 8(a)–8(c)) or occur as ~50 μm thick rims/halos surrounding plagioclase phenocrysts ("H" in Figures 8(d)–8(g)), range up to ~100 μm in thickness at the swells (e.g., Figure 8(c)), have relatively sharp curvilinear contacts with basaltic glass (i.e., smooth and sharp glass–devitrification interfaces: Figure 8(c)), and are composed primarily of microcrystalline K-feldspar (i.e., are K-, Al-, and especially Si-rich based on EDS analysis (by SEM): see Figure 8(j)) (±quartz or cristobalite?)—similar in composition to some other examples of devitrified volcanic glass (p. 418 in Cas and Wright [184], [185]).

Locally, these white (K-Al-Si)-rich devitrified zones exhibit a weakly fibrous internal microtexture, visible with transmitted light microscopy (Figure 8(b)) and define an overall axiolitic structure (i.e., with fibers growing outward from the observed linear fractures along the central axis of the pinch and swell structures)—similar in nature to the axiolitic devitrification textures commonly observed in rhyolitic glasses [185]. Similarly, the white devitrified zones that form halos around plagioclase phenocrysts also exhibit a fibrous internal microtexture, but in this case the fibers appear to radiate around the phenocrysts, maintaining perpendicularity to the plagioclase contact (Figures 8(d) and 8(e)). High resolution SEM imaging of these white devitrified zones does not reveal any evidence of this fibrous/axiolitic microstructure and instead shows a more mottled and even textured material (Figures 8(h) and 8(i)); however, this is in keeping with the definition of axiolitic structure defined by A. Allaby and M. Allaby [185], which states that such axiolitic fibers are typically only visible by petrographic microscope.

The white (K-Al-Si)-rich devitrified zones described here are similar in size, shape, geological context, and chemical composition to light coloured K-rich zones documented in a previous study of DSDP 418A basaltic glasses [186], where they were also found to occur in basaltic glass along some fractures and as rims around some plagioclase phenocrysts and interpreted as poorly crystalline, secondary K-feldspar.

Although the glass-devitrification interface is for the most part quite sharp and curvilinear (e.g., see boundary between "D" and "FG" in Figure 8(c)), close-up SEM imaging

reveals the presence of a minor amount of "granular textured" corrosion features (defined by ~0.5–1 μm wide granules) that extend outwards (typically <10 μm) into fresh basaltic glass ("gt" in Figure 8(i)), which we interpret as incipient corrosion of previously formed alpha-recoil tracks in the glass during devitrification. This interpretation is based on the similarity in size and form of these corrosion microtextures to other such "granular textures" described from the glass-palagonite interface that are considered to have formed by preferential corrosion of multitudes of randomly distributed alpha-recoil tracks (see Figure 10 and Section 3.3.2). However, because the formation of these white devitrified zones is presumed to have taken place relatively early on in the alteration history of these rocks (<~1 million years after pillow eruption—possibly during burial beneath the overlying volcanic pile—see below), only a minor amount of alpha-recoil tracks were likely to have been present during devitrification—resulting in only minor/sparse development of granular corrosion microtextures at the glass-devitrification interface (Figure 8(i)).

Aside from being the possible end product of the solid-state transformation of basaltic glass into poorly crystalline materials (i.e., secondary K-feldspar ± quartz or cristobalite?) during devitrification [185], the formation of these white axiolitic devitrified zones (and associated f_2 fractures) and halos in DSDP 418A basaltic glass might have been triggered as a diagenetic/low-grade-metamorphic response (cf. p. 418 in Cas and Wright [184]) to deep burial (to 408–461 m) of these glasses beneath the overlying volcanic pile (Figure 5(b)), late in the history of the spreading ridge. High temperature devitrification of submarine glasses (i.e., penecontemporaneous with eruption and quenching) commonly results in the formation of varioles (also known as "variolites" [184] or "spherulites"—in glasses with more felsic compositions [187, 188]), which are typically comprised of tiny radiating crystals of clinopyroxene and/or plagioclase (±quartz or cristobalite) [184, 187, 188], but this can be ruled out in the present case because the fractures along which these white devitrified zones occur (i.e., "f_2" in Figures 7 and 8) clearly truncate ("t" in Figures 7(a) and 8(a)) against an earlier-formed set of fractures ("f_1" in Figures 7(a) and 8(a)) along which "low temperature" palagonite had already formed (e.g., "ip" in Figures 7(a), 8(a), and 8(g)) and, moreover, the mere fact that these white devitrified zones occur along fractures suggests that they must have formed sometime after the glass had cooled below the glass transition temperature (i.e., ~600–700°C [189]) to allow the fracturing to occur in the first place. Furthermore, the composition of these white (K-Al-Si)-rich devitrified zones is consistent with poorly crystalline K-feldspar (Figure 8(j)) and not plagioclase or clinopyroxene, and they do not form masses of coalescing spheroidal bodies (of radiate fibers) that are typical of high temperature devitrification textures (e.g., Figures F14, F15, F23, and F31 in Shipboard Scientific Party [187]). However, some minor occurrences of varioles do exist locally in some of the basaltic glass pillow margin samples studied (e.g., "var." in Figure 9(m)), and although they also occur as halos around some plagioclase phenocrysts, they are distinctly different from the white (K-Al-Si)-rich devitrified zones in that they are dark brown in colour (i.e., resemble very dark brown basaltic glass: Figure 9(m)) and are interpreted as primary (high temperature) igneous quench features—similar in nature to the dark-coloured varioles described by Fisk and McLoughlin [103]. Another possible explanation for the origin of the white (K-Al-Si)-rich devitrified zones (Figure 8) in DSDP 418A glasses (i.e., aside from diagenetic/low-grade metamorphic response to deep burial—alluded to above) is that they are more externally linked to an episode of late alkalic hydrothermal alteration (i.e., metasomatism) affecting these pillow basalts, given that late-stage (i.e., renewed or continued) "off-axis" alkali fixation is known to occur in the alteration history of the oceanic crust [164] and that K-feldspar can be a product of hydrothermal alteration of glassy volcanic rocks [190].

3.2. Stepwise Development of Fracturing, White Devitrified Zones, and Palagonitization.

The textural relationships observed between different generations of fractures (f_1 and f_2), development of white (K-Al-Si)-rich devitrified zones, and stepwise/incremental encroachment of orange-brown palagonite alteration zones in these volcanic glasses are somewhat complex (e.g., Figures 7(a), 7(b), 8(a), 8(d), 8(g), and 8(h)). Initially (sometime soon after pillow eruption and quenching of basaltic glass), these glassy pillow margins underwent f_1 fracturing followed by infiltration of seawater and incipient palagonitization along these fractures, resulting in thin (typically 10–50 μm thick), straight, *initial* palagonitic layers ("ip" in Figures 7(a), 8(a), 8(g), 9(i), 9(n), and 9(o)) that are parallel to (and occur along) f_1 fractures. Some *later* "f_1" fractures probably formed during the initial stages of development of the volcanic pile (early burial) and are considered to have formed as a series of individual discrete curviplanar fractures (i.e., not occurring in parallel sets) that represent incremental "cracking" of glass in stages (e.g., see f_{1a} to f_{1e} in Figure 8(g)). The formation of initial orange-brown palagonitic layers along f_1 fractures ("ip" in Figures 7(a), 8(a), 8(g), 9(i), 9(n), and 9(o)) is interpreted to have occurred during the first several thousand years of glass alteration (palagonitization) by infiltrating seawater, during which radiation damage in the glass was quite minimal and therefore the glass-palagonite interface on either side of the f_1 fractures remained quite sharp and parallel to the f_1 fractures—essentially equivalent to the straight, fracture-parallel, palagonite alteration textures observed in young Icelandic basaltic glasses (see Figure 1 in Crovisier et al. [136]) and also in some submarine basaltic glasses (i.e., so-called "abiotic" palagonite alteration described in Section 4.1 of Furnes et al. [59], Figure 1 in Staudigel et al. [75], Figure 13(a) in Furnes et al. [11], Figure 1 in Furnes et al. [26], Figure 2 in McLoughlin et al. [31], and Figure 1 in Staudigel et al. [27]). These relict "sharp contacts" can still be seen ("ip" in Figures 7(a), 8(a), 8(g), 9(i), 9(n), and 9(o)), even though they have now been overgrown by a subsequent/ongoing stage of palagonitization ("P" in Figures 7(a), 8(a), 8(g), 9(i), 9(n), and 9(o)) discussed below (in Section 3.3). Such slow advancement of the glass-palagonite interface (i.e., on the order of microns over timescales of thousands or millions of years) is a well-known aspect of the palagonitization process [136] and is probably in part due to the protective effect of the alteration layer [136, 191] and has even prompted some to consider using palagonite rind thickness as an archaeological dating tool for certain basaltic glass/obsidian artefacts [40].

Subsequent to the development of these narrow incipient (i.e., initial) palagonitic layers along f_1 fractures in basaltic glass ("ip" in Figures 7(a), 8(a), 8(g), 9(i), 9(n), and 9(o)), a second stage of fracturing (f_2) of these glasses took place along which white (K-Al-Si)-rich devitrified zones developed (Figures 7(a), 7(b) and 8) and as highlighted above (in Section 3.1) this probably took place in response to deep burial of these pillow lavas beneath some 450 m of overlying lavas (perhaps 50 to 100 thousand years after their initial eruption, during the building up of ancient "axial volcanic ridges" (which may be up to ~600 m high [192]) on the floor of the median valley of the Early Cretaceous Mid-Atlantic Ridge). Where they intersect the earlier-formed (f_1) fractures (along which palagonite is rooted), these second-stage (f_2) fractures are observed to terminate against them ("t" in Figures 7(a) and 8(a))—thus indicating their "younger" relative age—which is consistent with late-stage (f_2) "cracking" (and associated devitrification) of the intervening basaltic glass that occurs between preexisting f_1 fractures (i.e., during deep burial).

Ongoing palagonitization of basaltic glass by infiltrating seawater (i.e., *still* rooted along f_1 fractures) then appears to have continued long after the formation of white (K-Al-Si)-rich devitrified zones and associated f_2 fractures, as evidenced by the advancement of orange-brown palagonite ("P" in Figure 7(a)) that has partially enveloped/overgrown preexisting white (K-Al-Si)-rich devitrified zones (oval labelled "e" in Figure 7(a)). Unlike the early-formed, narrow (<50 μm) layers of incipient palagonite ("ip" in Figures 7(a), 8(a), 8(g), 9(i), 9(n), and 9(o)), which seem to exhibit relict "sharp contacts" with preexisting glass, this next stage of continued (and ongoing) palagonitization of basaltic glass ("P" in Figures 7–10) extends for distances of up to several hundred microns away from the f_1 fractures and appears to have been advancing in the wake of an irregular and complex corrosion front ("cf" in Figures 7(a), 8(h), 9(a), 9(e), and 10(d)) that occurs at the present-day glass-palagonite interface. This corrosion front is locally characterized by a pronounced etch-tunnel zone (Figures 1(b), 1(d), 1(f), 7, and 9) that extends out in front of the palagonite zone into fresh basaltic glass for distances of up to a few hundred microns (e.g., Figure 7(b))—in addition to regions characterized by "granular" palagonite microtextures (Figures 7(f), 8(h), 9(i), 9(n), 9(o), and 10(a)–10(f)). The nature and origin of this complex corrosion front—and associated "etch-tunnel zone" and "granular palagonite microtexture"—is the main focus of the present study on DSDP 418A basaltic glass, because in many previous studies of basaltic glasses from this drill site, such alteration microtextures have classically been interpreted as evidence for microbial activity (i.e., biocorrosion/bioalteration) [11, 26–28, 31, 33, 59, 61, 93, 94, 100]—whereas we think that they instead represent evidence of "abiotic" corrosion (palagonitization and etch-tunnelling: by seawater) of dense concentrations of randomly distributed, radiation damaged sites in the glass (p. 10 in French and Muehlenbachs [107], [111–113, 139]).

Therefore as a first step, in the next few Sections we provide a range of detailed observations of complex corrosion microtextures that occur at the glass-palagonite interface in these studied samples of basaltic glass pillow margins from DSDP 418A.

3.3. Complex Corrosion Microtextures Associated with Ongoing Palagonitization.
Palagonitization is interpreted to have started early—soon after quenching (in the Early Cretaceous), in association with f_1 fracturing (as outlined in Section 3.2)—but also to have been episodic and ongoing in these rocks, possibly right up until the point of drilling and sample collection. The present-day glass-palagonite interface is very irregular and mottled in form due to the presence of a complex corrosion front that occurs there ("cf" in Figures 7(a), 8(h), 9(a), 9(e), and 10(d))—the formation of which seems to occur in advance of the encroaching orange-brown palagonite by some kind of initial corrosion/dissolution/etch-tunnelling process that takes place within fresh basaltic glass. Similar alteration microtextures have also been observed in the interior of some zircon grains caused by preferential corrosion of high U and Th (radiation damaged) regions during weathering (e.g., "cf" in Figure 7(a) (inset BSE image); Figure 9 in Lumpkin [193]; Appendix A in French [194]; Figure 1(d) in French [195]), and thus, by comparison, the irregular corrosion front observed here ("cf" in Figures 7(a), 8(h), 9(a), 9(e), and 10(d)) may also have been caused by preferential corrosion of radiation damaged regions of basaltic glass by seawater. In polished petrographic thin section, this corrosion front at the glass-palagonite interface in DSDP 418A basaltic glass can be subdivided into two distinct microtextural varieties including an "etch-tunnel zone" (i.e., "ARTETs" in Figures 1, 7, and 9) and a "granular palagonite alteration zone" (i.e., "granular palagonite ART alteration" or "GP" in Figures 2(b), 2(d), 7(f), 8(h), 9(i), 9(n), 9(o), and 10(a)–10(f)), which are described below in Sections 3.3.1 and 3.3.2, respectively.

3.3.1. The Etch-Tunnel Zone at the Glass-Palagonite Interface.
Palagonite at the glass-palagonite interface locally appears to be superficially much darker in colour (i.e., almost black)—when viewed under plane polarized light by petrographic microscope—because it grades into a complex network of nanoscopic etch-tunnels that scatter and absorb light (Figures 1(b), 7(a), 7(b), and 9(m)–9(o)). These etch-tunnels extend outwards into the fresh glass for up to several hundred microns past the palagonite zone (Figure 7(b)), tracing out intricate 3-dimensional curvilinear pathways through the glass (Figure 7(c)) that exhibit a high degree of tortuosity (i.e., highly anastamosing patterns), which allows only a small portion of them to be brought into focus at a given time under petrographic microscope (e.g., Figure 1(f)). A mosaic of many photomicrographic image fragments taken at 15 different focal depths shows a more complete "in focus" representation of the tunnel networks observed in a ~30 μm thick petrographic section (Figure 7(c)). SEM

imaging reveals that where they intersect the surface of the thin section, the tunnels imaged in these photomicrographs of sample *DSDP-418A-75-3-[120–123]* (Figures 1(f) and 7(a)–7(c)) are ~120 nm wide (see Figures 7(c) (inset secondary electron images) and 9(c), 9(d))—essentially identical in size as a typical alpha-recoil track (~120 nm in diameter [129]; see pink dots in Figure 7(c) (inset) and in Figure 9(d)). Similar SEM imaging of the etch-tunnel zone preserved at the glass-palagonite interface in a polished petrographic section of sample *DSDP-418A-68-3[40–43]* (Figures 9(e)–9(o)), ~50 m higher up in the volcanostratigraphic succession (Figure 5(b)) also reveals that these etch-tunnels at the glass-palagonite interface are typically ~120 nm wide (i.e., compare the diameters of hypothetical alpha-recoil tracks (pink dots) with those of the etch-tunnels—i.e., "ARTETs"—in Figures 9(h), 9(k), and 9(l)). The similarity in diameter between hypothetical alpha-recoil tracks and the observed etch-tunnels (i.e., circa 120 nm in both cases; Figures 9(d), 9(h), 9(k), and 9(l)) at the glass-palagonite interface in two different DSDP 418A pillow margin samples provides compelling evidence that this etch-tunnelling and corrosion which takes place in advance of the palagonitization front occurs primarily due to preferential dissolution (by seawater) of multitudes of randomly distributed alpha-recoil tracks in the glass—in contrast with previous biogenic (i.e., microbial trace fossil) interpretations (i.e., of "tubular" texture at DSDP 418A [11, 26–28, 31, 33, 59, 61, 93, 94, 100]). For that reason (also see Sections 3.4 and 5), we propose that the majority of etch-tunnels observed at the glass-palagonite interface in polished petrographic sections of DSDP 418A basaltic glass (e.g., Figures 1(a)–1(d), 1(f), 7(a)–7(c), 7(e), and 9) can now be interpreted as "alpha-recoil track etch-tunnels" (ARTETs).

Subsequent to their formation at the glass-palagonite interface, late secondary modification of some of these alpha-recoil track etch-tunnels has also locally taken place. For instance, postdissolutional "necking" of tunnel walls ("N" in Figure 7(c)) has healed/closed-off many portions of these alpha-recoil track etch-tunnels ("ARTETs" in Figure 7(c)) into what are now isolated, elongate (several microns long), narrow (~120 nm wide) fluid inclusions that presumably contain seawater. In addition, at many places along the glass-palagonite interface, encroachment of palagonite appears to have completely overprinted/destroyed preexisting alpha-recoil track etch-tunnels, preserving their elongate curvilinear form as "palagonite fingers" that extend outward into fresh glass (Figures 7(d) and 7(f)). Furthermore, prolonged overetching (perhaps caused by pressure solution etch-tunnelling—this idea is explored further in Section 5.3) has apparently widened some incipient alpha-recoil track etch-tunnels to significantly larger tunnels of myriad shapes (Figure 7(e)), including elongate wide tunnels (EWT), string-of-pearls (SOP) texture (cf. string-of-beads texture in Figure 1(f) of Fisk et al. [24], and string-of-pearls texture described in Figure F68 of Shipboard Scientific Party [60] and Banerjee and Muehlenbachs [64]), irregular bulbous cavities (IBC), and boudinaged tunnels (BT: i.e., shaped like sausage links—but not "stretched" as in other examples of boudinaged rocks in the field of structural geology). In the studied samples, these larger "overetched" tunnels (Figure 7(e)) are somewhat rare and atypical in comparison to the more common ~120 nm wide "incipient" alpha-recoil track etch-tunnels (Figures 1(b), 1(d), 1(f), 7(b), 7(c), 7(e), and 9), which represent >95% of all tunnels observed in the etch-tunnel zone at the glass-palagonite interface in basaltic glass pillow margins in this study.

3.3.2. Granular Palagonite ART Alteration Microtextures. In many places along the glass-palagonite interface, alpha-recoil track etch-tunnels are abundant (e.g., Figures 7 and 9)—but equally common along this interface is a distinctly different type of corrosion microtexture: granular palagonite ART (alpha-recoil track) alteration (Figure 10; also see Figures 2(b), 2(d), 7(f), 8(h), 9(i), 9(n), and 9(o)). For instance, along the f_1 fracture highlighted in Figures 9(e), 9(i), and 9(m)–9(o), a prominent alpha-recoil track etch-tunnel zone occurs at the glass-palagonite interface on one side (i.e., the bottom side: "ETZ" in Figures 9(f) and 9(i) and "ARTETs" in Figures 9(n) and 9(o)) in addition to abundant granular palagonite ART alteration ("GP" in Figures 9(i), 9(n), and 9(o)). In contrast, the glass-palagonite interface on the opposite side of this fracture exhibits only "granular palagonite ART alteration" microtextures (i.e., "GP," top side of the fracture in Figure 9(i), and "granular palagonite ART alteration" in Figure 9(o)).

Although such "granular" palagonite alteration microtextures may occur adjacent to alpha-recoil track etch-tunnels (as in the previous example), this style of corrosion/palagonitization of basaltic glass is not considered to form by any kind of prior dissolution (e.g., etch-tunnelling) process—in contrast with "palagonite fingers" (e.g., Figures 7(d) and 7(f)) which appear to form by overprinting of alpha-recoil track etch-tunnels by encroaching palagonite (Figures 11(a)–11(c)). Instead, "granular" palagonite seems to represent a distinct type of low temperature aqueous *alteration* (i.e., palagonitization) by seawater that takes place in a very spotty/mottled/granular fashion by preferential leaching, diffusion, hydrolysis, and reaction of chemical components at specific nucleation sites in the glass (i.e., numerous "point sources" of damage in the glass, which we attribute (below) to the presence of randomly distributed alpha-recoil tracks: Figures 10(e), 10(f), 10(h), and 10(i)). The individual palagonite spots (or granules) reported here range from about ~0.3 to ~1.0 μm in diameter but are most commonly ~0.6 μm across (Figures 10(e) and 10(f))—similar in size to other known examples of granular palagonite alteration in submarine glasses worldwide (i.e., those attributed to microbial activity: ~0.1–1.3 μm, but most commonly ~0.2–0.6 μm [11, 26, 28, 34, 59]; e.g., compare Figures 2(c) and 2(d))—and they tend to occur in dense constellations that form a spotty/mottled transition zone between fresh basaltic glass ("FG" in Figures 10(a)–10(d)) and glass that has been completely altered to palagonite ("P" in Figures 10(a)–10(d)). We interpret this pattern of alteration as evidence that the infiltration of seawater into the glass during palagonitization takes place not only by preferential etch-tunnelling through alpha-recoil track damaged sites in the glass (i.e., as described

in Section 3.3.1 and Figure 9), but also through "selective palagonitization" of multitudes of randomly distributed alpha-recoil tracks—which represent ideal point sources of radiation damage that are more amenable to chemical attack than surrounding glass—resulting in a characteristic "granular" palagonite microtexture (Figures 10(e), 10(f), 10(h), and 10(i)). Key evidence to support this claim comes from the observation that the areal density and distribution of palagonite granules imaged by SEM (e.g., Figures 10(e) and 10(f)) matches closely the calculated areal density and distribution of "model" alpha-recoil tracks in DSDP 418A basaltic glass (Figures 10(h) and 10(i))—predicted in our theoretical modelling study (Section 5). In addition, the characteristically small size of these palagonite "granules" (~600 nm) is only slightly larger than the diameter of these putative previously existing alpha-recoil tracks (typically ~120 nm [129]; see pink dots labelled "ART" in Figure 10(f)); that is, the ~120 nm alpha-recoil tracks provide ideal "point sources" for selective palagonitization that are inherent in these old glasses. However, during the process of selective palagonitization, some of the surrounding glass is evidently also palagonitized, resulting in a slightly larger size of ~600 nm for these preferentially palagonitized alpha-recoil tracks (i.e., "palagonite granules" in Figure 10(f)). Therefore, we propose that granular palagonite alteration microtextures found in submarine volcanic glasses worldwide (e.g., [11, 28, 34, 59] and this study) do not represent evidence of microbial activity/bioalteration or the presence of a global lithoautotrophic microbial community thriving at the glass-palagonite interface in submarine glasses (e.g., [11, 28, 34, 59]) but instead are a reflection of the preferential *abiotic* corrosion/palagonitization of multitudes of randomly distributed alpha-recoil tracks in submarine glasses by seawater—which appears to take place at the global scale because midocean ridge basaltic glasses are known to contain trace amounts of U and Th worldwide [146] and thus accumulate alpha-recoil tracks very quickly. Consequently, we suggest that the recently proposed microbial trace fossil (i.e., ichnospecies) *Granulohyalichnus vulgaris* and, in fact, the entire *Granulohyalichnus* ichnogenus [28] are now more or less in doubt in terms of biogenicity (see further discussion in Section 6.1).

3.4. High Resolution Scanning Electron Microscopy of Basaltic Glass "Chip Samples". Due to the polishing process used in the fabrication of polished petrographic thin sections, some of the more tiny (<0.1 μm) details preserved within corrosion microtextures and etch-tunnels have the potential to become slightly obscured/modified and thus difficult to image at high resolution by SEM. Therefore, in order to circumvent this problem in the present study, we also prepared hand-crushed (i.e., via mortar and pestle), sand-sized basaltic glass "chip samples" from *DSDP-418A-75-3-[120–123]* to allow for more detailed high resolution SEM imaging of corrosion microtextures (Figures 11–15)—to compliment the SEM studies of the surfaces of polished petrographic sections that we carried out above (Figures 8–10). In this way, fracture surfaces generated during crushing of the sample into small chips exposed "fresh surfaces" of the interior of the basaltic glass pillow margin for

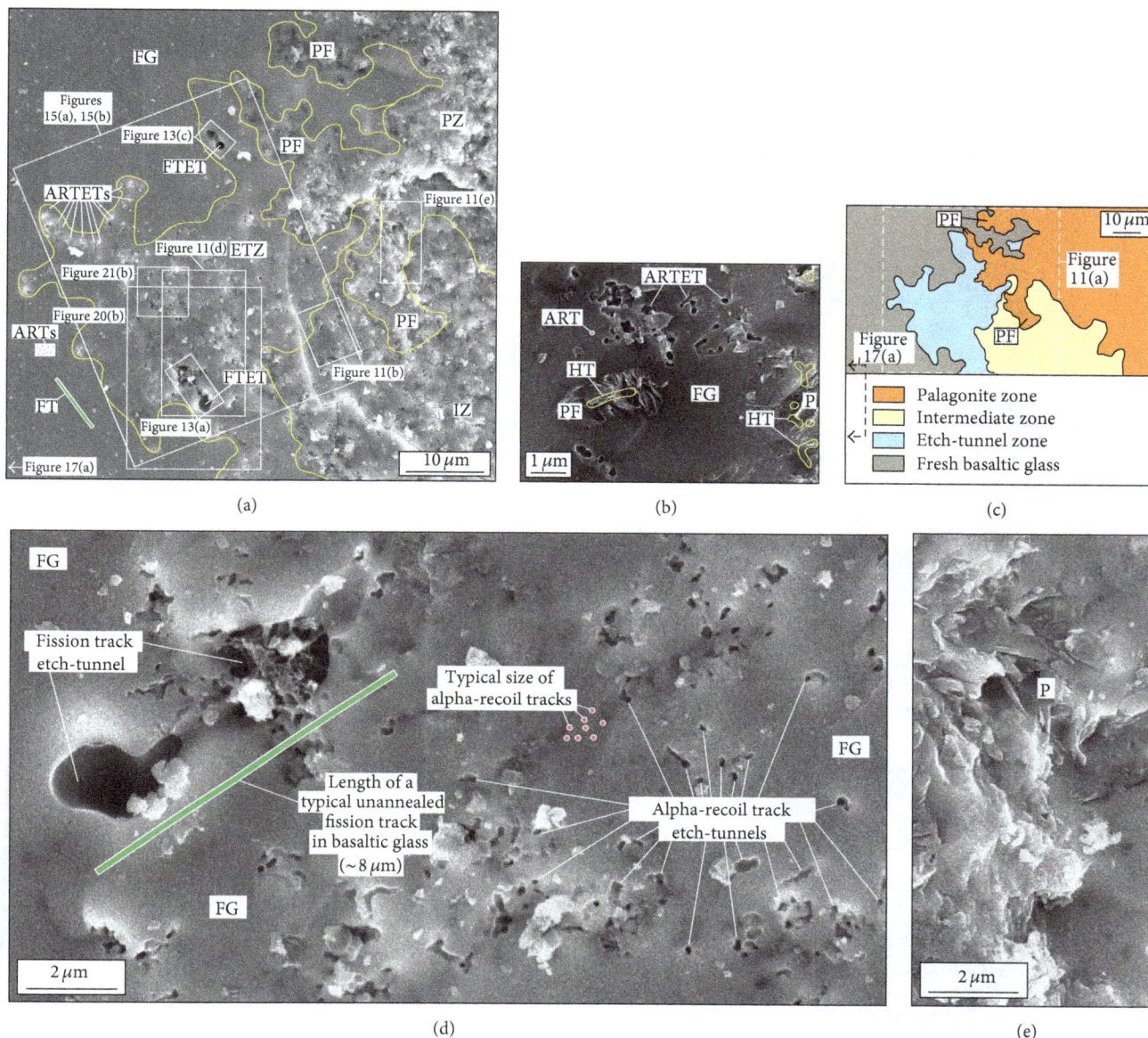

(a)

(b)

(c)

(d)

(e)

FIGURE 11: High resolution SEM mapping of the glass-palagonite interface and associated etch-tunnel zone (as observed on the freshly fractured surface of a basaltic glass "chip sample" of *DSDP-418A-75-3[120–123]*). (a, b, d, and e) Secondary electron images. (a) Overview of a representative region of the glass-palagonite interface, highlighting four microtextural domains (boundaries shown as yellow lines; also see (c)) that include two large FTETs and several hundred smaller ARTETs. (b) Close-up from (a) highlighting where the palagonite zone is encroaching upon (overprinting) the etch-tunnel zone. (c) Alteration map, showing the distribution of the four microtextural domains outlined in (a). (d) Close-up (from (a)) of the etch-tunnel zone, highlighting one FTET and several ARTETs. Note the similarity in size between the hypothetical fission track (green bar) and the FTET and between the hypothetical ARTs (pink dots) and the ARTETs. (e) Representative close-up (from (a)) of the palagonite zone. ARTs: alpha-recoil tracks; ARTETs: alpha-recoil track etch-tunnels; ETZ: etch-tunnel zone; FG: fresh basaltic glass; FT: fission track; FTET: fission track etch-tunnel; HT: hypothetical preexisting alpha-recoil track etch-tunnel; IZ: intermediate zone (between the ETZ and the PZ); P: palagonite; PZ: palagonite zone.

the high resolution SEM studies of corrosion microtextures (Figures 11–15; also see Figures 3(e) and 4(c)—left, and some images introduced in later Sections: Figures 17, 20(b), 21(b), 21(c), 21(e), and 21(f)). The sub-millimeter-sized glassy pillow margin fragments (i.e., chip samples) analysed by SEM in this study were sputtered with a 20 Å coating of iridium using a VCR group Inc IBS/TM200S Ion Beam Sputterer

and analysed using a JEOL 6301F field emission scanning electron microscope equipped with a PGT IMIX model X-ray analysis system (images presented in Figures 11(a), 11(d), 11(e), 13(a), 15(a) and some images introduced in later Sections: Figures 17, 20(b), 21(b), 21(c), 21(e), and 21(f)) or a Hitachi S-4000 scanning electron microscope (Figures 11(b), 12(d)–12(h), 13(b)–13(e), and 14(e)–14(h)), or alternatively coated

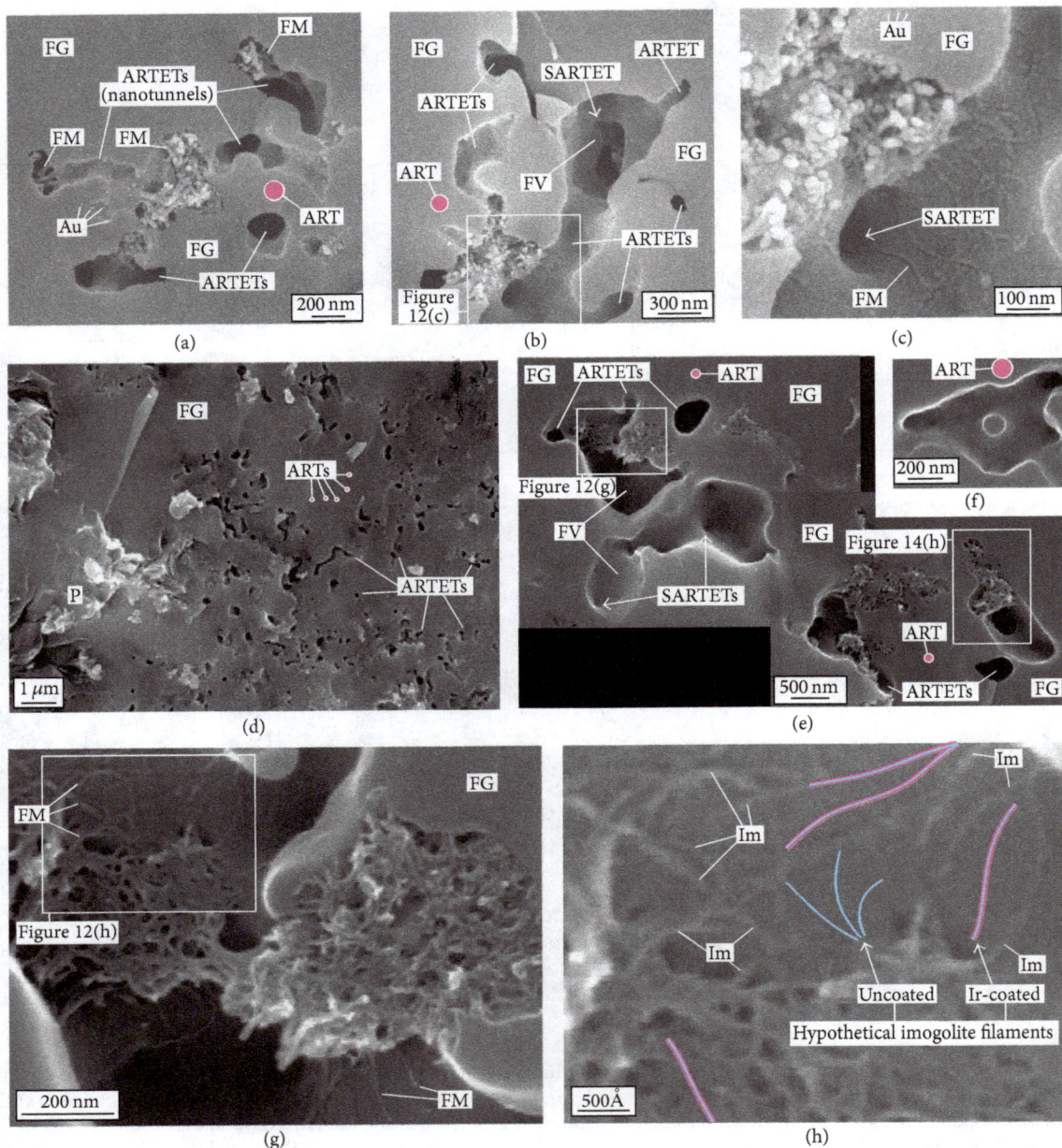

FIGURE 12: High resolution SEM (secondary electron) images of alpha-recoil track etch-tunnels (ARTETs) at the glass-palagonite interface (as observed on freshly fractured surfaces of basaltic glass "chip samples" from *DSDP-418A-75-3[120–123]*). (a–c) Representative close-up images of typical ARTETs found at the glass-palagonite interface. Note the meandering, branching nature of the etch-tunnels and the occurrence of FM within them, and note that tunnel diameters are about the same size as a typical alpha-recoil track (~120 nm pink dots labelled "ART"). Other features include "pockmarks" that represent new side-tunnels (SARTETs), and flare-out voids (interpreted as etch-tunnelling of locally dense regions of ARTs or as ARTETs affected by prolonged "overetching" ± pressure solution). (d) Overview image, showing a typical region of the glass-palagonite interface where abundant ARTETs occur, which are being encroached upon and overprinted by palagonite. (e) Overview of a region where numerous ARTETs, FVs, and SARTETs occur. (f) A region where an ARTET forms an etched-out "loop." (g) Close-up from (e), highlighting abundant FM lining tunnel walls and draping/bridging across tunnel void spaces. (h) Close-up from (g). For comparison, several hypothetical imogolite filaments are drawn to scale (in blue) (i.e., 20 Å wide, the exact thickness of a single strand of imogolite), some of which also depict the 20 Å thick coating of iridium (magenta). Note the identical thickness of these hypothetical, iridium-coated imogolite filaments and the real nanofilaments imaged by SEM. ART: alpha-recoil track; ARTETs: alpha-recoil track etch-tunnels; Au: granules of gold sputtering; FG: fresh basaltic glass; FM: filamentous material; FV: flare-out voids; Im: imogolite; P: palagonite; SARTETs: "starting" alpha-recoil track etch-tunnels.

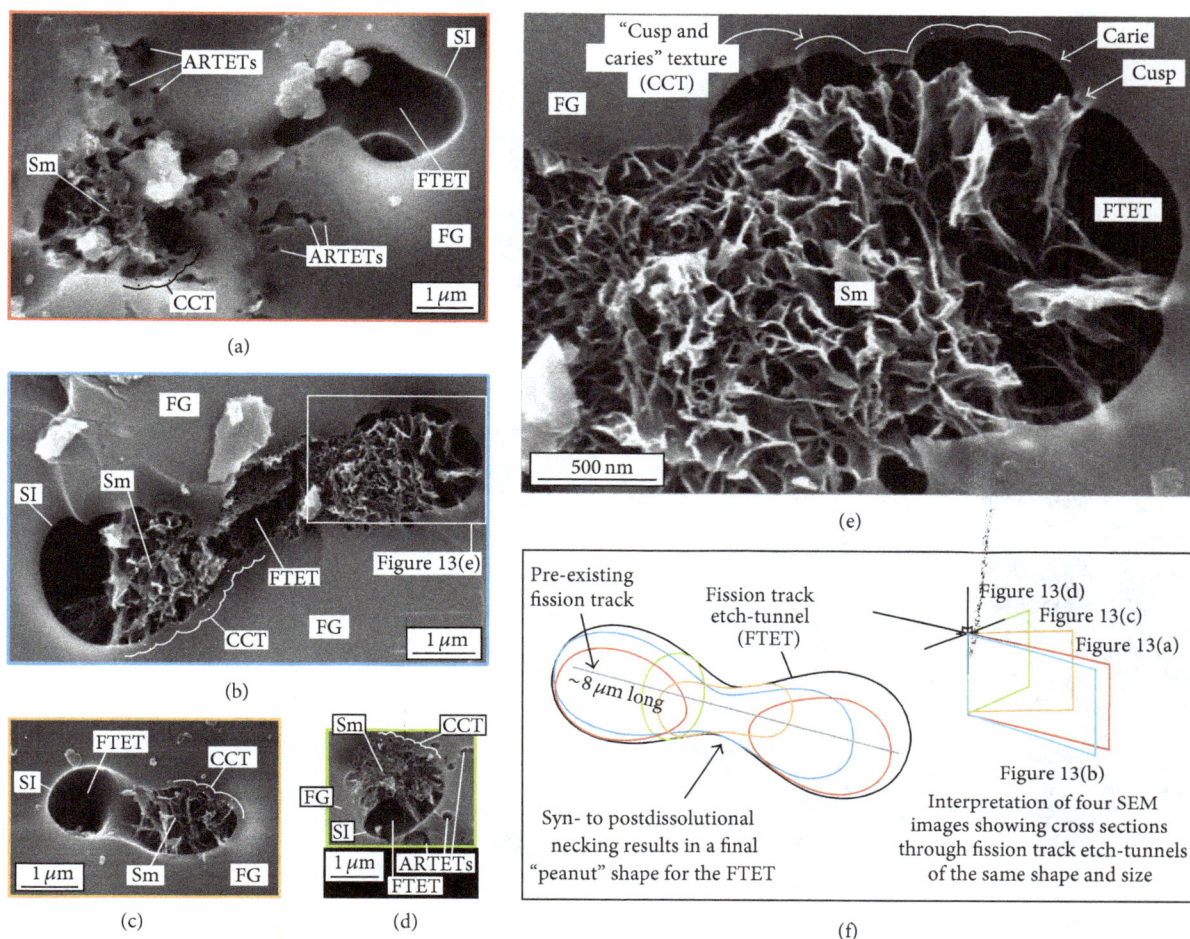

FIGURE 13: High resolution SEM (secondary electron) imaging of fission track etch-tunnels (FTETs) at the glass-palagonite interface (as observed on the freshly fractured surfaces of basaltic glass "chip samples" of *DSDP-418A-75-3[120–123]*). (a–d) Images of four FTETs shown at the same scale, all of which contain authigenic platy smectite. Note how two of these FTETs ((a) and (d)) are interconnected to several smaller ARTETs as a part of the same porosity/etch-tunnel network. In addition, all four FTETs exhibit two contrasting types of glass-dissolution interface: (1) a smooth interface (labelled "SI") at regions distal to secondary smectite growth and (2) a "cusp and caries" textured interface (labelled "CCT") at regions undergoing dissolution in the vicinity of secondary smectite growth (see Figure 14 for further explanation of CCT). (e) Close-up from (b), highlighting authigenic platy smectite within an FTET, as well as well-developed cusp and caries microtexture. (f) Interpretation of the four large etch-tunnels shown in (a–d) as variably oriented cross sections through FTETs of essentially the same shape and size (i.e., peanut-shaped and ~8 μm long). ARTETs: alpha-recoil track etch-tunnels; CCT: "cusp and caries" texture; FTET: fission track etch-tunnel; SI: smooth interface; Sm: platy smectite.

with ~150 Å of gold using a Nanotech Semprep2 and then analysed using the aforementioned JEOL 6301F instrument (Figures 12(a)–12(c) and one image introduced in a later Section: Figure 21(a)). During high resolution SEM imaging, the samples sputtered with a ~150 Å thick layer of gold showed a slight artificial granulation of the sample surface (see nanogranules of "Au" in Figures 12(a)–12(c)), which is why most of the chip samples in this study were coated with ~20 Å of iridium—allowing resolution of much finer details (as small as ~20 Å across: e.g., see the nanofilaments in Figure 12(h)).

3.4.1. *Overview of the Glass-Palagonite Interface and Associated "Etch-Tunnel Zone"*. We surveyed a large number of

fresh fracture surfaces of various basaltic glass chip samples (from *DSDP-418A-75-3-[120–123]*) and located partially palagonitized basaltic glass (and associated etch-tunnels) on several different grains, and a high resolution SEM overview of a representative region of the glass-palagonite interface is shown in Figure 11(a). Many of the corrosion/dissolution microtextures observed in polished petrographic thin section at the glass-palagonite interface (e.g., Figures 7 and 9) were also identified within this representative region (Figure 11(a)), which can be subdivided into four distinct textural domains, including fresh basaltic glass, an etch-tunnel zone, a palagonite zone, and an intermediate zone where the latter two coexist (Figures 11(a)–11(c)). The etch-tunnel zone occurs between the palagonite zone and fresh basaltic glass

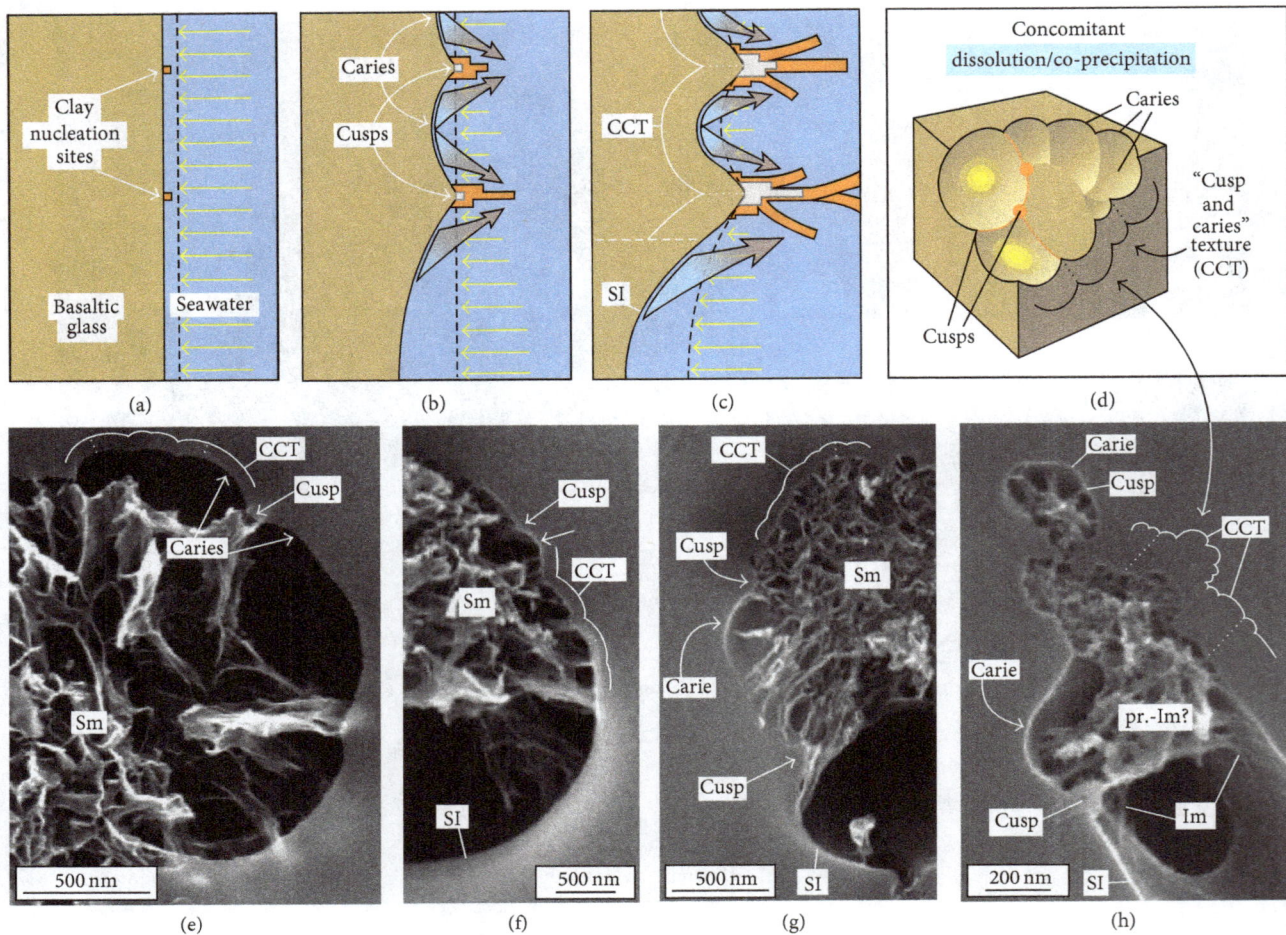

FIGURE 14: SEM images and a schematic model (a–d) explaining the origin of "cusp and caries" texture observed along the walls of fission track etch-tunnels (FTETs) and alpha-recoil track etch-tunnels (ARTETs) in basaltic glass. The secondary electron images in (e–h) are all close-up SEM images from other figures: (e) and (f) are close-up images from Figure 13(b); (g) is a close-up from Figure 13(d); and (h) is a close-up from Figure 12(e). Schematic model (a–d): at locations distal to the nucleation sites of secondary clays, the glass-water interface (i.e., dissolution front) is quite smooth and featureless (see "SI" in (c) and (f–h) and also in Figures 13(a)–13(d)). In contrast, where glass dissolution (yellow) is accompanied by the nearby nucleation and growth of secondary clays (orange), the protective effect of the newly formed clay minerals (±the kinetics of the dissolution/co-precipitation process—see orange/blue and yellow arrows in (b) and (c)) results in the formation of "cusps" along the dissolution front that are separated by "caries" (b)—the latter of which form by dissolution of glass as a concave front/incursion (slightly distal to secondary clay formation). In the case of FTETs, the secondary clay that forms during development of cusp and caries texture is platy smectite (e–g), whereas for ARTETs the secondary mineral is filamentous imogolite ((h); Figures 12(e), 12(g), and 12(h)). CCT: "cusp and caries" texture; Im: imogolite; pr.-Im?: proto-imogolite?; SI: smooth interface; Sm: platy smectite.

(Figures 11(a)–11(c)) and consists of fresh basaltic glass that is riddled with porosity (i.e., etch-tunnels: Figure 11(d)). These textural relationships indicate that advancement of the palagonite alteration front took place in the wake of a prominent dissolution/etch-tunnelling front caused by the infiltration of seawater into fresh basaltic glass (preferentially along radiation damaged regions—see Section 3.4.2). Furthermore, in some places the etch-tunnel zone has been completely overprinted by this encroachment of secondary palagonite, which locally forms palagonite fingers ("PF" in Figures 11(a)–11(c)) that extend outward into the etch-tunnel zone and fresh basaltic glass (akin to those shown in Figures 7(d) and 7(f)).

3.4.2. Nature of Alpha-Recoil Track Etch-Tunnels (ARTETs) and Fission Track Etch-Tunnels (FTETs).
Porosity in the etch-tunnel zone occurs in the form of a complex 3-dimensional network of anastamosing nanoscopic tunnels (i.e., "ARTETs" in Figures 11 and 12) that are typically ~120 nm in diameter (Figures 11(d), 12(a), 12(b), and 12(d)), exhibit a high degree of tortuosity (Figures 12(a), 12(b), and 12(d)), commonly branch (Figures 12(a) and 12(b)), occasionally flare out into larger irregular voids up to ~1 μm across ("FV" in Figures 12(b) and 12(e)) that have small "pockmarks" on their interior surface ("SARTETs" in Figures 12(b), 12(c), and 12(e)), and in one instance occur in the form of an etched-out loop

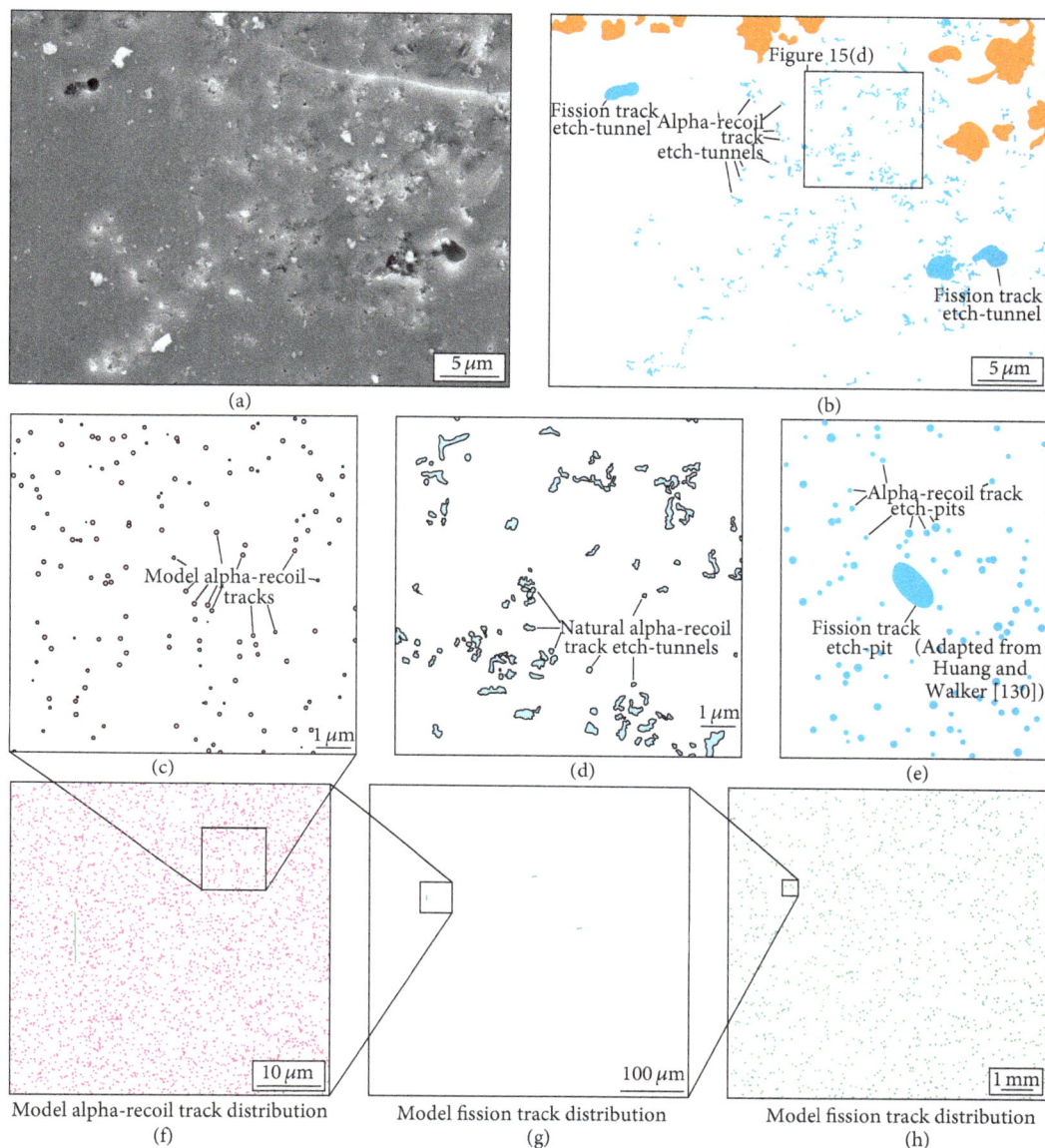

FIGURE 15: Comparison of theoretical modelling of radiation damage in DSDP 418A basaltic glass with the observed distribution of natural etch-tunnels at the glass-palagonite interface. (a) Representative SEM (secondary electron) image of the etch-tunnel zone (close-up from Figure 11(a)). (b) Porosity map (constructed from (a)) highlighting the distribution of fission track etch-tunnels (FTETs) and alpha-recoil track etch-tunnels (ARTETs) observed at the glass-palagonite interface. (c) Theoretical areal distribution of alpha-recoil tracks (ARTs) in DSDP 418A basaltic glass intersecting a hypothetical flat fracture plane through the glass (calculated using (1)). In this close-up theoretical plot (c), the individual model ARTs are plotted as randomly distributed 120 nm diameter spheres, resulting in pink circles of varying size (up to 120 nm), depending on their depth relative to the plane of the page (i.e., flat fracture surface). (d) Close-up porosity map from (b), highlighting a representative region of the etch-tunnel zone that is predominated by ARTETs. Note the similarity in size and areal distribution of the natural ARTETs (d) with the theoretically modelled ARTs (c), indicating a causal relationship. Many of the ARTETs in (d) exhibit elongate anastamosing shapes, the result of seawater "etching out" several nearby alpha-recoil tracks in cumulative succession (see "step 10a" in the lower left panel of Figure 18). (e) Etch-pit map, showing the observed distribution of experimentally etched ARTs and one etched-out fission track, on the cleavage surface of a mica crystal (adapted from Figure 1(b) of Huang and Walker [130], no scale bar available). (f–h) Theoretical models of the present-day distribution of randomly distributed ARTs (pink) and fission tracks (green) intersecting a hypothetical flat fracture plane through DSDP 418A basaltic glass, as determined using (1) and (2). For clarity, ARTs are not shown in (g) and (h) because of their exceptionally high numbers. Note how in (f), one single large fission track is surrounded by a multitude of smaller ARTs, similar in nature to the *bimodal* size versus population distributions of experimentally etched ARTs and fission tracks in mica (e) and naturally formed ARTETs and FTETs in DSDP 418A basaltic glass (b), a key argument for the "radiation damage" origin of the ARTETs and FTETs in DSDP 418A basaltic glass (see Sections 3.4.2, 5.2, and 6.1 for discussion). Colours in (b) and (d): blue = ARTETs and FTETs; white = fresh basaltic glass; orange = palagonite. Colours in (c), (f), (g), and (h): white = fresh basaltic glass not affected by radiation damage; pink = model ARTs; green = model fission tracks (randomly oriented lines in (f) and (g); dots in (h)). Colours in (e): white = fresh cleavage surface of mica; blue = ART etch-pits and one fission track etch-pit.

(Figure 12(f)). Rare larger chambers/tunnels that are ~1-2 μm in diameter and typically peanut-shaped (i.e., dumbbell shaped) and up to ~8 μm long (i.e., "FTETs" in Figures 11(a), 11(d) and 13) are also interconnected with this complex network of anastamosing nanoscopic tunnels. In cross section (i.e., on a large, freshly formed "chip"/fracture surface of basaltic glass pillow margin), these two contrasting types of exposed etch-tunnels show a bimodal size versus population distribution (Figures 11(a), 11(d), 15(a), and 15(b)) and also contain different types of infilling authigenic minerals. The typically peanut-shaped larger tunnels (i.e., fission track etch-tunnels or "FTETs") are much less abundant than the smaller tunnels (i.e., compare: 2 FTETs versus 379 ARTETs in Figures 15(a) and 15(b)) and contain platy material that is interpreted based on morphology, EDS analysis by SEM, geological context, and textural relationships with surrounding glass to be authigenic platy smectite ("Sm" in Figures 13 and 14). SEM images of four cross sections through the larger variety of etch-tunnel (Figures 13(a)–13(d)) may all be interpreted to represent variably oriented planar sections through a single type of peanut-shaped void space (i.e., fission track etch-tunnel) with the same overall shape and size in each case (Figure 13(f)). The smaller etch-tunnels (i.e., alpha-recoil track etch-tunnels or "ARTETs") predominate in the etch-tunnel zone (e.g., Figures 11(d) and 12(d)) and commonly contain filamentous material that forms cobweb-like bundles that drape across the tunnel void spaces and line tunnel walls ("FM" in Figures 12(a), 12(c), and 12(g)). On the basis of morphology, size, geological context, and textural relationships with surrounding glass, this filamentous material is interpreted to be authigenic imogolite (Figures 12(e), 12(g), 12(h), and 14(h); also see Figure 17(c) and 17(f)—introduced in a later Section—and [111, 139]). For instance, imogolite is typically described as the initial weathering product of glassy volcanic ash [196], and so it is logical that this mineral should also be forming here at the glass-water interface in submarine glasses that are undergoing corrosion/dissolution by seawater. In a close-up view of many imogolite filaments draped across the wall of a larger flare-out void that is a part of the smaller (alpha-recoil track) etch-tunnel network (Figure 12(h)), cross sections through hypothetical imogolite filaments are shown to scale as 20 Å thick (the known outer diameter of natural imogolite tubes [196]) filaments (blue) with an additional 20 Å of thickness on either side in order to represent the iridium coating (magenta). In this comparison, it is clear that the thickness of a hypothetical imogolite filament matches the size of the filaments draped across the tunnel walls in the SEM image given that they are coated with 20 Å of iridium (Figure 12(h); also see Figures 17(c) and 17(f)—introduced in a later Section). In SEM images of a different basaltic glass "chip" (also from sample *DSDP-418A-75-3-[120–123]*) coated in gold (Figures 12(a)–12(c)), similar-sized filaments also show through the relatively thicker and irregular coating formed on the sample during gold sputtering ("FM" in Figures 12(a)–12(c)).

The authigenicity of these imogolite filaments and platy smectite occurring within small versus large etch-tunnels, respectively (i.e., ARTETs versus FTETs), is constrained by the occurrence of pronounced "cusp and caries" texture (CCT) at the glass-water interface ("CCT" in Figures 13 and 14). This distinctive microtexture occurring at the glass-water interface (i.e., "cusp and caries texture" along etch-tunnel walls) is interpreted to arise from two concomitant processes taking place simultaneously during etch-tunnelling: dissolution (of glass) and coprecipitation (of secondary clays—i.e., smectite within FTETs and imogolite within ARTETs). According to this model (Figures 14(a)–14(d)), "cusps" form along the etch-tunnel walls at the sites of nucleation and growth of secondary clays (imogolite or smectite), whereby the protective effect of the secondary clays prevents dissolution from taking place directly at those regions, while "caries" (or "incursions" into the glass) form slightly adjacent to these sites of nucleation and growth (Figure 14(b)) through dissolution of nearby glass in a concave fashion. In regions sufficiently distal to the sites of clay nucleation and growth, glass dissolution proceeds without hindrance, resulting in a smooth interface ("SI" in Figures 13 and 14). This type of "cusp and caries" microtexture is a term traditionally used by microscopists in the study of ore deposits, in cases where a primary mineral is being replaced by a secondary (alteration) mineral, such that the new mineral forms concave incursions into the host, "as if the secondary mineral had bitten into the host" (p. 141 in Guilbert and Park Jr. [197]) and actually originates from a dental analogy—with the "caries" representing the concave incursions in a tooth cavity and the relict protuberences between them being "cusps" (p. 141 in Guilbert and Park Jr. [197]).

The smaller (~120 nm diameter) and more abundant variety of etch-tunnels (i.e., "ARTETs" in Figures 11(d) and 12) observed on the surfaces of these "chip" samples are essentially identical in size to the etch-tunnels identified at the glass-palagonite interface in polished petrographic thin sections in this study (Figure 9) and, therefore, also have incidentally the same size as a typical alpha-recoil track (~120 nm in diameter [129]—see "ARTETs" and pink dots labelled "ART" in Figures 9(d), 9(h), 9(k), 9(l), 11(a), 11(b), 11(d), 12(a), 12(b), and 12(d)–12(f)). This adds further support for our conclusion in Section 3.3.1 that the etch-tunnel zone at the glass-palagonite interface forms in advance of the palagonitization front primarily due to preferential etch-tunnelling (by seawater) of randomly distributed alpha-recoil tracks in the glass resulting from radioactive decay of U and Th—again, in contrast with previous biogenic (i.e., microbial trace fossil) interpretations (i.e., of "tubular" texture at DSDP 418A [11, 26–28, 31, 33, 59, 61, 93, 94, 100]). Additional support for a "radiation damage origin" for this etch-tunnel zone comes from the coincidence in size of the rare, larger (~1-2 μm wide by up to ~8 μm long) etch-tunnels ("FTETs" in Figures 11(a), 11(d) and 13) with the typical size of a "pristine" fission track in volcanic/impact glass (i.e., ~8 μm in length—see a direct comparison in Figure 3(e); "Pristine" means naturally occurring *spontaneous* fission tracks in volcanic/impact glass originating from the natural radioactive decay (spontaneous fission) of ^{238}U that have not been affected by subsequent thermal annealing/track-shortening or in the case of *induced* fission tracks in volcanic glass, originating from the induced fission of ^{235}U triggered by irradiating the sample with thermal neutrons in a nuclear reactor—fission tracks which are also typically ~8 μm in length: see Figure 1 in Sandhu et al. [198], Figure 8(b) in Westgate and Naeser [199], Figure 3 in this study, and descriptions in Arias et al. [126] and Sandhu and Westgate [128]). As such, we interpret the etch-tunnel zone at the glass-palagonite interface (Figures 7, 9, 11–13, 15(a), 15(b), and 15(d); also see Figures 1(b), 1(d), 1(f), 3(e), 4(c)—left) to have formed by preferential dissolution and etch-tunnelling (by seawater) through radiation damaged regions of basaltic glass (i.e., randomly distributed alpha-recoil tracks and spontaneous fission tracks)—in advance of the encroaching palagonitization front that formed along fractures during low temperature subaqueous alteration of these glassy margins of midocean ridge pillow lavas—long after their initial eruption and quenching.

It is also important to highlight at this point that the *bimodal* distributions observed for "etch-tunnel size" versus "etch-tunnel population/areal density" in this study (i.e., a multitude (379) of small (~120 nm wide) etch-tunnels versus only a few (2) large (~8 μm long) etch-tunnels: Figures 11(a), 15(a), and 15(b)) also provides strong "microtextural" evidence that the etch-tunnel zone at the glass-palagonite interface is abiotic in origin and the result of preferential etching (by seawater) of radiation damaged regions of glass. For instance, because fossil fission tracks in most geological materials are produced by the spontaneous fission of ^{238}U, they should occur amongst a much larger population of relatively smaller alpha-recoil damage tracks caused by the eight alpha-recoil events (Figure 4(a)) that are a part of the ^{238}U–^{206}Pb radioactive decay chain (e.g., [130]; Figures 4(b): left and 15(e)). A contrast in relative abundance of the two types of radiation damage occurs (i.e., several orders of magnitude more alpha-recoil tracks than fission tracks) mostly because of the difference in half-lives for spontaneous fission of ^{238}U and the ^{238}U–^{206}Pb alpha/beta decay chain, which are 8.2×10^{15} and 4.468×10^{9} y, respectively [200, 201]. But this is also in part because a significant number of additional alpha-recoil tracks are caused by the radioactive decay of other isotopes present such as ^{235}U to ^{207}Pb and ^{232}Th to ^{208}Pb, which for these particular decay chains involves a total of seven and six alpha-recoil events, respectively [202]. This bimodal size/population distribution for alpha-recoil tracks versus fission tracks was recognized early on in the studies of experimentally etched fission tracks and alpha-recoil tracks on the cleavage surfaces of micas ([130]; i.e., one large fission track etch-pit surrounded by a multitude of tiny alpha-recoil track etch-pits: Figures 4(b): left and 15(e)). Further discussions on the significance of these "bimodal distributions" of corrosion microtextures—linked to the differences in size and areal density of fission tracks versus alpha-recoil tracks—can be found in Sections 5.2, 6.1, and 6.2.2, including the observation of bimodal distributions of "granular" palagonite textures in glasses from the Costa Rica Rift (Section 6.1).

In summary, there are three lines of microtextural evidence identified thus far—from the "etch-tunnel zone" at the glass-palagonite interface in DSDP-418A basaltic glass—that

support a *radiation damage origin* for these etch-tunnels and *not a microbial origin*. (1) The smaller variety of etch-tunnels (~120 nm diameter "ARTETs" in Figures 9(d), 9(h), 9(k), 9(l), 11(a), 11(b), 11(d), 12, 13(a), 13(d), 15(a), 15(b), and 15(d)) are typically about the same size as an alpha-recoil track (~120 nm in diameter [129]). (2) The larger variety of etch-tunnels ("FTETs" in Figures 11(a), 11(d), 13, 15(a), and 15(b)) range in length up to ~8 μm long, that is, the same size as a typical unannealed fission track in volcanic glass (as well as in tektite glasses, that is, ~8 μm in length; see Figures 3(b) and 3(e) (this study), Figure 1 in Sandhu et al. [198], Figure 8(b) in Westgate and Naeser [199], and descriptions in Arias et al. [126] and Sandhu and Westgate [128]). (3) The smaller (~120 nm diameter) variety of etch-tunnels are several orders of magnitude more abundant than the larger (up to ~8 μm long) variety (e.g., 379 "ARTETs" versus 2 "FTETs" in Figures 15(a) and 15(b))—consistent with the well-known concept (outlined above) that the areal density of alpha-recoil tracks (i.e., #tracks/cm^2) accumulates several orders of magnitude faster than fission track areal densities.

In light of these similarities between natural etch-tunnels in DSDP-418A basaltic glass and alpha-recoil tracks and fission tracks, it is important to evaluate precisely how much radiation damage should actually be present in these glasses (i.e., in the form of randomly distributed fission tracks and alpha-recoil tracks) based on the known age of these pillow lavas (~120.6 Ma; for age constraints, see Figures 5(c) and 5(d) and Section 2.2) and the measured concentrations of U and Th present in fresh basaltic glass, and that is the subject of the following two Sections (4 and 5).

4. Determination of U and Th Concentrations in Fresh Basaltic Glass by ICP-MS

For theoretical modelling of radiation damage in the glassy margins of pillow basalts in this study, the concentrations of U and Th in fresh glass from sample *DSDP-418A-75-3-[120–123]* were measured by ICP-MS. A few fragments of this basaltic glass pillow margin sample (each several mm across) were crushed in ethanol to achieve a smaller grain size using an agate mortar and pestle. The resulting rock powder was then sieved in ethanol using a silkscreen mesh to remove the fines (material less than ~50 μm in grain size). Crushing and sieving were necessary to isolate numerous fragments of only fresh basaltic glass that were of adequate size. The grains were then hand-picked in ethanol under binocular microscope, using a custom built pipette made of Tygon and Teflon. Selecting fresh material required sorting through and picking grains of basaltic glass that contained no alteration (e.g., palagonitized or devitrified zones), no evidence of still attached fragments of mineral inclusions (e.g., clinopyroxene and plagioclase), and only those grains that were devoid of vesicles.

In all, 827 individual fragments of pristine basaltic glass were picked for U and Th analysis, ranging in size from ~50 to ~300 μm (Table 1). They are described as fresh shards of basaltic glass that are light tan brown (smallest shards) to dark brown (largest shards) in colour. Grain surfaces invariably

TABLE 1: ICP-MS results for U and Th concentrations in DSDP 418A basaltic glass.

Sample name	Sample grain size; {# grains}	Sample description	Sample weight (mg)	Th (ppm)	U (ppm)	Th/U
DSDP-418A-75-3-[120–123]-ICP–MS #1	~0.3–1.0 mm; {13}	(a)	5.3	0.097	0.037	2.66
DSDP-418A-75-3-[120–123]-ICP–MS #2	~0.3–1.0 mm; {42}	(a)	10.4	0.129	0.037	3.51
DSDP-418A-75-3-[120–123]-ICP–MS #3	~1.5 × 3 × 4 mm; {1}	(a)	17.8	0.164	0.053	3.10
DSDP-418A-75-3-[120–123]-ICP–MS #4	~1 × 1 × 1 mm; {1}	(a)	1.6	0.112	0.038	2.97
DSDP-418A-75-3-[120–123]-ICP–MS #5	~1 × 1 × 1 mm; {1}	(a)	1.1	0.107	0.040	2.67
DSDP-418A-75-3-[120–123]-ICP–MS #6	~50–300 μm; {266}	(b)	0.9	0.132*	0.042*	3.16*
DSDP-418A-75-3-[120–123]-ICP–MS #7	~50–300 μm; {263}	(b)	0.9	0.117	0.035	3.31
DSDP-418A-75-3-[120–123]-ICP–MS #8	~50–300 μm; {298}	(b)	1.2	0.108	0.032	3.35

For Th and U determinations (ppm), the external reproducibility is 5–10% (2σ level) of the quoted abundances.
Sample descriptions:
(a) Test fractions: they comprise predominantly fresh basaltic glass, which also contains numerous small (<100 μm) vesicles and their inclusions (e.g., vesicles commonly contain numerous ~1 μm sized Fe-sulfide spherules) and possibly some small domains (~a few % by volume) that comprise altered (palagonitized/devitrified) zones and/or phenocryst inclusions such as clinopyroxene and plagioclase.
(b) Pristine, fresh basaltic glass, devoid of alteration, inclusions, or vesicles: the shards of glass are transparent and light tan brown (thinnest shards) to dark brown (thickest shards) in colour. Surfaces invariably show a vitreous luster and comprise curviplanar to conchoidal fractures that impart a myriad of shapes to the grains. All of these grains were found to be isotropic when viewed under petrographic microscope with crossed nicols.
*values used for theoretical modelling of alpha-recoil track (Figures 10(h), 10(i), 15(c), and 15(f)) and fission track (Figures 10(g), 15(g), and 15(h)) areal densities in DSDP 418A basaltic glass in this study.

show a vitreous luster and because they are curviplanar to conchoidal fractures they impart a myriad of shapes to the grains from elongate blades with razor sharp edges to more equant shapes. As an additional precautionary measure after picking, all of the grains were then examined with crossed polars using a petrographic microscope to ensure that they were isotropic. This was also done to exclude any fragments that might still contain attached splinters of other minerals (e.g., plagioclase) that were difficult to see under binocular microscope. It was found that all of the grains were indeed isotropic, as expected for basaltic glass, although ~2% of the grains had tiny inclusions, or fragments on their edges, which were readily visible as bright birefringent material (likely phenocryst fragments). These grains were removed leaving the final population of 827 fresh basaltic glass fragments that was split into three 0.9 to 1.2 mg fractions for trace element analysis (Table 1).

To ensure that these fractions would comprise an adequate amount of material for successful analysis by ICP-MS, a series of five test fractions was run first. These also originate from sample DSDP-418A-75-3-[120–123] and ranged in weight from 1.1 to 17.8 mg (Table 1). These fractions comprised larger grains (one to several mm in size) of pillow margin that contained predominantly fresh basaltic glass, which also contained a significant amount of vesicles, altered domains, and mineral inclusions (Table 1).

Before dissolution, the grains were transferred into Savillex beakers in MilliQ H_2O (18.6 MΩ) and placed into an ultrasonic bath for several minutes. The water was decanted and the grains rinsed again in MilliQ H_2O. This was followed by addition of four parts 49% HF and one part 68% HNO_3 into the Savillex beakers for dissolution on a hot plate for 48 hours at 135°C. Solutions were then evaporated to dryness and then reequilibrated in 1 mL of 2% HNO_3, and an internal standard solution (Bismuth) was added before analysis.

For determination of U and Th concentrations, solutions were analysed on a Perkin Elmer Elan 6000 quadrupole

ICP-MS. Operating conditions are as outlined in Simonetti et al. [203], with the exception that dwell times for Th and U were 120 and 60 ms, respectively. Each sample analysis consisted of 35 sweeps/reading (3 replicates) for a total analysis time of 21 s per sample, and the external reproducibility is 5–10% (2σ level) of the quoted abundances (Table 1).

The five test samples yielded positive results and so the three small fractions of fresh basaltic glass were also analysed by ICP-MS. The results for these eight samples were quite similar overall, and they are shown in Table 1. For the test fractions, Th concentration ranged from 0.097 to 0.164 ppm and U concentrations from 0.037 to 0.053, with Th/U values between 2.66 and 3.51. The fresh glass showed a more narrow range of trace element concentrations, with 0.108–0.132 ppm Th, 0.032–0.042 ppm U, and an exceptionally narrower range of Th/U values of 3.16–3.35. The broader range of values for the test fractions likely resulted from the inclusion of minor amounts of altered glass, mineral inclusions, and vesicles. Because—as highlighted in Section 2—all three basaltic glass pillow margin samples in this study belong to the same chemical (i.e., glass) type "J" [162] of lithologic unit "13" of volcanic eruptive unit "Vb" [161], it is reasonable to make the geochemical prediction that the U and Th contents of fresh basaltic glass from samples DSDP-418A-68-3[40–43] and DSDP-418A-72-4[13–15] in this study are probably quite similar to the values reported here for sample DSDP-418A-75-3[120–123] (i.e., U = 0.032–0.042 ppm; Th = 0.108–0.132 ppm). Consequently, we consider our theoretical modelling study (Section 5) of the present-day distribution of radiation damage in fresh basaltic glass—which utilizes these U and Th concentrations determined here by ICP-MS—to apply to all three pillow margin samples.

The Th/U ratios and U and Th contents in DSDP-418A basaltic glass determined in this study are similar to other previous results for midocean ridge basaltic glass. For example, fresh basaltic glass from globally widespread, dredged samples of normal MORB have U concentrations that range from 0.0215 to 0.129 ppm, with Th/U rising smoothly from 1.4 to 3.1 as Th concentrations increase from 0.03 to 0.399 ppm [146]. The measured U concentrations in basaltic glasses from other pillow margins higher up in the volcanostratigraphic sequence at DSDP-418A are also quite similar (0.018–0.037 ppm [125]).

5. Theoretical Modelling of Present-Day Radiation Damage in Fresh Basaltic Glass

In order to evaluate in a more quantitative way the role of radiation damage on microtextural development during the natural corrosion/palagonitization of DSDP-418A basaltic glasses by seawater, we carried out a theoretical modelling study of the present-day distribution of radiation damage in these glasses (i.e., areal densities of alpha-recoil tracks and spontaneous fission tracks intersecting a hypothetical flat fracture surface in the glass), based on the known age of these pillow lavas (~120.6 Ma; for age constraints, see Figures 5(c) and 5(d) and Section 2.2) and the measured concentrations of U and Th in fresh basaltic glass determined by ICP-MS (Table 1).

5.1. Modelling Methodology. As they grow—for example, during the eight successive alpha-recoil events during the complete radioactive decay chain of ^{238}U to ^{206}Pb—alpha-recoil tracks typically "zigzag" through their host material (Figure 4(a)), ~30–50 nm in new random directions with each successive alpha-recoil event (e.g., Figure 3 in Jonckheere and Gögen [131]; Table 1 in Stübner and Jonckheere [132]), resulting in a final composite cluster of radiation damage measuring about 120 nm across (e.g., estimates for the mean diameter of alpha-recoil tracks derived from the complete decay of ^{238}U to stable ^{206}Pb in micas range from ~110 nm [132] to ~120 nm [129] and to ~125 nm [131]). In this theoretical modelling study of the distribution of alpha-recoil tracks in DSDP 418A basaltic glass, we consider each alpha-recoil track to be the end result of complete U- or Th-series decay. For example, even though a total of eight alpha-recoil events take place during the complete radioactive decay of ^{238}U to ^{206}Pb, because they are all interconnected they are considered as a "single" composite ~120 nm wide alpha-recoil track (Figure 4(a) [129, 131, 132]). In addition, for simplicity in this high spatial resolution theoretical modelling study (Figures 10(i) and 15(c)) each alpha-recoil track is modelled as a ~120 nm diameter sphere (instead of as myriad different geometric varieties of zigzagging structures) and this is in keeping with the observation that when fully etched, alpha-recoil tracks tend to form relatively equant etch-pits (e.g., see Figure 4(b); Figure 1 in Stübner et al. [133]); that is, the theoretical nanoscopic "zigzagging" three-dimensional structure of alpha-recoil tracks is effectively destroyed upon etching. Furthermore, because these DSDP 418A basaltic glasses are inferred to have only experienced relatively low temperatures throughout their entire geological history (since quenching), we assume that (a) no ^{222}Rn diffusion has occurred (which can "split" the ^{238}U-^{206}Pb composite alpha-recoil track in half—see Figure 4(a)) AND (b) no alpha-recoil track "fading/shortening" has occurred.

The areal density of alpha-recoil tracks intersecting a hypothetical flat fracture plane through DSDP-418A basaltic glass (Figures 10(h), 10(i), 15(c), and 15(f)) was calculated using (1), which is adapted from (4) and (5) of Gögen and Wagner [129] with two additional exceptions. (i) The effects of partitioning of Th/U between mineral and melt are excluded in the present case because we are dealing with a quenched glass and (ii) we are calculating a theoretical alpha-recoil track areal density (ρ_a), not a volume density (ρ_v), and therefore multiply by R_e (the etchable range of an alpha-recoil track).

Accordingly, in (1), ρ_{ART} is the areal density of etchable alpha-recoil tracks (i.e., # of alpha-recoil tracks per cm^2), ^{238}U$_g$ is the weight concentration of ^{238}U in DSDP 418A basaltic glass (4.17 × 10^{-8} g/g, calculated from the ^{235}U/^{238}U isotopic abundance ratio of 0.00725 [204] and the U concentration of 42 ppb determined by ICP-MS: DSDP-418A-75-3[120–123]-ICP-MS #6, Table 1), n_s is the density of basaltic glass (3.0 g/cm^3: cf. [205]), N_A is Avogadro's number (6.02 × 10^{23} atoms/mol), M_U is the molar mass of uranium (238.0289 g/mol), λ_{238} is the decay constant for ^{238}U (1.55125 × 10^{-10} y^{-1} [204]), λ_{230} is the decay constant for ^{230}Th (9.158 × 10^{-6} y^{-1} [206]), λ_{235} is the decay constant for ^{235}U (9.8485 × 10^{-10} y^{-1} [204]), λ_{234} is the decay constant for ^{234}U (2.8262 × 10^{-6} y^{-1} [206]), λ_{232} is the decay constant for ^{232}Th (4.9475 × 10^{-11} y^{-1} [204]), t is the time since quenching of the glass (120,600,000 y; for age constraints see Figures 5(c), 5(d) and Section 2.2), (Th/U)$_{glass}$ is the Th/U ratio of the glass (3.16, determined from the ICP-MS data: DSDP-418A-75-3-[120–123]-ICP-MS #6, Table 1), I is the ^{235}U/^{238}U isotopic abundance ratio (0.00725 [204]), and R_e is the etchable range of an alpha-recoil track (0.000012 cm; i.e., 120 nm [129, 131]). The value for the total efficiency coefficient for alpha-recoil track revelation (η_{tot}) in conventional alpha-recoil track dating studies is assumed to be 1 [129], which holds true in the present case because the tracks are hypothetical and therefore automatically revealed. Consider

$$
\rho_{ART} = \left\{ \frac{\left[\left(^{238}U_g \right) (n_s) (N_A) \right]}{(M_U)} \right\} \left(e^{\lambda_{238}t} \right) \left\{ \left(1 - e^{-\lambda_{238}t} \right) \right.
$$

$$
+ \left(\frac{\lambda_{238}}{\lambda_{234}} \right) \left(1 - e^{-\lambda_{234}t} \right) + \left(\frac{\lambda_{238}}{\lambda_{230}} \right) \left(1 - e^{-\lambda_{230}t} \right)
$$

$$
\left. + I \left(1 - e^{-\lambda_{235}t} \right) + \left(\frac{Th}{U} \right)_{glass} \left(1 - e^{-\lambda_{232}t} \right) \right\} (\eta_{tot}) \tag{1}
$$

$$
\cdot \left(R_e \right).
$$

Note that one particular subpart of this equation (and the same goes for (2) below) is equal to the atomic frequency of ^{238}U in basaltic glass (i.e., ^{238}U$_a$ in: #atoms of ^{238}U/cm^3), whereby ^{238}U$_a$ = {[(^{238}U$_g$)(n_s)(N_A)]/(M_U)} (i.e., after (5) of Gögen and Wagner [129]).

The areal density of fossil fission tracks intersecting a hypothetical flat fracture plane through DSDP-418A basaltic glass (Figures 10(g), 15(g), and 15(h)) is calculated here using (2) (adapted from Bigazzi [207] and Galbraith and Laslett [208]), where ρ_{FT} is the areal density of etchable fossil fission tracks, λ_{238} is the decay constant for ^{238}U (1.55125 × 10^{-10} y^{-1} [204]), t is the time since quenching of the glass (120,600,000 y; for age constraints see Figures 5(c), 5(d) and Section 2.2), λ_F is the decay constant for spontaneous fission of ^{238}U (8.5 × 10^{-17} y^{-1} calculated from the half-life value of 8.2 × 10^{15} y reported for the spontaneous fission of ^{238}U by Holden and Hoffman [201] using the equation $T_{1/2}$ = ln 2/λ, that is, (4.10) of Faure [202]), $^{238}U_g$ is the weight concentration of ^{238}U in g/g (4.17 × 10^{-8} g/g, calculated using the $^{235}U/^{238}U$ isotopic abundance ratio of 0.00725 [204] and the U concentration of 42 ppb determined by ICP-MS: *DSDP-418A-75-3[120–123]-ICP-MS #6*, Table 1), n_s is the density of basaltic glass (3.0 g/cm^3: cf. [205]), N_A is Avogadro's number (6.02 × 10^{23} atoms/mol), M_U is the molar mass of uranium (238.0289 g/mol), and R_F is the etchable length of a fission track in volcanic glass (0.0008 cm; i.e., ~8 μm: see Figures 3(b) and 3(e) (this study); Figure 1 in Sandhu et al. [198]; Figure 8(b) in Westgate and Naeser [199]; and descriptions in Arias et al. [126]; Sandhu and Westgate [128]). Consider

$$\rho_{FT} = \left(\frac{1}{2}\right) \left[\frac{\left(e^{\lambda_{238}t} - 1\right)}{\lambda_{238}} \right] (\lambda_F)$$

$$\cdot \left\{ \frac{\left[\left(^{238}U_g\right)(n_s)(N_A)\right]}{(M_U)} \right\} (R_F). \tag{2}$$

The 1/2 in (2) falls out of the Poisson line-segment model "in which the orientation distribution of a track is uniform with respect to solid angle, and the joint distribution of length and orientation of a track is independent of its location" [208], which holds true for basaltic glass because it is isotropic with a liquid like structure.

5.2. Modelling Results. From (1), we calculate the present-day alpha-recoil track areal density in DSDP-418A basaltic glass to be very high at 148,000,000 alpha-recoil tracks/cm^2, which indicates that these glasses are absolutely riddled with alpha-recoil track damage (Figures 10(h), 10(i), 15(c), and 15(f)) amenable to preferential dissolution/etch-tunnelling (e.g., Figures 1(b), 1(d), 1(f), 7(a)–7(c), 9, 11, 12, 15(a), 15(b), and 15(d)) and preferential palagonitization (e.g., Figure 10) during corrosion/alteration by infiltrating seawater. In our theoretical model, U and Th contributed about equally to the accumulation of alpha-recoil tracks, with the majority of tracks originating from radioactive decay of ^{238}U (accounting for ~71,700,000 alpha-recoil tracks/cm^2) and ^{232}Th (accounting for ~72,700,000 alpha-recoil tracks/cm^2). Comparatively, few alpha-recoil tracks originated from radioactive decay of ^{235}U (3,140,000 alpha-recoil tracks/cm^2), ^{234}U (~212,000 alpha-recoil tracks/cm^2), and ^{230}Th (~65,500 alpha-recoil tracks/cm^2).

A direct 1:1 scale comparison of the resultant map of theoretical alpha-recoil track distribution in DSDP 418A basaltic glass (Figure 15(c)) to the map of natural porosity in the etch-tunnel zone (Figure 15(d)) at the same scale reveals two very important similarities. Firstly, the model alpha-recoil tracks are about the same width as the natural etch-tunnels (circa 120 nm). Secondly, the model alpha-recoil track areal density (148 alpha-recoil tracks in a 10 × 10 μm region: Figures 10(i) and 15(c)) is high and quite close to the observed areal density of natural nanoscopic etch-tunnels (94 nanotunnels observed in a representative 10 × 10 μm region: Figure 15(d)). This indicates that the numerically predicted alpha-recoil track areal density is more than sufficient to account for the observed areal density of ~120 nm wide etch-tunnels. From these two observations, we conclude that the complex networks of nanoscopic etch-tunnels observed at the glass-palagonite interface in DSDP 418A basaltic glass (Figures 1(b), 1(d), 1(f), 7, 9, 11, 12, and 15(a)) are in fact naturally formed alpha-recoil track etch-tunnels (ARTETs) caused by the infiltration of seawater into the glass through preferential dissolution/etch-tunnelling along multitudes of randomly distributed alpha-recoil track damaged sites in the glass and therefore not the result of microbial activity (i.e., in the case of "tubular" microtextures in DSDP 418A basaltic glasses [11, 26–28, 31, 33, 59, 61, 93, 94, 100]).

Similarly, direct 1:1 scale comparisons of the model alpha-recoil track distribution in DSDP-418A basaltic glass (Figures 10(h) and 10(i)) with the observed distribution of "granular palagonite" (Figures 10(e) and 10(f)) reveal very similar "spotty" patterns of randomly distributed submicroscopic bodies; that is, the areal density of model alpha-recoil tracks (Figures 10(h) and 10(i)) is very similar to the observed areal density of palagonite granules (Figures 10(e) and 10(f)). This provides strong microtextural evidence to suggest that such granular palagonite microtextures at the glass-palagonite interface (both in DSDP-418A basaltic glass and probably in submarine basaltic glasses worldwide; see further discussion in Sections 3.3.2 and 6.1) originate by preferential alteration/corrosion (i.e., palagonitization) of randomly distributed alpha-recoil track damaged sites in the glass and, therefore, not by microbial activity/bioalteration as previously thought (e.g., [11, 26–28, 31]).

As expected (see Section 3.4.2), the model fission track areal density calculated from (2) is several orders of magnitude smaller at 1,310 fission tracks/cm^2 (Figure 15(h); also see Figures 10(g), 15(f), and 15(g)). This indicates that although they are larger, fission track etch-tunnels should be comparatively few with respect to the number of alpha-recoil track etch-tunnels found (e.g., compare track distributions in Figures 15(f) and 15(g)), and this is consistent with our observations (2 FTETs versus 379 ARTETs in Figures 15(a) and 15(b)), and indeed only 4 were found in our search for fission track etch-tunnels (Figures 13(a)–13(d)). This type of bimodal population versus size distribution of etched fission tracks versus etched alpha-recoil tracks was recognized early on in etching studies of micas (i.e., Figure 1 in Huang and Walker [130]; Figures 4(b): left and 15(e) in this study) and is reminiscent of the pattern of etch-tunnelling that we have observed here in DSDP 418A basaltic glass (i.e., compare

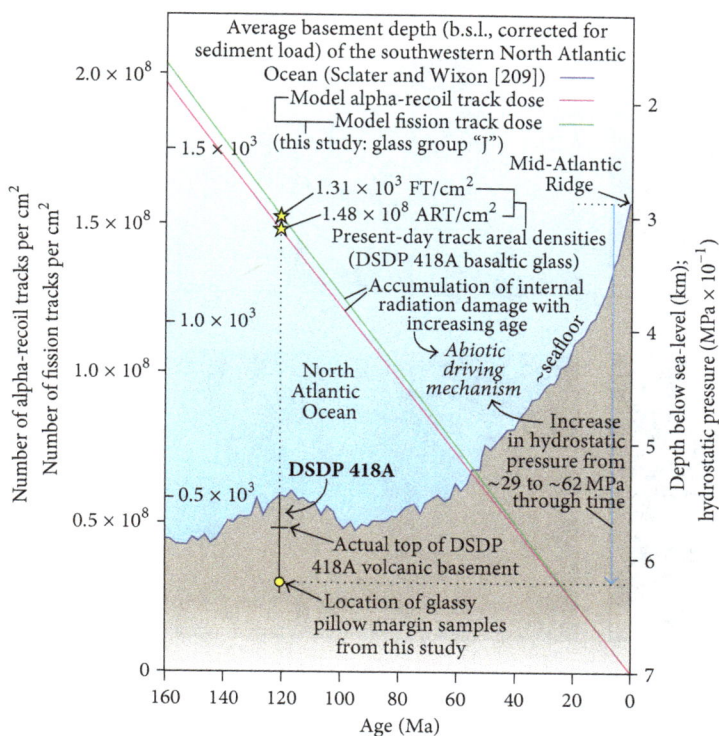

FIGURE 16: Diagram summarizing the "abiotic driving mechanism" for the formation of complex corrosion microtextures (i.e., microscopic etch-tunnels and granular palagonite) in DSDP 418A basaltic glass: the concomitant increase in hydrostatic pressure and accumulation of radiation damage in basaltic glass with aging of the oceanic crust. The following is depicted in this diagram: (i) the depth-age relationships for volcanic basement in the oceanic crust—depicted as the average depth to basement below sea level (b.s.l.) (corrected for sediment load) for the southwestern North Atlantic Ocean (jagged blue line; from Sclater and Wixon [209]); (ii) the location of DSDP Hole 418A (note that the actual top of volcanic basement in this drill hole is slightly lower than the southwestern North Atlantic average (jagged blue line)); (iii) the inferred hydrostatic pressure in the water column (which can be read at right, along with ocean depth); and (iv) curves (although they are nearly straight lines in this plot) depicting the accumulation (dose) of radiation damage in basaltic glass group "J" (g.g. "J" in Figure 5(b)) through time—that is, the systematic increase (with age) in model fission track and alpha-recoil track areal densities intersecting a hypothetical flat fracture plane through DSDP 418A basaltic glass (calculated using (1) and (2)).

Figures 15(b) with 15(e)). This similarity between the pattern of natural etch-tunnelling in DSDP 418A basaltic glass (e.g., Figure 15(b)) and etch-pit formation on the cleavage surface of experimentally etched micas (e.g., Figure 15(e)—adapted from Huang and Walker [130]) is taken as further micro-textural evidence to support a "radiation damage origin" for the etch-tunnels observed at the glass-palagonite interface in DSDP 418A basaltic glass—vis-à-vis natural preferential etch-tunnelling of alpha-recoil tracks and spontaneous fission tracks in the glass by infiltrating seawater.

5.3. The Abiotic Driving Mechanism for the Development of Complex Corrosion Microtextures in Submarine Glasses. The "abiotic" driving mechanism for the formation of micro-scopic etch-tunnels and granular palagonite microtextures at the glass-palagonite interface in submarine glasses is really quite simple (Figure 16). As the oceanic crust ages and moves away from the spreading ridge, it subsides (e.g., [209]) caus-ing the ambient hydrostatic pressure in volcanic basement rocks to rise incrementally with deepening of the overlying ocean, and this happens concomitantly as radiation damage

in the rocks accumulates with time (Figure 16). Consequently, as basaltic glass ages and becomes more and more amenable to preferential corrosion along alpha-recoil track and fission track damaged regions, it may also become increasingly susceptible to the effects of pressure solution etch-tunnelling associated with increasing hydrostatic pressure. Classically, pressure solution in rocks—for example, during diagenesis—takes place in response to increasing *lithostatic* pressure (e.g., under "nonhydrostatic" stress conditions at grain-to-grain contacts, or between sedimentary layers in the case of stylolites [210]). As outlined in Section 3.1, the basaltic glass in this study appears to have undergone an episode of (f_2) fracturing and concomitant devitrification during the formation of white (K-Al-Si)-rich devitrified zones, and this appears to have taken place as a direct consequence of deep burial beneath the overlying volcanic pile (i.e., in response to increasing *lithostatic* pressure). Hence, the diagenetic/low-grade metamorphic response of volcanic glasses to increasing lithostatic load is not necessarily a form of pressure solu-tion, but rather, the solid-state transformation of glass into microcrystalline K-feldspar (±quartz or cristobalite) during the process of "devitrification" (p. 418 in Cas and Wright [184],

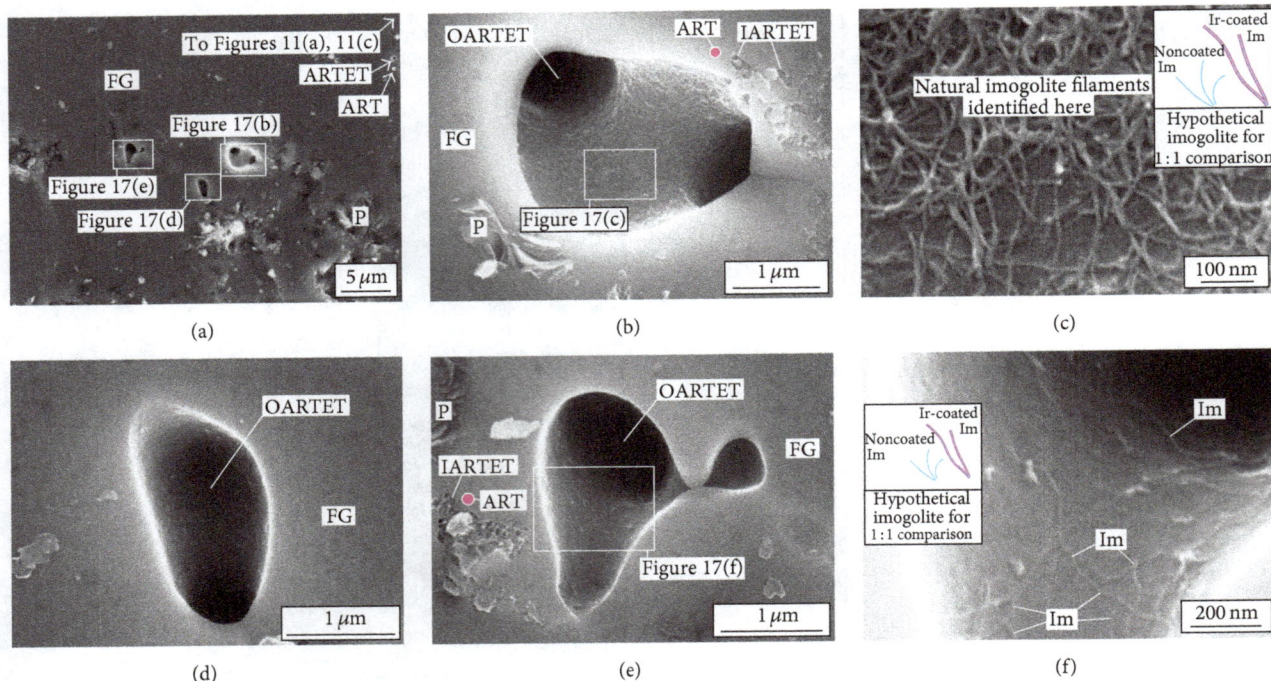

FIGURE 17: Examples of a peculiar class of etch-tunnels at the glass-palagonite interface that formed either by (i) prolonged overetching of alpha-recoil track etch-tunnels (i.e., are overetched ARTETs or "OARTETs") and/or (ii) by the advance of etch-tunnelling primarily through pressure solution (caused by increasing hydrostatic pressure through time: Figure 16). All images (a–f) are SEM (secondary electron) images, obtained from the freshly fractured surface of a basaltic glass "chip sample" from *DSDP-418A-75-3[120–123]*. Regardless of their origin, these etch-tunnels contain authigenic imogolite filaments within them (c, f) (similar in nature to the imogolite filaments found within ARTETs— see Figure 12), which seems to indicate possible links with alpha-recoil track etch-tunnelling. Note that the etch-tunnels in this region (a) actually occur along the same glass-palagonite interface and etch-tunnel zone shown in Figure 11 (which contains both FTETs and ARTETs) and occur nearby to some ARTETs (e.g., see ARTET at top right in (a)). Furthermore, some of the alteration features in the immediate vicinity of these tunnels are close to the same size as an alpha-recoil track (~120 nm) and are therefore interpreted as corroded or "infilled ARTETs" (see "IARTETs" at top right in (b) and at left in (e)). Although these comparatively large etch-tunnels (a, b, d, e) are closer in size to FTETs (~1-2 μm wide)—hinting at a possible "etched out" fission track cluster—they do not contain any authigenic platy smectite, which is observed ubiquitously within FTETs (see Figure 13). Therefore, these observations point to an origin that is different than fission track etch-tunnelling (i.e., (i) and/or (ii), above). Nevertheless, this peculiar class of etch-tunnels (OARTETs?) (b, d, e), is interpreted to be more or less equivalent to the variety of overetched microtunnels shown in Figure 7(e) that were imaged by transmitted light microscopy.

[185]). Nevertheless, here we suggest that some localized pressure solution of basaltic glass may actually be taking place—not in response to increasing lithostatic pressure— but rather, increasing *hydrostatic* pressure as the overlying Atlantic Ocean has deepened systematically with the passage of geologic time (Figure 16).

According to this hypothesis, the etch-tunnels in DSDP 418A basaltic glass have formed not only through preferential dissolution of radiation damage, but also by a process of pressure solution etch-tunnelling (e.g., enhancement or runaway dissolution of ARTETs)—as if the ever deepening Atlantic Ocean has been incrementally squeezing or "injecting" its way into the glass with microscopic etch-tunnelling "needles" of seawater that propagate preferentially through radiation damaged regions. In this regard, this type of "hydrostatic" pressure solution may be an important compounding factor in controlling pattern formation during etch-tunnelling (e.g., by connecting up nearby tracks during etch-tunnelling: see step 10a of Figure 18 and curvy blue arrows in Figure 19 (bottom), or by etch-tunnel "widening," as seen in Figures

7(e) and 17, and step 12d of Figure 18). Consequently, some etch-tunnels might start out by preferential etch-tunnelling of a single alpha-recoil track by seawater but then continue to propagate through the glass via the process of pressure solution etch-tunnelling (i.e., "runaway etch-tunnelling"), which could explain the occurrence of some relatively long and straight etch-tunnels (e.g., Figure 7(c)). Although testing this hypothesis regarding *hydrostatic* "pressure solution etch-tunnelling" in DSDP 418A basaltic glass is beyond the scope of this study, it does seem like a reasonable proposition given that (a) Since pillow eruption at the Mid-Atlantic ridge ~120.6-million-years-ago, the ambient hydrostatic pressure has more than doubled from ~29 MPa to ~62 MPa, and (b) "Classic" pressure solution is known to take place at around these conditions of pressure and temperature (e.g., during burial of quartzose sandstones: ~20–60°C and ~18–30 MPa— albeit "lithostatic" pressure [210]).

Therefore, we suggest that microscopic etch-tunnels and granular palagonite microtextures in submarine glasses are not microbial trace fossils, but rather, artefacts of three

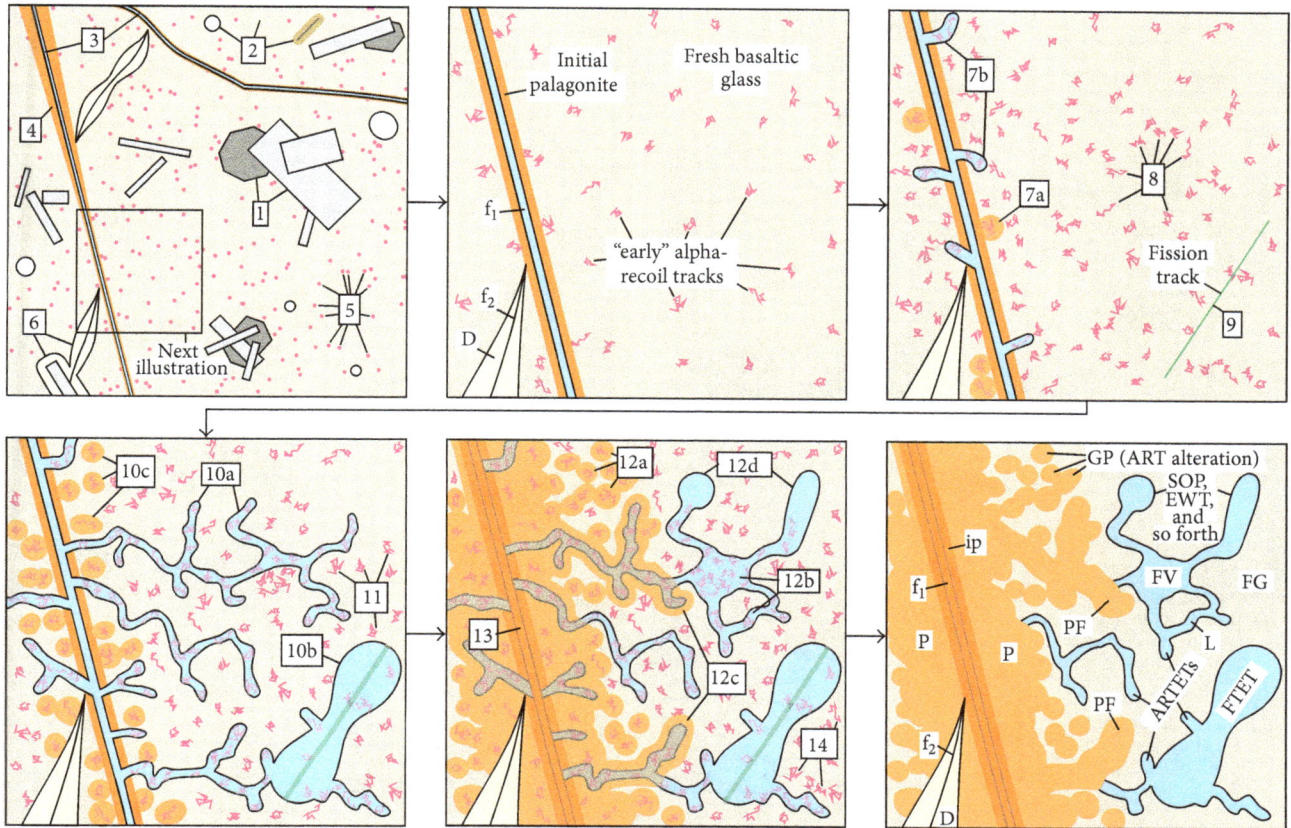

FIGURE 18: Schematic illustrations depicting the abiotic, stepwise development of complex petrographic microtextures in DSDP 418A basaltic glass (with emphasis on the corrosion of radiation damage). The final illustration (at lower right) highlights the key dissolution/palagonitization microtextures in DSDP 418A basaltic glass that are readily observable by transmitted light microscopy and SEM. Steps in microtextural/petrographic development: (1) nucleation and growth of plagioclase and clinopyroxene phenocrysts in the parent magma; (2) formation of basaltic glass, vesicles, and varioles, upon pillow eruption and quenching of glassy pillow margins; (3) early (f_1) fracturing of glass, allowing the infiltration of seawater into pillow margin interiors; (4) formation of inicipient (initial) palagonite—that is, before the onset of radiation damage—resulting in a sharp glass-palagonite interface; (5) the accumulation of significant numbers of alpha-recoil tracks (ARTs) in basaltic glass (originating from radioactive decay of U and Th) begins soon after pillow eruption (~23,500 ARTs/cm^2 after only the first 10,000 years); (6) deep burial of the pillow lavas being studied to depths of ~408–461 m beneath the overlying volcanic pile causes (f_2) fracturing and development of white (K-Al-Si)-rich devitrified zones, both along f_2 fractures and as halos surrounding plagioclase phenocrysts; (7) early-formed ARTs in the vicinity of f_1 fractures undergo preferential palagonitization (7a) and/or preferential dissolution/etch-tunnelling (7b); (8) ARTs continue to accumulate inside fresh basaltic glass; (9) eventually, fission tracks begin to accumulate in significant numbers within the glass, albeit in a much more sluggish fashion than ARTs (i.e., only ~11 fission tracks/cm^2 after the first 1,000,000 years); (10) over the passage of many tens of millions of years, basaltic glass becomes "riddled" with radiation damage. Meanwhile, the local oceanic crust ages and subsides under a deepening ocean (Figure 16), causing large incremental increases in hydrostatic pressure. The combination of these two processes leads to more advanced corrosion (dissolution and palagonitization) of basaltic glass, including development of a complex etch-tunnel network at the glass-palagonite interface defined by immense numbers of ARTETs (step 10a) interconnected with comparatively sparse FTETs (step 10b). In addition, multitudes of additional ARTs undergo "selective palagonitization" that collectively results in the development of a "granular palagonite ART alteration microtexture" (step 10c). (11-12) More ARTs and fission tracks would continue to form with the passage of time (step 11), ultimately resulting in an increase in complexity of both "granular palagonite ART alteration microtextures" (step 12a) and the ARTET-FTET network (step 12b). Eventually, some previously existing ARTETs/FTETs would become completely overprinted by the advancing palagonite zone, resulting in the formation of "palagonite fingers" that extend outwards into glass (step 12c; Figures 7(d) and 7(f)). Incremental increases in hydrostatic pressure could lead to "pressure solution" etch-tunnelling (step 12d), resulting in more peculiar etch-tunnel structures such as string-of-pearls (SOP), elongate wide tunnels (EWT), and overetched ARTETs (Figures 7(e) and 17). (13) Locally, the permeability of f_1 fractures could be reduced to zero by infilling of fractures with clays/palagonite, terminating the corrosion process. (14) Even after the termination of etch-tunnelling and alteration of the glass by seawater, additional ARTs (and fission tracks) would continue to form within fresh glass. ART: alpha-recoil track; ARTETs: alpha-recoil track etch-tunnels; D: devitrified zone; f_1: early fractures (associated with incipient/initial and ongoing palagonitization); f_2: late fractures (associated with devitrification); FTET: fission track etch-tunnel; FG: fresh glass; FV: flare-out void (formed by dissolution of multiple nearby ARTs); GP: granular palagonite; ip: initial palagonite; L: loop; P: palagonite; PF: palagonite fingers; SOP, EWT, and so forth: string-of-pearls texture, elongate wide tunnels, and other peculiar etch-tunnel varieties caused by prolonged overetching of alpha-recoil tracks and/or pressure solution etch-tunnelling; see Figures 7(e) and 17.

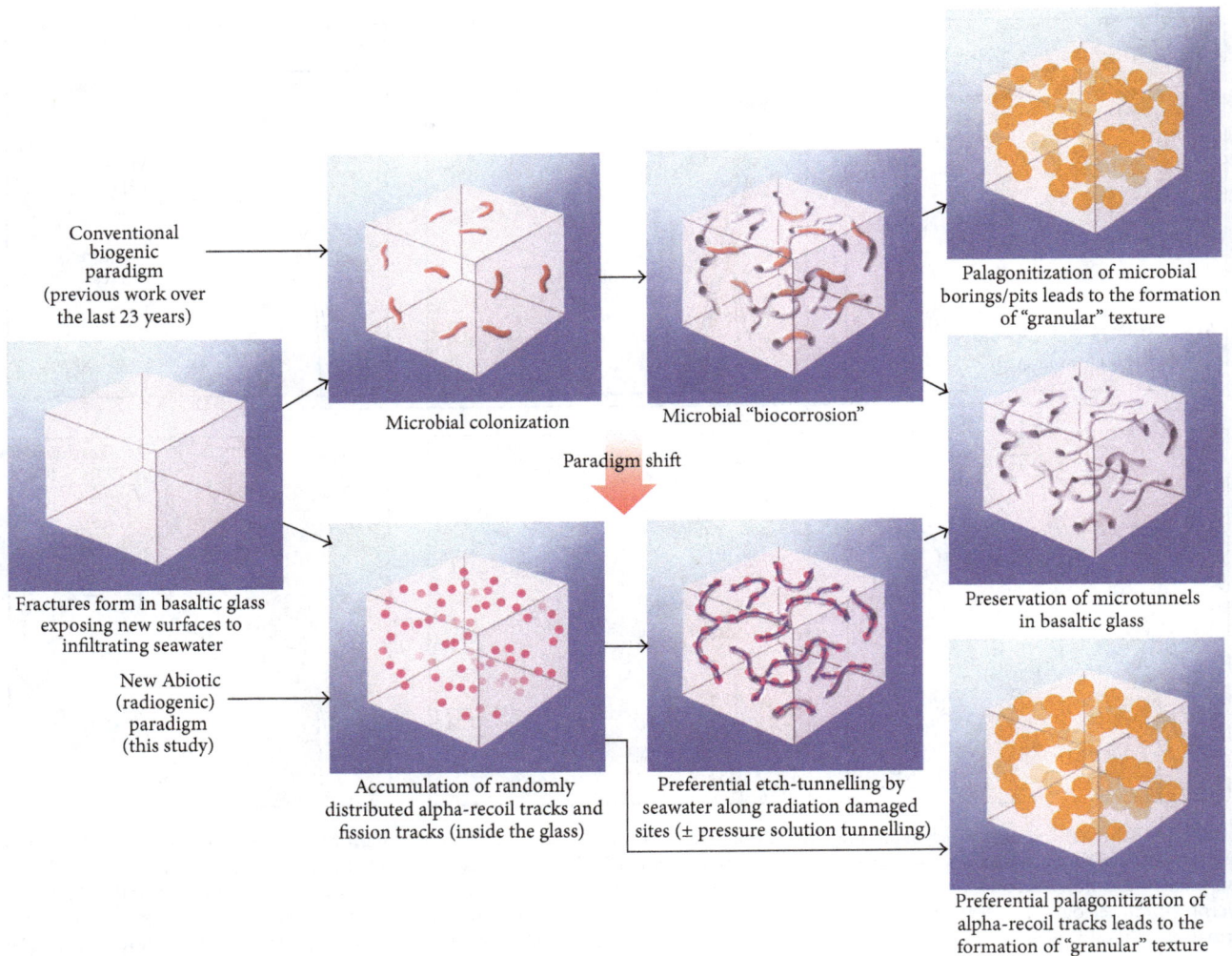

FIGURE 19: Diagram outlining a paradigm shift in scientific perspectives on the origin of complex corrosion microtextures in submarine basaltic glass—constrained in this study with our model for the preferential corrosion of radiation damage (by seawater) in basaltic glass pillow margins at DSDP 418A (i.e., new abiotic "Radiogenic" paradigm). The conventional "biogenic" paradigm is outlined in previous studies on microbial bioalteration of basaltic glass (e.g., see reviews in Furnes et al. [11], Furnes et al. [26], Staudigel et al. [27], McLoughlin et al. [28], and Furnes et al. [59]).

compounding "abiotic" geological processes: (1) accumulation of radiation damage (i.e., fission tracks and alpha-recoil tracks) in the glass by radioactive decay of U and Th; (2) preferential corrosion (i.e., palagonitization and dissolution) of radiation damage in the glass by seawater; (3) subsidence of the oceanic crust during the cooling and thickening of oceanic lithosphere as it moves away from the spreading ridge and the associated incremental increases in hydrostatic pressure under a deepening ocean—leading to localized pressure solution of glass (and alpha-recoil track etch-tunnel widening/enhancement).

As with the study of mm-scale pressure solution structures in fossiliferous shales by Lescinsky and Benninger [211]—which may be readily mistaken for predator traces (i.e., are *pseudo*ichnofossils or *pseudo*borings)—the pressure solution microstructures (i.e., enhanced alpha-recoil track

etch-tunnels) described here may also be readily mistaken for microbial trace fossils (e.g., compare SOP texture in Figure 7(e) of this study with string-of-beads texture in Figure 1(f) of Fisk et al. [24]).

6. Discussion and Implications

6.1. Discovery of Naturally Etched Nuclear Tracks in Basaltic Glass: Implications for Geomicrobiology. The worldwide corrosion of basaltic glass throughout geological history on planet Earth has resulted in the development of a variety of distinct and complex alteration/dissolution microtextures preserved in submarine glasses of the *in situ* oceanic crust, ophiolites, and greenstone belts dating back to ~3.5 Ga, and, for about the last 20 years, the consensus has been that these globally widespread, complex petrographic microtextures in volcanic glass (i.e., "tubular" etch-tunnels and "granular" palagonite) originate strictly through microbial activity [11, 23, 24, 26–28, 59, 104]). Therefore, our discovery in the present study that microscopic etch-tunnels and granular palagonite microtextures in DSDP-418A basaltic glass originate primarily by preferential corrosion of radiation damage (i.e., fission tracks and alpha-recoil tracks) has immediate implications for geomicrobiology, microbial ecology, and studies aimed at identifying microscopic morphological biomarkers in volcanic/impact glasses on planet Earth (or even Mars). For instance, alpha-recoil track etch-tunnels and granular palagonite "ART" alteration identified at the glass-palagonite interface in DSDP-418A basaltic glass in this study are basically indistinguishable from many previous reports of "tubular" and "granular" bioalteration (i.e., compare "tubular bioalteration" in Figures 1(a), 1(c), and 1(e), to "alpha-recoil track etch-tunnels" in Figures 1(b), 1(d), and 1(f), and compare "granular bioalteration" in Figures 2(a) and 2(c) to "granular palagonite ART alteration" in Figures 2(b) and 2(d)). What this means is that probably in many cases in the past, such "abiotic" corrosion microtextures in submarine basaltic glass have likely been mistaken for signs of microbial activity/bioalteration. In this regard, naturally etched fission tracks and alpha-recoil tracks in submarine glasses—including microscopic "fission track etch-tunnels" (e.g., Figures 11, 13, 15(a), and 15(b)), "alpha-recoil track etch-tunnels" (e.g., Figures 7, 9, 11, 12, 15(a), and 15(b)), and "granular palagonite ART alteration" (e.g., Figures 7(f) and 10)—can also be regarded as *pseudomicrobial trace fossils*—that is, abiotic microstructures that look conspicuously biogenic, but which are not (akin to previously described *pseudo*borings/predator traces" documented in fossiliferous shales that are simply artefacts of pressure-dissolution [211]). Consequently, the depth of the biosphere (in the oceanic crust), the distribution of microbes and microbial habitat and the possible preservation of microbial trace fossils in volcanic glass within the *in situ* oceanic crust, ophiolites, and greenstone belts dating back to ~3.5 Ga now need to be completely reevaluated in light of our abiotic (i.e., U–Th–Pb radiogenic) model for the development of microscopic etch-tunnels (i.e., tubular textures) and granular palagonite microtextures in submarine basaltic glass. Therefore, our study marks the second "paradigm shift" in scientific perspectives on the origin of complex corrosion microtextures in submarine volcanic glass (Figure 19), from an early "abiotic" paradigm (i.e., before Thorseth et al. [23] [39], "mist zone" of Morgenstein and Riley [40], [41–44]), to the popular "biotic" paradigm which has dominated the scientific literature over the last 23 years (e.g., [11, 12, 23–28, 34, 49, 59, 64, 96, 104])—and now back to an "abiotic" (i.e., U–Th–Pb radiogenic) paradigm with the present study (Figure 19).

Probably in most previously documented cases, the notion of "granular bioalteration" of basaltic glass is now more or less in doubt—in light of our simple "abiotic" model of granular palagonite formation by the preferential palagonitization of multitudes of alpha-recoil tracks (Figures 2, 10, 18, and 19). Therefore, we suggest that the recently proposed trace fossil names for granular bioalteration of basaltic glass, including the *Granulohyalichnus vulgaris* ichnospecies and

the *Granulohyalichnus* ichnogenus [28], can now essentially be challenged. These granular palagonite microtextures are quite clearly the result of preferential palagonitization of multitudes of randomly distributed alpha-recoil tracks in the glass (Figures 10 and 18), and they are basically identical (Figure 2) to previous reports of granular bioalteration microtexture—most notably those microtextures presented as "classic examples" of granular bioalteration (Figures 2(a) and 2(c) [11, 28, 59]).

We did not observe any examples of preferentially palagonitized fission tracks in DSDP 418A basaltic glass in this study (i.e., "granular palagonite FT alteration" as opposed to "granular palagonite ART alteration"), and this is probably due to the relatively sparse distribution of fission tracks at the scale of observation of granular palagonite (i.e., compare Figures 10(d) and 10(g), shown at similar scales). However, some previous reports of granular palagonite microtextures in submarine glasses from the Costa Rica Rift (*Plate 1 [3]* in Furnes et al. [50]), interpreted as evidence of microbial etching—but reinterpreted here as "abiotic" corrosion microtextures (Figure 20(c))—do provide compelling textural evidence to suggest that preferential palagonitization of *both alpha-recoil tracks and fission tracks* does take place during the alteration of submarine glasses. In particular, the bimodal distribution of a few large (ca. 5 μm) "fission track granules" surrounded by multitudes of smaller (ca. 0.5–1 μm) "alpha-recoil track granules" (Figure 20(c)) is consistent with the expected size and population distributions of fission tracks versus alpha-recoil tracks. Therefore, we suggest that in addition to the ca. 0.6 μm variety of granular palagonite (i.e., "granular palagonite ART alteration"; Figure 10), some larger (ca. 5 μm) varieties of granular palagonite (i.e., possibly representing "granular palagonite FT alteration"; Figure 20(c)) may also be considered to represent pseudo*microbial trace fossils*.

Similarly, it seems clear at this point that many previously documented examples of "tubular bioalteration" of basaltic glass (i.e., microscopic tunnels at the glass-palagonite interface in submarine glasses)—can now be reinterpreted as alpha-recoil track etch-tunnels. For instance, tubular textures in submarine basaltic glasses presented as "classic examples" of tubular bioalteration are essentially identical in size, form, and geological setting to the alpha-recoil track etch-tunnels documented at the glass-palagonite interface in DSDP-418A basaltic glass in the present study (e.g., compare at the same scale: Figures 1(a), 1(c), and 1(e) with Figures 1(b), 1(d), and 1(f), resp.). Therefore, we suggest that the biogenicity of most "tubular bioalteration" microtextures documented in submarine glasses worldwide (e.g., [11, 26], including the recently proposed *Tubulohyalichnus ichnogenus* and related ichnospecies, particularly the ichnospecies *Tubulohyalichnus simplus* of McLoughlin et al. [28]) can probably in many (if not most) cases be effectively refuted by our simple model of alpha-recoil track and fission track etch-tunnelling (possibly combined with etch-tunnel enhancement by pressure solution). For example, compare at the same scale, representative examples of *Tubulohyalichnus simplus* presented in Figures 2(a)–2(c) of McLoughlin et al. [28] with representative examples of alpha-recoil track etch-tunnels in Figures 1(b), 1(d), 1(f), and 7 in the present study.

(a)

(b)

(c)

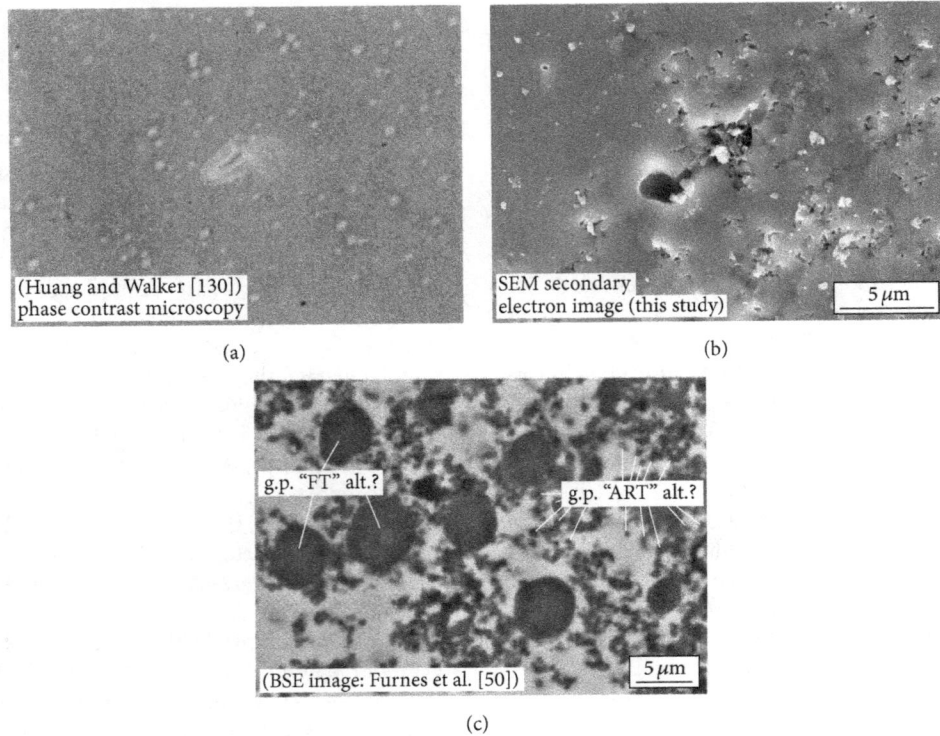

FIGURE 20: Three examples of "bimodal" distributions of microscopic corrosion features in geological samples—in each case, most likely related to the bimodal "size versus population (i.e., areal density)" distribution of fission tracks versus alpha-recoil tracks. (a) Phase contrast photomicrograph showing one large fission track etch-pit surrounded by a multitude of much smaller alpha-recoil track etch-pits on the experimentally etched cleavage surface of mica (from Huang and Walker [130]—no scale bar available). (b) SEM (secondary electron) image of one large fission track etch-tunnel surrounded by several much smaller alpha-recoil track etch-tunnels in *DSDP-418A-75-3-[120–123]* basaltic glass (close-up from Figure 11(a)). (c) SEM (BSE) image of several large palagonite granules surrounded by a multitude of much smaller palagonite granules (from Plate 1 in Furnes et al. [50] of sample *148–896A-11R-1*, 111–113 cm—from the Costa Rica Rift). The original interpretation is that both varieties of palagonite granules are biogenic in origin (i.e., "produced by the etching of microbes": Furnes et al. [50]). However, in light of our newly proposed "abiotic" paradigm for interpreting corrosion microtextures in submarine glasses (Figure 19), we reinterpret the bimodal distribution of palagonite granules in (c) to have more likely arisen due to the concomitant, preferential palagonitization of several large fission tracks (granular palagonite "FT" alteration—labelled as "g.p. 'FT' alt.?") and a multitude of tiny alpha-recoil tracks (granular palagonite "ART" alteration—labelled as "g.p. 'ART' alt.?"), during infiltration of seawater.

It then follows that "older" examples of titanite-mineralized tubular textures attributed to microbial activity that are preserved along healed fractures in the metamorphosed margins of pillow lavas in Archean greenstone belts [25, 35, 37, 91, 94] might also simply represent the preserved (titanite-mineralized) relicts of preferential corrosion of radiation damage and/or pressure solution etch-tunnelling (i.e., "abiotic" microtextures) that formed during the encroachment of seawater into these basaltic glass pillow margins in Archean times.

Consequently, we suggest that the observed worldwide distribution of "tubular" and "granular" microtextures in submarine glasses from >40 geological sites worldwide (e.g., [26, 27, 103]) might not necessarily be a reflection of a global lithoautotrophic microbial community causing the biocorrosion/bioalteration of volcanic glass, but instead it may simply be a reflection of the worldwide occurrence of U- and Th-bearing submarine glasses [146] and the associated "abiotic" corrosion/dissolution of radiation damage by encroaching seawater combined with the possible effects of

pressure solution etch-tunnelling, as these glasses age under a deepening ocean (Figure 16).

At this point, it is important to highlight that there is also a much larger variety of other more peculiar morphological types of microscopic tunnels documented in submarine glasses around the globe (but not found in the present study), whose origin needs to be explained, whether "biotic" or "abiotic." This includes spiral/helical microtunnels, annulated microtunnels, and dendritic microtunnels (Figures 4–6 in McLoughlin et al. [28]), along with several other intricate microtunnel varieties documented more recently in submarine glasses such as flattened "petal shaped" tunnels exhibiting honey-comb or ribbed textures, crowned and palmate tunnel varieties, and tunnels exhibiting septae [103]. It has been suggested that some of these other intricate microtunnel varieties might be the product of microbial activity/microboring [28, 103] and some are currently desig-nated as *bona fide* microbial trace fossils in basaltic glass (e.g., the recently erected ichnofossil species *Tubulohyalichnus spiralis*, *Tubulohyalichnus annularis*, and *Tubulohyalichnus*

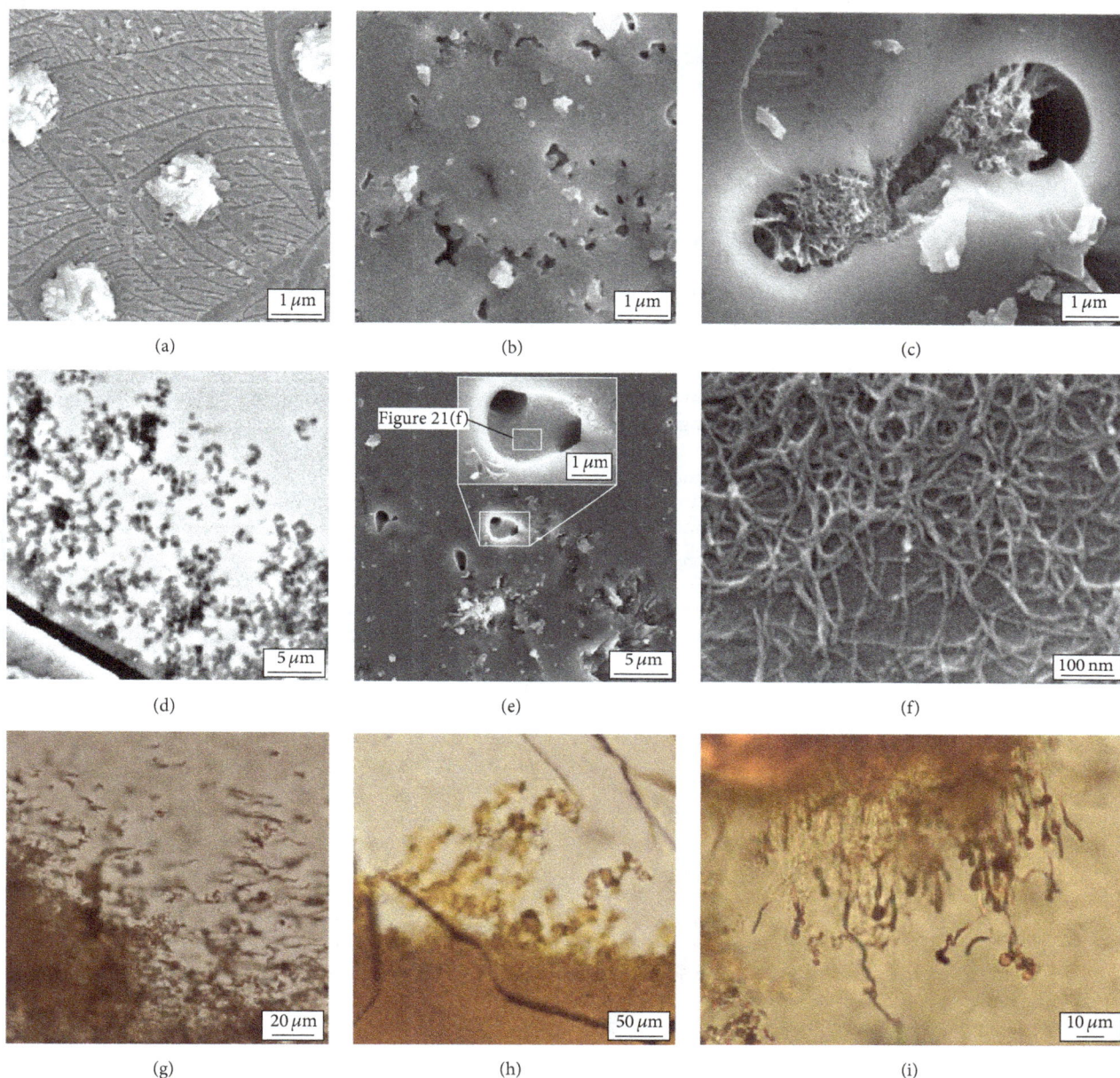

FIGURE 21: Summary of complex "abiotic" microtextures found in DSDP 418A basaltic glass that have the potential to be mistaken for signs of microbial activity (i.e., microbial etchings, groovings, borings, alteration, or remains). (a–c, e, f, h, and i) are from sample *DSDP-418A-75-3-[120–123]* and (d, g) are from sample *DSDP-418A-68-3[40–43]*. (a–f) are SEM (secondary electron) images ((a–c, e, and f) are from basaltic glass "chip samples" and (d) is from a polished petrographic section) and (g–i) are transmitted light photomicrographs of polished sections taken in plane polarized light (uncrossed polars). (a) Dendritic nanogrooves on a vesicle wall (close-up image from Figure 1(b) in French and Muehlenbachs [107]) that represent frozen viscous fingers of magmatic fluid injected into the vesicle wall upon quenching of the glass. (b) Alpha-recoil track etch-tunnels (ARTETs). (c) A fission track etch-tunnel. (d) Granular palagonite ART alteration (close-up from Figure 2(b)). (e) ARTETs at the glass-palagonite interface affected by prolonged overetching and/or are "pressure solution enhanced" (from Figure 17 in this study; Figure 1 in French [139])—akin to the tunnels shown in Figures 7(e) and 21(i)—also note that this region (e) occurs along the same glass-palagonite interface that is shown in Figure 11(a) (i.e., in the vicinity of dense concentrations of other *bona fide*/incipient ARTETs). (f) Authigenic imogolite filaments (close-up from (e); Figure 17; Figure 1 in French [139]). (g) ARTETs at the glass-palagonite interface. (h) Palagonite fingers that have overprinted previously existing ARTETs. (i) ARTETs—most of which have been affected by prolonged overetching and/or are "pressure solution enhanced," resulting in string-of-pearls texture and elongate wide tunnels (among other forms—see Figure 7(e)).

stipes of McLoughlin et al. [28]). Here, we suggest that these other morphological varieties of microtunnels in volcanic glass might also originate through some kind of pressure solution etch-tunnelling process (similar to that outlined in Section 5.3) as opposed to microbial activity, such that they could instead represent spiral/helical, annulated, and dendritic pressure solution etch-tunnels (e.g., that form in the absence of radiation damage?). Note that the idea of the "injection" of microscopic channels of geological fluids into basaltic glass is not that far-fetched, because dendritic patterns of nanoscopic grooves found on vesicle walls in submarine basaltic glass have recently been shown to form by the fluid mechanical process of viscous fingering, resulting essentially in the "injection" of narrow (~50–75 nm wide) fingers of magmatic vapour into hot basaltic glass upon quenching of the glass during pillow eruption (see companion study on nanogrooves within vesicles in DSDP-418A basaltic glass [107]). At any rate, the experimental etch-tunnelling of quartz using HF has also resulted in the formation of micron-scale spiral/helical microtunnels [116] that are similar in nature to those found in basaltic glass (e.g., those shown in Figure 5 of McLoughlin et al. [28]), and therefore nonbiological explanations such as microscopic etch-tunnelling, pressure solution etch-tunnelling, injection, and possibly even viscous fingering (cf. dendritic microchannels documented in French and Muehlenbachs [107]) should also be sought for the origin of these other (more peculiar) varieties of microtunnel in basaltic glass (i.e., spiral/helical, annulated, and dendritic, among other forms).

The discovery of naturally etched fission tracks and alpha-recoil tracks in submarine basaltic glass in this study has important implications for other fields of research as well, including geochronology (alpha-recoil track and fission track dating), the influence of radiation damage on weathering and corrosion processes in rocks, minerals, and glasses, and the astrobiological exploration of Mars (e.g., evaluating the origin and possible biogenicity/abiogenicity of corrosion microtextures found in Martian glasses), and provides an ideal natural laboratory for understanding the long-term breakdown and corrosion of nuclear waste glasses stored in deep geological repositories, all of which we address in the following four Sections 6.2–6.5.

6.2. Identification of Naturally Etched Fission Tracks in Submarine Basaltic Glass: Context within the Broader Field of Fission Track Dating

6.2.1. Background on Experimentally Etched Fission Tracks in Minerals and Glasses. The *experimental* etch-tunnelling of fission tracks is routinely carried out in fission track dating studies of a variety of geological materials that contain trace U (Figures 3(b) and 3(c)), and this includes minerals such as zircon [212], apatite [213], monazite [149], sphene [148], titanite [214], and muscovite [207], a variety of natural glasses including the glassy margins of midocean ridge pillow basalts [125, 147], glassy shards in marine sediments [140] and tephra beds [128], basaltic glass inclusions in volcanic quartz [215], tektites [216–219], and obsidian artefacts [220]—as well as in

studies of borosilicate nuclear waste glasses [138, 144] and other synthetic materials used as fission track "detectors" such as diallyl phthalate resin [221] and CR-39 plastic [222]. Exposing a polished or cleavage surface to an etchant preferentially dissolves the damaged regions in the mineral or glass that result from the spontaneous fission of ^{238}U (Figure 3) or by the induced fission of ^{235}U caused by irradiating the sample with thermal neutrons from a nuclear reactor (Figure 3(b)—right) [202, 223, 224] (note: strong etchants such as HF, HNO_3, HCl, or NaOH solutions are typically used in fission track dating studies, although weak etchants—such as deionized water—have also been used to reveal fission tracks in silicate glasses (e.g., Figure 3(d)) [138]). Fission fragments cause linear damage tracks (Figure 3(a)) that range in etchable length depending on numerous factors including host mineralogy and composition [225, 226], fission fragment mass [225, 227], crystallographic orientation [228], etching time and efficiency [222], and the degree of thermal annealing which may have occurred [229]. For example, the mean etchable length of pristine fission tracks (e.g., unannealed or induced tracks) is ~8-9 μm in volcanic glass (Figure 3(b); Figure 1 in Sandhu et al. [198]; Figure 8(b) in Westgate and Naeser [199]; Arias et al. [126]; Sandhu and Westgate [128]) and tektites (Figures 1 and 2 in Storzer and Wagner [218], ~11 μm in zircon [212, 230], ~16 μm in apatite [213], and ~20 μm in micas [207, 229], although experimentally etched "fossil" (i.e., spontaneous) fission tracks may be substantially shorter in each case, especially due to annealing and track fading [126, 128, 207, 212, 213, 229, 230]. For natural (and synthetic) silicate glasses exhibiting a wide range of chemical compositions, the lengths of fully revealed (i.e., etched) fission tracks are consistently reported as being ~6-9 μm long (e.g., Figures 3(b), 3(d), and 3(e))—including those in hydrated silicic volcanic glass shards [128], tektites (including: australite [216, 231], bediasite [218], indochinite [216], and moldavite [128, 219]), obsidian archaeological artefacts [220], rhyolite glass [232], dredged and drilled samples of basaltic glass pillow margins [125, 127], and glasses from volcanic flows and glassy shards interlayered within sediments [126]—and they etch much faster in basaltic glass than in glasses with higher SiO_2 contents [144]. As highlighted above and in Section 3.4.2, the typical size of fully etched "pristine" fission tracks in volcanic glass (and tektites) is ~8-9 μm long (Figure 3(b) [126, 128]; Figure 1 in Sandhu et al. [198]; Figure 8(b) in Westgate and Naeser [199]; Figures 1 and 2 in Storzer and Wagner [218]).

During etch-tunnelling of fission tracks in the laboratory, the etched track widths grow linearly with time [212, 222], and when the full track length is revealed with minimal overetching they are on the order of ~1 μm wide [212, 221]. Thus, at ~1-2 μm wide and up to ~8 μm long, the naturally formed fission track etch-tunnels identified in DSDP-418A basaltic glass in this study (Figures 11, 13, and 15) are consistent with the expected dimensions of fully etched fission tracks in volcanic glass (and tektites) (see 1:1 scale comparison in Figure 3(e)), which adds support for our interpretation that these peanut-shaped cavities at the glass-palagonite interface (e.g., Figure 13) are in fact naturally etched fission tracks. Fission tracks in volcanic/impact glasses that are etched in the laboratory, typically exhibit a pointed "conical" shape during etching of the track (at first), but once the etch-pit reaches the end of the track (i.e., when fully etched), the etch-pits then progressively take on a more rounded (prolate spheroidal to spherical) shape (Figure 2.6 in Wagner and Van den Haute [219]), such that fission tracks oriented perpendicular to the polished surface look like circles when fully etched, while those at oblique angles look like elongate ovals (Figure 3(b); Figures 2.7 and 6.7 in Wagner and Van den Haute [219]). Similarly, the fission track etch-tunnels identified here in DSDP-418A basaltic glass (Figure 13) have rounded tips and are considered to be "fully etched/revealed" fission tracks, which may also exhibit a circular appearance when they are intersected at a high angle (i.e., when viewed end-on: Figures 13(d) and 13(f)) and more elongate shapes when intersected at a shallow angle (e.g., Figures 13(a)–13(c), and 13(f)). However—instead of an oval shape—in this geological environment, the naturally etched fission tracks take on an overall "dumbbell" or "peanut" shape (Figures 3(e) and 13)—and this is interpreted to reflect syn- to postdissolutional necking (Figure 13(f)) that takes place in response to the conditions of high confining pressure that these glasses are subjected to during fission track etching (i.e., the elongate etch-tunnel has attempted—but failed—to pinch itself off as two smaller spheres that would ultimately have less surface energy and thus be more stable—a process that commonly takes place in tubular or planar fluid inclusions within minerals [122, 233]).

6.2.2. Adding to the Known Record of Naturally Etched Fission Tracks in Minerals and Glasses.

It was noted early on that despite their possible widespread occurrence, reports of *naturally* etched fission tracks are rare [138, 234], prompting some to duly note that given the abundance of latent fission tracks in ancient U-bearing minerals and glasses "one can wonder why such a preferential etching of fission tracks in natural systems has never been reported." [138]. Some possible reasons put forth to explain this apparent lack of naturally etched fission tracks include the difficulty in recognizing naturally etched fission tracks due to the rough nature of mineral surfaces at the micron scale, the low concentrations of fissionable elements in many natural glasses, and the low dissolution rates of many ancient U-bearing minerals that contain abundant fission tracks (e.g., zircon and monazite). Nevertheless, at this point in time, there are actually a few documented cases of naturally etched fission tracks in geological samples, including the weathered outer surfaces of sphene grains from the ca. 450 Ma Coleraine granite, Western Australia [148], occurrences within apatite crystals of the Late Cretaceous Kunon pluton, Zhangzhou Igneous Complex, southeast China [150], and those formed by natural etching caused by fluids circulating along fractures within ~1 Ma monazite of unknown provenance [149] and possibly also within quartz grains picked on the borders of the natural fossil nuclear reactors at Oklo, Gabon [138, 234]. Therefore, the ~1-2 μm wide, up to ~8 μm long, fission track etch-tunnels identified in DSDP 418A basaltic glass in this study (Figures 11, 13, and 15) add to this relatively sparse record and represent the first documented occurrence of naturally etched fission

tracks in basaltic glass (or for that matter, any other type of natural glass). This discovery provides new evidence that naturally etched fission tracks could be quite widespread on planet Earth (i.e., within a wide variety of possible U-bearing minerals and glasses that have been subjected to subaqueous weathering and dissolution); it is just that we have been relatively slow to recognize them thus far. Further evidence to support this idea comes from the bimodal distribution of "granular" palagonite alteration observed in some partially palagonitized submarine glasses (Figure 20(c) [50]), which may arise from the preferential palagonitization of relatively few large fission tracks along with multitudes of smaller alpha-recoil tracks (Figure 20(c)) during the infiltration of seawater.

6.3. Identification of Naturally Etched Alpha-Recoil Tracks in Submarine Basaltic Glass: Context within the Broader Field of Alpha-Recoil Track Dating

6.3.1. Adding to the Record of Naturally Etched Alpha-Recoil Tracks in Minerals and Glasses.

Similarly, there is much direct (and indirect) evidence which suggests that the natural etching of alpha-recoil tracks in minerals is a widespread phenomenon on Earth [141], including, for example, irregular corrosion fronts associated with preferential alteration/weathering of high U- and Th-bearing metamict domains in zircon (e.g., Figure 7(a)—inset BSE image; Figure 9 in Lumpkin [193]; Appendix A in French [194]; Figure 1(d) in French [195]) or the leaching of certain radionuclides (e.g., ^{234}U and radiogenic Pb) from alpha-recoil track damaged sites within zircon (e.g., [141, 235]). Therefore, the discovery of alpha-recoil track etch-tunnels at the glass-palagonite interface in submarine basaltic glass in this study (readily observable by petrographic microscope: Figures 1(b), 1(d), 1(f), 7, 9(n), and 9(o); and by SEM: Figures 9, 11–13, and 15) provides new "microtextural evidence" for this likely globally widespread corrosion process and represents the first ever documented occurrence of naturally formed alpha-recoil track etch-tunnels (ARTETs) in any type of geological material. Furthermore, the observation that *granular palagonite ART alteration* at the glass-palagonite interface in DSDP 418A basaltic glass can be explained by the preferential palagonitization of multitudes of alpha-recoil tracks (Figure 10—also see Section 3.3.2) provides further evidence that alpha-recoil track etching/corrosion is an important process taking place within submarine basaltic glass—possibly at a global scale given the widespread distribution of U- and Th-bearing submarine glasses in the oceanic crust [146] and the widespread occurrence of microtunnels and granular alteration textures in submarine volcanic glasses (e.g., [26, 27, 103]).

6.3.2. Implications for Alpha-Recoil Track Dating of Natural Glasses.

As with fission tracks, the areal density of alpha-recoil tracks in geological samples can be exploited as a geochronometer, when the tracks are revealed by chemical etching in the laboratory and then counted—although this geochronological technique has so far only been applied to very young volcanic micas (i.e., <1 Ma phlogopites) due to the relatively rapid accumulation of alpha-recoil tracks and the simplicity of etching those which intersect the exposed cleavage surfaces of micas (Figure 4(b)) [129–134]. Therefore, the discovery of naturally etched alpha-recoil tracks in ~120.60 Ma basaltic glass in this study opens up for the first time, the possibility of carrying out geochronological studies of volcanic/impact glasses and obsidian archaeological artefacts using alpha-recoil tracks—originally suggested as being worthy of pursuit if alpha-recoil tracks were ever found in volcanic glass [130]. As mentioned above, the revelation of alpha-recoil tracks by experimental chemical etching has only ever been applied to micas [129], and this is in part because the perfectly flat cleavage surfaces of mica provide ideal surfaces for chemical etching of such faint and small (~120 nm diameter) nuclear tracks—avoiding the need for generating flat surfaces by means of polishing and mechanical abrasion—as is carried out in fission track dating studies of minerals such as apatite, zircon, or sphene, and notably volcanic/impact glasses. Therefore, if experimental chemical etching of latent alpha-recoil tracks in volcanic glass is ever attempted (it is beyond the scope of the present study), we suggest that it should be carried out on freshly broken fracture surfaces (i.e., generated in the laboratory) that are suitably flat and fresh, to avoid the possible destructive impact that polishing with fine abrasive powders would have on the revelation of such tiny damage tracks on an artificially "polished" surface (i.e., see polishing "scratches" in Figure 3(d)). It is also important to highlight that the naturally formed alpha-recoil track etch-tunnels identified in DSDP-418A basaltic glass in this study were "naturally etched" under conditions of considerably high hydrostatic pressure (~29–62 MPa), and so the "natural" revelation of alpha-recoil tracks in these volcanic glasses could in part have been facilitated by the immense weight of the overlying water column (Atlantic Ocean). This might also explain why, despite decades of fission track dating of natural glasses (e.g., [125, 126, 128, 140, 144, 147, 216, 219, 232]), no studies have ever hinted at the possible presence of etched alpha-recoil tracks in laboratory etching studies of polished samples of volcanic/impact glasses. Moreover, when comparing the etching acceleration (i.e., preferential etching ratio: $(V_T - V_G)/V_G$, where V_T is the "track" etch rate and V_G is the "general" etch rate) of alpha-recoil tracks (0.0015; i.e., V_T is 0.15% higher than V_G) versus fission tracks (3000; i.e., V_T is 3000 times higher than V_G!) in muscovite [145], it again becomes clear why alpha-recoil tracks may have been overlooked (or simply not revealed at all) in many previous fission track dating studies of natural glasses.

6.4. Implications for Long-Term Storage of Nuclear Waste Glass.

The envisaged final steps in the nuclear fuel cycle are solidifying the high level radioactive waste, generally by vitrifying it and subsequently safe storage of this material in a deep dry geological repository such as salt domes, granite, or clay [227, 236]. Nuclear waste glasses are typically comprised of 40–50 wt.% SiO_2, 10–15 wt.% B_2O_3, 8–20 wt.% Na_2O, some MgO, Al_2O_3, Fe_2O_3, CaO, and TiO_2, and 10–15 wt.% high

level radioactive waste [227]. Basaltic glass is considered to represent a suitable natural analog for the breakdown of borosilicate nuclear waste glasses, because both have similar SiO_2 contents and related corrosion rates and mechanisms [135, 191].

Most of the radiation damage in nuclear waste glasses comprises alpha-recoil tracks [227, 236–238]. As with alpha decay in rocks and minerals [132, 134], the kinetic energy of each recoiling daughter nucleus (~0.7 keV/nucleon) is lost through elastic collisions with the host glass, forming a ~30–50 nm long damage trail that comprises on the order of ~1000 atomic displacements [129, 237]. Some radiation damage in nuclear waste glasses also originates from other forms of self-radiation including alpha particles themselves, which cause ~200 atomic displacements over their travel range of ~10–20 μm, spontaneous fission tracks, which are very rare relative to alpha-recoil tracks but cause about ~40,000–60,000 atomic displacements per fission fragment, and beta particles which cause on average less than one atomic displacement per beta decay event [227, 236, 238].

Extrapolation of laboratory assessments of nuclear waste glass behaviour to longer time periods (i.e., 10,000 years or more) is currently one of the most important problems in nuclear waste disposal [193]. In the event that ground water breaches a repository sometime in the future, it is particularly important to understand the interactions of water with radiation damaged regions in old glass because leaching will occur preferentially in those regions [227]. Of special interest in this problem are the long-lived actinides (e.g., Np, Pu, Am, and Cm) in nuclear waste glasses that will cause abundant alpha-recoil damage during radioactive decay over the realistically long periods of storage time (i.e., some 10,000 years) in the repository [193, 236]. To some degree, this problem can be investigated by doping of synthetic nuclear waste glasses with short-lived actinides such as ^{238}Pu (half-life of 87.7 years) which leads to accelerated alpha-recoil damage during self-radiation and then assessing the stability of these glasses [238].

The natural system provided by DSDP 418A basaltic glass pillow margins described here represents an exceptional analog for this type of long-term investigation of nuclear waste glass stability, because the glass is ~120.6 million years old (for age constraints see Figures 5(c), 5(d) and Section 2.2), contains abundant alpha-recoil tracks (148,000,000 alpha-recoil tracks/cm^2) and fission tracks (1,310 fission tracks/cm^2), contains regions of fresh unaltered glass—as well as glass that has been affected by preferential corrosion of radiation damage (e.g., Figures 7 and 9–12), and has been exposed to seawater—possibly in a continual fashion since the Early Cretaceous. The antiquity of these basaltic glass pillow margins also compensates to some degree for the comparatively low concentrations of U and Th in the glass (e.g., 0.032–0.042 ppm U and 0.108–0.132 ppm Th: Table 1) when compared with the ~10–15 wt.% high level radioactive waste typically present in borosilicate nuclear waste glasses [227], and this is revealed by the observation that (in addition to the pattern of fracturing in the glass) the distributions of fission tracks and especially alpha-recoil tracks were primary factors in controlling the textural development of the palagonite alteration/dissolution (i.e., corrosion) front (Figures 10 and 15). Fission tracks and alpha-recoil tracks (simulated by ion implantation) are known to drastically increase the etch rates of borosilicate nuclear waste glasses in water (e.g., Figure 3(d)) [137, 138]. Therefore, the discovery of naturally formed fission track etch-tunnels (Figure 13), alpha-recoil track etch-tunnels (Figure 12), and granular palagonite ART alteration (Figure 10) caused by the advancement of encroaching seawater into basaltic glass at DSDP 418A is particularly relevant to understanding the long-term behaviour of nuclear waste glass, especially given the relatively low (about a million times less) alpha-recoil dose of these natural submarine glasses (ca. 10^{18} alpha decay/m^3) when compared to that observed in accelerated long-term studies of synthetic nuclear waste glasses doped with short-lived actinides (e.g., ca. 10^{24} alpha decay/m^3 [238]).

Consequently, based on our observations of pronounced natural corrosion (i.e., etch-tunnelling and palagonitization) of radiation damaged sites in fresh basaltic glass at DSDP 418A, we highlight that deep burial of high level nuclear waste glasses in geological repositories should probably be avoided altogether to prevent dangerous radionuclides from leaking out into Earth's biosphere during the aging and associated physical breakdown of such nuclear waste glasses—should a repository be breached by subsurface groundwaters in the future.

6.5. Implications for the Astrobiological Exploration of Mars.
Recent studies have suggested that the complex alteration microtextures commonly found within volcanic glass on Earth (i.e., granular and tubular bioalteration textures) should be sought as possible target biosignatures to look for in Martian glasses during future astrobiology missions to Mars, and possibly even other solar system bodies such as Europa [12]. Although it is true that "tubular" etch-tunnels and "granular" palagonite microtextures in submarine glasses do coincide with the sizes of typical microbes (which is cited as the main reason why they are thought to be biogenic in origin: e.g., [26]), here we highlight that these corrosion microtextures in basaltic glass are also, incidentally, the same size as fission tracks and alpha-recoil tracks (e.g., Figure 11(d)) and/or exhibit similar areal density and distribution (e.g.; Figures 10(e), 10(f), 10(h), 10(i), 15(c), and 15(d)). Therefore, the complex "abiotic" corrosion microtextures described from DSDP Hole 418A basaltic glass pillow margins in this study (i.e., alpha-recoil track etch-tunnels: Figures 7, 9, and 12; fission track etch-tunnels: Figure 13; and granular palagonite ART alteration: Figure 10) are important terrestrial analogs for Martian samples, because, if present, these abiotic corrosion microtextures might be readily mistaken for signs of biological activity (i.e., microbial borings/bioalteration) on Mars (as highlighted by French and Blake [111], French and Blake [112], and French [113]). Consequently, if such complex corrosion microtextures are found in samples of Martian glasses obtained during future sample return, robotic rover, or manned missions to Mars, by comparison with the present study, they would *not* constitute signs of past microbial life on Mars (i.e., microbial "trace fossils").

The possibility that alpha-recoil track etch-tunnels, fission track etch-tunnels, and granular palagonite ART alteration might occur in volcanic/impact glasses on the surface of Mars is considered to be very likely. This owes in part to the abundant remote sensing evidence for Hawaiian-style shield volcanism, flood volcanism, and giant dyke swarms in the Tharsis and Elysium regions [239, 240] and the evidence for widely distributed ground ice and past action of liquid water at the surface including development of valley systems, outflow channels, and possible oceans [158, 159]. Spectral measurements of Mars made by orbiting spacecraft and Earth based telescopes have long revealed a surface composition consistent with weathered mafic igneous rocks, and which are similar, for instance, to the corresponding spectra for Hawaiian palagonitic soils from Mauna Kea [151, 156, 241]. Impact cratering, which took place around the planet throughout much of the geological history of Mars [242], also provides a viable mechanism of forming abundant, widely distributed impact glasses through the formation and scattering of impact spherules. It has also been suggested that the interaction of basaltic magma with ground ice or glaciers could have resulted in the formation of abundant basaltic glass tuff deposits on Mars, ultimately supplying a large quantity of palagonite to the Martian soil through subsequent alteration and weathering [151, 243]. The surface of Mars has also been subdivided into two contrasting petrological domains along the planetary dichotomy on the basis of spectral data, including a basaltic composition dominated by plagioclase and clinopyroxene and an andesitic composition dominated by plagioclase and volcanic glass [153]. Remote sensing data also indicates that, in middle Mars history, the North Polar and Utopia basins in the northern hemisphere may have contained an ocean potentially as deep as ~ 1680 m [159]. Therefore, it is also possible that basaltic/impact glass may have existed within deep water on Mars and consequently been subjected to the effects of high hydrostatic pressure during preferential etch-tunnelling along radiation damaged sites, akin to the conditions of formation of abiotic corrosion microtextures in basaltic glass from DSDP 418A (Figure 16).

Consequently, in future astrobiology missions to Mars, evaluating the possible biogenicity of nano- to microscopic tunnels or nanoscopic filaments identified in Martian glasses should be carried out first by (a) assessing the possibility that the tunnels are alpha-recoil track etch-tunnels or fission track etch-tunnels; (b) whether or not pressure solution (or some other kind of abiotic etching process) may have played a role in forming them; and (c) whether or not the nanoscopic filaments are authigenic imogolite tubes, as we have suggested here for the nanofilaments observed within several etch-tunnels in DSDP 418A basaltic glass (Figures 12(g), 12(h), 17(c), and 17(f)).

It is also important to highlight at this point that a completely different variety of abiotically produced branching nanoscopic channel has also been described in a companion study on basaltic glass pillow margins from DSDP 418A (Figure 21(a) [107]). In that study, dendritic patterns of branching nanoscopic grooves are described from the interior surface of some vesicle walls of sample 418A-75-3-[120–123] (i.e., the same pillow margin sample as the present study), which appear to have formed by the nonbiological process of viscous fingering. These grooves are ~50 nm deep, 50–75 nm in width, and individual branches end as slightly larger terminal bulbs that measure 150–300 nm across, which makes them similar in size to alpha-recoil track etch-tunnels from this study (see a scale comparison in Figures 21(a) and 21(b)). In particular, these dendritic nanoscopic grooves represent frozen viscous fingers of magmatic fluid that were injected into the hot walls of some vesicles upon cooling through the glass transition during pillow eruption [107], and they represent another form of abiotic dendritic microchannel that might be found in Martian glasses, which could also potentially be mistaken as another form of microbial trace fossil on Mars. Therefore, during the astrobiological exploration of Mars, any branching nanoscopic channels found on vesicle walls in Martian glasses should also be compared with those nonbiological ones documented in our companion study on DSDP 418A basaltic glass (Figure 21(a) [107]).

A summary of all of these different varieties of complex "abiotic" microtextures found in DSDP 418A basaltic glass pillow margins—that look conspicuously biogenic but which are not—is shown in Figure 21, including dendritic nanogrooves on vesicle walls (Figure 21(a)), alpha-recoil track etch-tunnels (Figures 21(b) and 21(g)), fission track etch-tunnels (Figure 21(c)), granular palagonite ART alteration (Figure 21(d)), alpha-recoil track etch-tunnels affected by prolonged overetching (Figures 21(e) and 21(i)), authigenic imogolite filaments (Figure 21(f)), and palagonite fingers (overprinted ARTETs; Figure 21(h)). Note that authigenic imogolite filaments (Figure 21(f)) have good potential to be mistaken for certain biofilaments of similar size and form, such as the nanofilaments found within dessicated exopolysaccharide mucus produced by bacteria (see cover photo of Barker et al. [244]), or even filamentous strands of DNA (see Figure 1 in Anselmetti et al. [245]), which coincidentally have the same diameter (20 Å) and long, flexible, filamentous nature as imogolite [111, 139, 246]. Consequently, we highlight that all of these various "abiotic" microtextures found in submarine glasses at DSDP 418A (Figure 21) might also be found in volcanic/impact glasses on Mars, and so, in future astrobiology missions to Mars, special care should be taken to evaluate possible nonbiological origins for such features if analogous microtextures are eventually found on that planet.

Acknowledgments

Karlis Muehlenbachs (Department of Earth and Atmospheric Sciences, University of Alberta) is thanked for his comments and advice during collection of scientific data and for providing financial support for this project through the Natural Sciences and Engineering Research Council of Canada. The authors thank George Braybrook (Department of Earth and Atmospheric Sciences, University of Alberta) and Kathy Kato (NASA Ames Research Center) for assistance with SEM studies and Antonio Simonetti (University of Notre Dame) for the help with ICP-MS analyses carried out at the Radiogenic Isotope Facility (Department of Earth and Atmospheric Sciences, University of Alberta). Jason E. French would like to thank Thomas Chacko and Jeremy P. Richards (Department of Earth and Atmospheric Sciences, University of Alberta) for providing access to petrographic microscope and camera (photomicrographic) equipment during petrographic thin section work. In addition, Jason E. French would like to acknowledge the NASA Planetary Biology Internship Program for enabling SEM studies at the NASA Ames Research Center under the cosponsorship of David F. Blake and David J. Des Marais in July/August 1999.

References

[1] D. F. Blake, A. H. Treiman, H. E. F. Amundsen, S. J. Mojzsis, and T. Bunch, "Carbonate globules, analogous to those in ALH84001, from Spitsbergen, Norway: formation in a hydrothermal environment," in *Proceedings of the Abstracts: 30th Lunar and Planetary Science Conference*, Houston, Tex, USA, March 1999.

[2] J. P. Bradley, R. P. Harvey, H. Y. McSween Jr., E. Gibson Jr., K. Thomas-Keprta, and H. Vali, "No 'nanofossils' in martian meteorite," *Nature*, vol. 390, no. 6659, pp. 454–456, 1997.

[3] D. S. McKay, E. K. Gibson Jr., K. L. Thomas-Keprta et al., "Search for past life on Mars: possible relic biogenic activity in martian meteorite ALH84001," *Science*, vol. 273, no. 5277, pp. 924–930, 1996.

[4] D. S. McKay, E. Gibson Jr., K. Thomas-Keprta, and H. Vali, "Reply to: no 'nanofossils' in martian meteorite," *Nature*, vol. 390, pp. 455–456, 1997.

[5] K. L. Thomas-Keprta, D. S. McKay, S. J. Wentworth et al., "Bacterial mineralization patterns in basaltic aquifers: implications for possible life in martian meteorite ALH84001," *Geology*, vol. 26, no. 11, pp. 1031–1034, 1998.

[6] A. H. Treiman, H. E. F. Amundsen, D. F. Blake, and T. Bunch, "Hydrothermal origin for carbonate globules in Martian meteorite ALH84001: a terrestrial analogue from Spitsbergen (Norway)," *Earth and Planetary Science Letters*, vol. 204, no. 3-4, pp. 323–332, 2002.

[7] L. J. Rothschild, "Earth analogs for Martian life. Microbes in evaporites, a new model system for life on Mars," *Icarus*, vol. 88, no. 1, pp. 246–260, 1990.

[8] M. R. Walter and D. J. Des Marais, "Preservation of biological information in thermal spring deposits: developing a strategy for the search for fossil life on Mars," *Icarus*, vol. 101, no. 1, pp. 129–143, 1993.

[9] P. T. Doran, R. A. Wharton Jr., D. J. Des Marais, and C. P. McKay, "Antarctic paleolake sediments and the search for extinct life on Mars," *Journal of Geophysical Research E: Planets*, vol. 103, no. 12, pp. 28481–28493, 1998.

[10] P. J. Boston, M. V. Ivanov, and C. P. McKay, "On the possibility of chemosynthetic ecosystems in subsurface habitats on Mars," *Icarus*, vol. 95, no. 2, pp. 300–308, 1992.

[11] H. Furnes, N. R. Banerjee, H. Staudigel et al., "Comparing petrographic signatures of bioalteration in recent to Mesoarchean

pillow lavas: tracing subsurface life in oceanic igneous rocks," *Precambrian Research*, vol. 158, no. 3-4, pp. 156–176, 2007.

[12] M. R. M. Izawa, N. R. Banerjee, R. L. Flemming, N. J. Bridge, and C. Schultz, "Basaltic glass as a habitat for microbial life: implications for astrobiology and planetary exploration," *Planetary and Space Science*, vol. 58, no. 4, pp. 583–591, 2010.

[13] J. Farmer, D. Des Marais, R. Greeley, R. Landheim, and H. Klein, "Site selection for Mars exobiology," *Advances in Space Research*, vol. 15, no. 3, pp. 157–162, 1995.

[14] L. J. Rothschild and D. Des Marais, "Stable carbon isotope fractionation in the search for life on early Mars," *Advances in Space Research*, vol. 9, no. 6, pp. 159–165, 1989.

[15] D. E. Schwartz and R. L. Mancinelli, "Bio-markers and the search for extinct life on Mars," *Advances in Space Research*, vol. 9, no. 6, pp. 155–158, 1989.

[16] D. E. Schwartz, R. L. Mancinelli, and E. S. Kaneshiro, "The use of mineral crystals as bio-markers in the search for life on Mars," *Advances in Space Research*, vol. 12, no. 4, pp. 117–119, 1992.

[17] K. L. Thomas-Keprta, D. A. Bazylinski, J. L. Kirschvink et al., "Elongated prismatic magnetite crystals in ALH84001 carbonate globules: potential Martian magnetofossils," *Geochimica et Cosmochimica Acta*, vol. 64, no. 23, pp. 4049–4081, 2000.

[18] J. L. Bada and G. D. McDonald, "Amino acid racemization on Mars: implications for the preservation of biomolecules from an extinct Martian Biota," *Icarus*, vol. 114, no. 1, pp. 139–143, 1995.

[19] C. C. Allen, F. G. Albert, H. S. Chafetz et al., "Microscopic physical biomarkers in carbonate hot springs: implications in the search for life on Mars," *Icarus*, vol. 147, no. 1, pp. 49–67, 2000.

[20] B. L. Kirkland, F. L. Lynch, M. A. Rahnis, R. L. Folk, I. J. Molineux, and R. J. C. McLean, "Alternative origins for nannobacteria-like objects in calcite," *Geology*, vol. 27, no. 4, pp. 347–350, 1999.

[21] J. W. Schopf, "Microfossils of the early Archean apex chert: new evidence of the antiquity of life," *Science*, vol. 260, no. 5108, pp. 640–646, 1993.

[22] M. D. Brasier, O. R. Green, A. P. Jephcoat et al., "Questioning the evidence for Earth's oldest fossils," *Nature*, vol. 416, no. 6876, pp. 76–81, 2002.

[23] I. H. Thorseth, H. Furnes, and M. Heldal, "The importance of microbiological activity in the alteration of natural basaltic glass," *Geochimica et Cosmochimica Acta*, vol. 56, no. 2, pp. 845–850, 1992.

[24] M. R. Fisk, S. J. Giovannoni, and I. H. Thorseth, "Alteration of oceanic volcanic glass: textural evidence of microbial activity," *Science*, vol. 281, no. 5379, pp. 978–980, 1998.

[25] N. R. Banerjee, A. Simonetti, H. Furnes et al., "Direct dating of Archean microbial ichnofossils," *Geology*, vol. 35, no. 6, pp. 487–490, 2007.

[26] H. Furnes, N. McLoughlin, K. Muehlenbachs et al., "Oceanic pillow lavas and hyaloclastites as habitats for microbial life through time—a review," in *Links Between Geological Processes, Microbial Activities & Evolution of Life*, Y. Dilek, H. Furnes, and K. Muehlenbachs, Eds., vol. 4 of *Modern Approaches in Solid Earth Sciences*, pp. 1–68, Springer, Dordrecht, The Netherlands, 2008.

[27] H. Staudigel, H. Furnes, N. McLoughlin, N. R. Banerjee, L. B. Connell, and A. Templeton, "3.5 billion years of glass bioalteration: volcanic rocks as a basis for microbial life?" *Earth-Science Reviews*, vol. 89, no. 3-4, pp. 156–176, 2008.

[28] N. McLoughlin, H. Furnes, N. R. Banerjee, K. Muehlenbachs, and H. Staudigel, "Ichnotaxonomy of microbial trace fossils in volcanic glass," *Journal of the Geological Society*, vol. 166, no. 1, pp. 159–169, 2009.

[29] K. Muehlenbachs, H. Furnes, and M. de Wit, "Thermophile inhabitants of the Archean seafloor basalts," in *Proceedings of the Goldschmidt Conference Abstracts*, p. A911, Davos, Switzerland, June 2009.

[30] H. M. Sapers, G. R. Osinski, N. R. Banerjee, and L. J. Preston, "Enigmatic tubular features in impact glass," *Geology*, vol. 42, no. 6, pp. 471–474, 2014.

[31] N. McLoughlin, H. Furnes, N. Banerjee et al., "Micro-bioerosion in volcanic glass: extending the ichnofossil record to Archean basaltic crust," in *Current Developments in Bioerosion*, M. Wisshak and L. Tapanila, Eds., Erlangen Earth Conference Series, pp. 372–396, Springer, Berlin, Germany, 2008.

[32] C. Heberling, R. P. Lowell, L. Liu, and M. R. Fisk, "Extent of the microbial biosphere in the oceanic crust," *G-Cubed*, vol. 11, no. 8, pp. 1–15, 2010.

[33] H. Furnes and H. Staudigel, "Biological mediation in ocean crust alteration: how deep is the deep biosphere?" *Earth and Planetary Science Letters*, vol. 166, no. 3-4, pp. 97–103, 1999.

[34] H. Staudigel, B. Tebo, A. Yayanos et al., "The oceanic crust as a bioreactor," in *The Subseafloor Biosphere at Mid-Ocean Ridges*, W. S. Wilcock, E. F. DeLong, D. S. Kelley, J. A. Baross, and S. Craig Cary, Eds., vol. 144 of *AGU Geophysical Monograph Series*, pp. 325–341, AGU, 2004.

[35] H. Furnes, N. R. Banerjee, K. Muehlenbachs, H. Staudigel, and M. de Wit, "Early life recorded in archean pillow lavas," *Science*, vol. 304, no. 5670, pp. 578–581, 2004.

[36] R. A. Kerr, "New biomarker proposed for earliest life on Earth," *Science*, vol. 304, no. 5670, article 503, 2004.

[37] N. R. Banerjee, H. Furnes, K. Muehlenbachs, H. Staudigel, and M. De Wit, "Preservation of ~3.4–3.5 Ga microbial biomarkers in pillow lavas and hyaloclastites from the Barberton Greenstone Belt, South Africa," *Earth and Planetary Science Letters*, vol. 241, no. 3-4, pp. 707–722, 2006.

[38] D. Fliegel, J. Kosler, N. McLoughlin et al., "In-situ dating of the Earth's oldest trace fossil at 3.34 Ga," *Earth and Planetary Science Letters*, vol. 299, no. 3-4, pp. 290–298, 2010.

[39] W. G. Melson and G. Thompson, "Glassy abyssal basalts, Atlantic Sea floor near St. Paul's rocks: petrography and composition of secondary clay minerals," *Geological Society of America Bulletin*, vol. 84, no. 2, pp. 703–716, 1973.

[40] M. Morgenstein and T. J. Riley, "Hydration-rind dating of basaltic glass: a new method for archaeological chronologies," *Asian Perspectives*, vol. 17, no. 2, pp. 145–162, 1974.

[41] A. R. Geptner, "Palagonite and the process of palagonitization," *Litologiya i Poleznye Iskopaemye*, vol. 5, pp. 113–130, 1977.

[42] A. G. Kossowskaya, V. V. Petrova, and V. D. Shutov, "Mineral associations of palagonitization of oceanic basalts and aspects of extraction of mineral components," *Litologiya i Poleznye Iskopaemye*, vol. 4, pp. 10–31, 1982.

[43] H. Staudigel and S. R. Hart, "Alteration of basaltic glass: mechanisms and significance for the oceanic crust-seawater budget," *Geochimica et Cosmochimica Acta*, vol. 47, no. 3, pp. 337–350, 1983.

[44] Z. Zhou and W. S. Fyfe, "Palagonitization of basaltic glass from DSDP Site 335, Leg 37: textures, chemical composition, and mechanism of formation," *American Mineralogist*, vol. 74, no. 9-10, pp. 1045–1053, 1989.

[45] K. A. Ross and R. V. Fisher, "Biogenic grooving on glass shards," *Geology*, vol. 14, no. 7, pp. 571–573, 1986.

[46] K. Ross and R. V. Fisher, "Comment and reply on 'biogenic grooving on glass shards': reply," *Geology*, vol. 15, no. 5, p. 471, 1987.

[47] H. Staudigel, R. A. Chastain, A. Yayanos, and W. Bourcier, "Biologically mediated dissolution of glass," *Chemical Geology*, vol. 126, no. 2, pp. 147–154, 1995.

[48] I. H. Thorseth, H. Furnes, and O. Tumyr, "Textural and chemical effects of bacterial activity on basaltic glass: an experimental approach," *Chemical Geology*, vol. 119, no. 1–4, pp. 139–160, 1995.

[49] I. H. Thorseth, T. Torsvik, H. Furnes, and K. Muehlenbachs, "Microbes play an important role in the alteration of oceanic crust," *Chemical Geology*, vol. 126, no. 2, pp. 137–146, 1995.

[50] H. Furnes, I. H. Thorseth, O. Tumyr, T. Torsvik, and M. R. Fisk, "13. Microbial activity in the alteration of glass from pillow lavas from Hole 896A," in *Proceedings of the Ocean Drilling Program, Scientific Results*, J. C. Alt, H. Kinoshita, L. B. Stokking, and P. J. Michael, Eds., vol. 148, pp. 191–206, US Government Printing Office, 1996.

[51] S. J. Giovannoni, M. R. Fisk, T. D. Mullins, and H. Furnes, "14. Genetic evidence for endolithic microbial life colonizing basaltic glass/seawater interfaces," in *Proceedings of the Ocean Drilling Program, Scientific Results*, J. C. Alt, H. Kinoshita, L. B. Stokking, and P. J. Michael, Eds., vol. 148, pp. 207–214, Texas A&M University, College Station, Tex, USA, 1996.

[52] H. Staudigel, A. Yayanos, R. Chastain et al., "Biologically mediated dissolution of volcanic glass in seawater," *Earth and Planetary Science Letters*, vol. 164, no. 1-2, pp. 233–244, 1998.

[53] T. Torsvik, H. Furnes, K. Muehlenbachs, I. H. Thorseth, and O. Tumyr, "Evidence for microbial activity at the glass-alteration interface in oceanic basalts," *Earth and Planetary Science Letters*, vol. 162, no. 1–4, pp. 165–176, 1998.

[54] H. Furnes, K. Muehlenbachs, O. Tumyr, T. Torsvik, and I. H. Thorseth, "Depth of active bio-alteration in the ocean crust: Costa Rica Rift (Hole 504B)," *Terra Nova*, vol. 11, no. 5, pp. 228–233, 1999.

[55] J. C. Alt and P. Mata, "On the role of microbes in the alteration of submarine basaltic glass: a TEM study," *Earth and Planetary Science Letters*, vol. 181, no. 3, pp. 301–313, 2000.

[56] M. R. Fisk, I. H. Thorseth, U. Erbach, and S. J. Giovannoni, "14. Investigation of microorganisms and DNA from subsurface thermal water and rock from the east flank of Juan de Fuca Ridge," in *Proceedings of the Ocean Drilling Program, Scientific Results*, A. Fisher, E. E. Davis, and C. Escutia, Eds., vol. 168, pp. 167–174, Texas A&M University, College Station, Tex, USA, 2000.

[57] H. Furnes, K. Muehlenbachs, T. Torsvik, I. H. Thorseth, and O. Tumyr, "Microbial fractionation of carbon isotopes in altered basaltic glass from the Atlantic Ocean, Lau Basin and Costa Rica Rift," *Chemical Geology*, vol. 173, no. 4, pp. 313–330, 2001.

[58] H. Furnes, K. Muehlenbachs, O. Tumyr, T. Torsvik, and C. Xenophontos, "Biogenic alteration of volcanic glass from the Troodos ophiolite, Cyprus," *Journal of the Geological Society*, vol. 158, no. 1, pp. 75–84, 2001.

[59] H. Furnes, H. Staudigel, I. H. Thorseth, T. Torsvik, K. Muehlenbachs, and O. Tumyr, "Bioalteration of basaltic glass in the oceanic crust," *G-Cubed*, vol. 2, pp. 1–30, 2001.

[60] Shipboard Scientific Party, "Site 1184," in *Proceedings of the Ocean Drilling Program, Initial Reports*, J. J. Mahoney, J. G. Fitton, P. J. Wallace et al., Eds., vol. 192, pp. 1–131, Ocean Drilling Program, College Station, Tex, USA, 2001.

[61] H. Furnes, K. Muehlenbachs, T. Torsvik, O. Tumyr, and L. Shi, "Bio-signatures in metabasaltic glass of a Caledonian ophiolite, West Norway," *Geological Magazine*, vol. 139, no. 6, pp. 601–608, 2002.

[62] H. Furnes, I. H. Thorseth, T. Torsvik, K. Muehlenbachs, H. Staudigel, and O. Tumyr, "Identifying bio-interaction with basaltic glass in oceanic crust and implications for estimating the depth of the oceanic biosphere: a review," in *Volcano-Ice Interaction on Earth and Mars*, J. L. Smellie and M. G. Chapman, Eds., pp. 407–421, The Geological Society, London, UK, 2002.

[63] W. Bach and K. J. Edwards, "Iron and sulfide oxidation within the basaltic ocean crust: implications for chemolithoautotrophic microbial biomass production," *Geochimica et Cosmochimica Acta*, vol. 67, no. 20, pp. 3871–3887, 2003.

[64] N. R. Banerjee and K. Muehlenbachs, "Tuff life: bioalteration in volcaniclastic rocks from the Ontong Java Plateau," *G-Cubed*, vol. 4, no. 4, pp. 1–22, 2003.

[65] M. R. Fisk, M. C. Storrie-Lombardi, S. Douglas, R. Popa, G. McDonald, and C. Di Meo-Savoie, "Evidence of biological activity in Hawaiian subsurface basalts," *G-Cubed*, vol. 4, no. 12, pp. 1–24, 2003.

[66] H. Furnes and K. Muehlenbachs, "Bioalteration recorded in ophiolitic pillow lavas," in *Ophiolites in Earth's History*, Y. Dilek and P. T. Robinson, Eds., vol. 218 of *Geological Society of London Special Publications no. 218*, pp. 415–426, Geological Society of London, 2003.

[67] I. H. Thorseth, R. B. Pedersen, and D. M. Christie, "Microbial alteration of 0–30-Ma seafloor and sub-seafloor basaltic glasses from the Australian Antarctic Discordance," *Earth and Planetary Science Letters*, vol. 215, no. 1-2, pp. 237–247, 2003.

[68] A. W. Walton and P. Schiffman, "Alteration of hyaloclastites in the HSDP 2 phase 1 drill core 1. Description and paragenesis," *Geochemistry, Geophysics, Geosystems*, vol. 4, no. 5, pp. 1–31, 2003.

[69] H. Staudigel and H. Furnes, "Microbial mediation of oceanic crust alteration," in *Hydrogeology of the Oceanic Lithosphere*, E. Davis and H. Elderfield, Eds., pp. 606–626, Cambridge University Press, 2004.

[70] M. C. Storrie-Lombardi and M. R. Fisk, "Elemental abundance distributions in suboceanic basalt glass: evidence of biogenic alteration," *Geochemistry, Geophysics, Geosystems*, vol. 5, no. 10, Article ID Q10005, 15 pages, 2004.

[71] H. Furnes, N. R. Banerjee, K. Muehlenbachs, and A. Kontinen, "Preservation of biosignatures in metaglassy volcanic rocks from the Jormua ophiolite complex, Finland," *Precambrian Research*, vol. 136, no. 2, pp. 125–137, 2005.

[72] J. Einen, C. Kruber, L. Øvreås, I. H. Thorseth, and T. Torsvik, "Microbial colonization and alteration of basaltic glass," *Biogeosciences Discussions*, vol. 3, no. 2, pp. 273–307, 2006.

[73] M. R. Fisk, R. Popa, O. U. Mason, M. C. Storrie-Lombardie, and E. P. Vicenzi, "Iron-magnesium silicate bioweathering on Earth (and Mars?)," *Astrobiology*, vol. 6, no. 1, pp. 48–68, 2006.

[74] H. Furnes, Y. Dilek, K. Muehlenbachs, and N. R. Banerjee, "Tectonic control of bioalteration in modern and ancient oceanic crust as evidenced by carbon isotopes," *Island Arc*, vol. 15, no. 1, pp. 143–155, 2006.

[75] H. Staudigel, H. Furnes, N. R. Banerjee, Y. Dilek, and K. Muehlenbachs, "Microbes and volcanoes: a tale from the oceans, ophiolites, and greenstone belts," *GSA Today*, vol. 16, no. 10, pp. 4–10, 2006.

[76] K. Benzerara, N. Menguy, N. R. Banerjee, T. Tyliszczak, G. E. Brown Jr., and F. Guyot, "Alteration of submarine basaltic glass

from the Ontong Java Plateau: a STXM and TEM study," *Earth and Planetary Science Letters*, vol. 260, no. 1-2, pp. 187–200, 2007.

[77] H. Furnes, N. R. Banerjee, H. Staudigel, and K. Muehlenbachs, "Pillow lavas as a habitat for microbial life," *Geology Today*, vol. 23, no. 4, pp. 143–146, 2007.

[78] N. McLoughlin, M. D. Brasier, D. Wacey, O. R. Green, and R. S. Perry, "On biogenicity criteria for endolithic microborings on early Earth and beyond," *Astrobiology*, vol. 7, no. 1, pp. 10–26, 2007.

[79] C. S. Cockell and A. Herrera, "Why are some microorganisms boring?" *Trends in Microbiology*, vol. 16, no. 3, pp. 101–106, 2008.

[80] A. Herrera, C. S. Cockell, S. Self et al., "Bacterial colonization and weathering of terrestrial obsidian in Iceland," *Geomicrobiology Journal*, vol. 25, no. 1, pp. 25–37, 2008.

[81] M. Ivarsson, J. Lausmaa, S. Lindblom, C. Broman, and N. G. Holm, "Fossilized microorganisms from the emperor seamounts: implications for the search for a subsurface fossil record on earth and Mars," *Astrobiology*, vol. 8, no. 6, pp. 1139–1157, 2008.

[82] C. Kruber, I. H. Thorseth, and R. B. Pedersen, "Seafloor alteration of basaltic glass: textures, geochemistry, and endolithic microorganisms," *Geochemistry, Geophysics, Geosystems*, vol. 9, no. 12, 2008.

[83] A. W. Walton, "Microtubules in basalt glass from Hawaii Scientific Drilling Project #2 phase 1 core and Hilina slope, Hawaii: evidence of the occurrence and behavior of endolithic microorganisms," *Geobiology*, vol. 6, no. 4, pp. 351–364, 2008.

[84] C. S. Cockell, K. Olsson-Francis, A. Herrera, and A. Meunier, "Alteration textures in terrestrial volcanic glass and the associated bacterial community," *Geobiology*, vol. 7, no. 1, pp. 50–65, 2009.

[85] C. S. Cockell, K. Olsson, F. Knowles et al., "Bacteria in weathered basaltic glass, Iceland," *Geomicrobiology Journal*, vol. 26, no. 7, pp. 491–507, 2009.

[86] C. R. Cousins, J. L. Smellie, A. P. Jones, and I. A. Crawford, "A comparative study of endolithic microborings in basaltic lavas from a transitional subglacial—marine environment," *International Journal of Astrobiology*, vol. 8, no. 1, pp. 37–49, 2009.

[87] E. G. Grosch, N. McLoughlin, M. de Wit, and H. Furnes, "Drilling for the archean roots of life and tectonic earth in the barberton mountains," *Scientific Drilling*, no. 8, pp. 24–28, 2009.

[88] G. F. Slater, "International year of planet earth 6. Biosignatures, interpreting evidence of the origins and diversity of life," *Geoscience Canada*, vol. 36, no. 4, pp. 170–178, 2009.

[89] N. R. Banerjee, M. R. M. Izawa, H. M. Sapers, and M. J. Whitehouse, "Geochemical biosignatures preserved in microbially altered basaltic glass," *Surface and Interface Analysis*, vol. 43, no. 1-2, pp. 452–457, 2011.

[90] N. J. Bridge, N. R. Banerjee, W. Mueller, K. Muehlenbachs, and T. Chacko, "A volcanic habitat for early life preserved in the Abitibi Greenstone belt, Canada," *Precambrian Research*, vol. 179, no. 1-4, pp. 88–98, 2010.

[91] D. Fliegel, R. Wirth, A. Simonetti et al., "Septate-tubular textures in 2.0-Ga pillow lavas from the Pechenga Greenstone Belt: a nano-spectroscopic approach to investigate their biogenicity," *Geobiology*, vol. 8, no. 5, pp. 372–390, 2010.

[92] M. R. M. Izawa, N. R. Banerjee, R. L. Flemming, and N. J. Bridge, "Preservation of microbial ichnofossils in basaltic glass by titanite mineralization," *Canadian Mineralogist*, vol. 48, no. 5, pp. 1255–1265, 2010.

[93] N. Mcloughlin, H. Staudigel, H. Furnes, B. Eickmann, and M. Ivarsson, "Mechanisms of microtunneling in rock substrates: distinguishing endolithic biosignatures from abiotic microtunnels," *Geobiology*, vol. 8, no. 4, pp. 245–255, 2010.

[94] N. McLoughlin, D. J. Fliegel, H. Furnes et al., "Assessing the biogenicity and syngenicity of candidate bioalteration textures in pillow lavas of the ~2.52 Ga Wutai greenstone terrane of China," *Chinese Science Bulletin*, vol. 55, no. 2, pp. 188–199, 2010.

[95] H. Staudigel and D. A. Clague, "The geological history of deep-sea volcanoes: biosphere, hydrosphere, and lithosphere interactions," *Oceanography*, vol. 23, no. 1, pp. 58–71, 2010.

[96] D. Fliegel, R. Wirth, A. Simonetti, A. Schreiber, H. Furnes, and K. Muehlenbachs, "Tubular textures in pillow lavas from a Caledonian west Norwegian ophiolite: a combined TEM, LA–ICP–MS, and STXM study," *Geochemistry, Geophysics, Geosystems*, vol. 12, no. 2, Article ID Q02010, 21 pages, 2011.

[97] N. McLoughlin, "Archean traces of life," in *Encyclopedia of Astrobiology*, M. Gargaud, R. Amils, J. Cernicharo Quintanilla et al., Eds., pp. 74–84, Springer, Berlin, Germany, 2011.

[98] N. McLoughlin, D. Wacey, C. Kruber, M. R. Kilburn, I. H. Thorseth, and R. B. Pedersen, "A combined TEM and NanoSIMS study of endolithic microfossils in altered seafloor basalt," *Chemical Geology*, vol. 289, no. 1-2, pp. 154–162, 2011.

[99] L. J. Preston, M. R. M. Izawa, and N. R. Banerjee, "Infrared spectroscopic characterization of organic matter associated with microbial bioalteration textures in basaltic glass," *Astrobiology*, vol. 11, no. 7, pp. 585–599, 2011.

[100] D. Fliegel, E. Knowles, R. Wirth et al., "Characterization of alteration textures in Cretaceous oceanic crust (pillow lava) from the N-Atlantic (DSDP Hole 418A) by spatially-resolved spectroscopy," *Geochimica et Cosmochimica Acta*, vol. 96, pp. 80–93, 2012.

[101] E. Knowles, R. Wirth, and A. Templeton, "A comparative analysis of potential biosignatures in basalt glass by FIB-TEM," *Chemical Geology*, vol. 330-331, pp. 165–175, 2012.

[102] N. McLoughlin, E. G. Grosch, M. R. Kilburn, and D. Wacey, "Sulfur isotope evidence for a Paleoarchean subseafloor biosphere, Barberton, South Africa," *Geology*, vol. 40, no. 11, pp. 1031–1034, 2012.

[103] M. Fisk and N. McLoughlin, "Atlas of alteration textures in volcanic glass from the ocean basins," *Geosphere*, vol. 9, no. 2, pp. 317–341, 2013.

[104] N. McLoughlin, H. Furnes, H. Staudigel, and E. Hanski, "7.8.4. Seeking textural evidence of a Palaeoproterozoic sub-seafloor biosphere in pillow lavas of the pechenga greenstone belt," in *Reading the Archive of Earth Oxygenation, Vol. 1: The Palaeoproterozoic of Fennoscandia as a Context for the Fennoscandian Arctic Russia—Drilling Early Earth Project*, V. A. Melezhik, A. R. Prave, A. E. Fallick et al., Eds., pp. 1371–1394, Springer, 2013.

[105] H. Staudigel, H. Furnes, and M. Smits, "Deep biosphere record of in situ oceanic lithosphere and ophiolites," *Elements*, vol. 10, no. 2, pp. 121–126, 2014.

[106] B. P. Glass, "Comment and reply on 'Biogenic grooving on glass shards': comment," *Geology*, vol. 15, no. 5, pp. 470–471, 1987.

[107] J. E. French and K. Muehlenbachs, "The origin of nanoscopic grooving on vesicle walls in submarine basaltic glass: implications for nanotechnology," *Journal of Nanomaterials*, vol. 2009, Article ID 309208, 14 pages, 2009.

[108] K. Lepot, K. Benzerara, and P. Philippot, "Biogenic versus metamorphic origins of diverse microtubes in 2.7 Gyr old volcanic ashes: Multi-scale investigations," *Earth and Planetary Science Letters*, vol. 312, no. 1-2, pp. 37–47, 2011.

[109] E. G. Grosch and N. McLoughlin, "Reassessing the biogenicity of Earth's oldest trace fossil with implications for biosignatures in the search for early life," *Proceedings of the National Academy of Sciences of the United States of America*, vol. 111, no. 23, pp. 8380–8385, 2014.

[110] A. Drief and P. Schiffman, "Very low-temperature alteration of sideromelane in hyaloclastites and hyalotuffs from Kilauea and Mauna Kea volcanoes: implications for the mechanism of palagonite formation," *Clays and Clay Minerals*, vol. 52, no. 5, pp. 622–634, 2004.

[111] J. E. French and D. F. Blake, "Abiotic corrosion microtextures in volcanic glass: reevaluation of a putative biosignature for Earth and Mars," in *2012 Conference on Life Detection in Extraterrestrial Samples, San Diego, California. Program and Abstract Volume*, LPI Contribution no. 1650, pp. 34–35, LPI, 2012.

[112] J. E. French and D. F. Blake, "Role of self-incurred radiation damage in development of tubular and granular microtextures in submarine volcanic glass: implications for Mars exploration," in *Proceedings of the Geological Association of Canada/Mineralogical Association of Canada Annual Meeting*, vol. 35 of *Abstracts*, p. 45, St. John's, Canada, May 2012.

[113] J. E. French, "Questioning the textural evidence for microbial bioalteration of volcanic glass and its robustness as a terrestrial analog biosignature for Mars astrobiology," in *Proceedings of the Geological Association of Canada/Mineralogical Association of Canada Annual Meeting*, vol. 36, p. Abstracts, Winnipeg, Canada, May 2013.

[114] L. R. Pedersen, N. McLoughlin, P. E. Vullum, and I. H. Thorseth, "Abiotic and candidate biotic micro-alteration textures in sub-seafloor basaltic glass: a high-resolution in-situ textural and geochemical investigation," *Chemical Geology*, vol. 410, pp. 124–137, 2015.

[115] J. W. Nielson and F. G. Foster, "Unusual etch pits in quartz crystals," *American Mineralogist*, vol. 45, pp. 299–310, 1960.

[116] T. Hanyu, "Dislocation etch tunnels in quartz crystals," *Journal of the Physical Society of Japan*, vol. 19, no. 8, p. 1489, 1964.

[117] H. Carstens, "The lineage structure of quartz crystals," *Contributions to Mineralogy and Petrology*, vol. 18, no. 4, pp. 295–304, 1968.

[118] S. Takasu and S. Shimanuki, "Tunnel-like defects of flux grown magnetic garnets," *Journal of Crystal Growth*, vol. 24-25, pp. 641–645, 1974.

[119] F. Iwasaki, "Line defects and etch tunnels in synthetic quartz," *Journal of Crystal Growth*, vol. 39, no. 2, pp. 291–298, 1977.

[120] S. L. Brantley, S. R. Crane, D. A. Crerar, R. Hellmann, and R. Stallard, "Dissolution at dislocation etch pits in quartz," *Geochimica et Cosmochimica Acta*, vol. 50, no. 10, pp. 2349–2361, 1986.

[121] A. C. Lasaga and A. E. Blum, "Surface chemistry, etch pits and mineral-water reactions," *Geochimica et Cosmochimica Acta*, vol. 50, no. 10, pp. 2363–2379, 1986.

[122] T. N. Tingle, E. Roedder, and H. W. Green II, "Formation of fluid inclusions and etch tunnels in olivine at high pressure," *American Mineralogist*, vol. 77, no. 3-4, pp. 296–302, 1992.

[123] T. Lu, J. E. Shigley, J. I. Koivula, and I. M. Reinitz, "Observation of etch channels in several natural diamonds," *Diamond and Related Materials*, vol. 10, no. 1, pp. 68–75, 2001.

[124] K. Lepot, P. Philippot, K. Benzerara, and G.-Y. Wang, "Garnet-filled trails associated with carbonaceous matter mimicking microbial filaments in Archean basalt," *Geobiology*, vol. 7, no. 4, pp. 393–402, 2009.

[125] D. Storzer and M. Sélo, "Fission track age of magnetic anomaly M-zero and some aspects of sea-water weathering," in *Initial Reports of the Deep Sea Drilling Project*, LI, LII, LIII, Part 2, pp. 1129–1133, US Government Publishing Office, Washington, DC, USA, 1980.

[126] C. Arias, G. Bigazzi, and F. P. Bonadonna, "Size corrections and plateau age in glass shards," *Nuclear Tracks*, vol. 5, no. 1-2, pp. 129–136, 1981.

[127] M. Sélo and D. Storzer, "Uranium distribution and age pattern of some deep-sea basalts from the Entrecasteaux area, south-western pacific: a fission-track analysis," *Nuclear Tracks*, vol. 5, no. 1-2, pp. 137–145, 1981.

[128] A. S. Sandhu and J. A. Westgate, "The correlation between reduction in fission-track diameter and areal track density in volcanic glass shards and its application in dating tephra beds," *Earth and Planetary Science Letters*, vol. 131, no. 3-4, pp. 289–299, 1995.

[129] K. Gögen and G. A. Wagner, "Alpha-recoil track dating of Quaternary volcanics," *Chemical Geology*, vol. 166, no. 1-2, pp. 127–137, 2000.

[130] W. H. Huang and R. M. Walker, "Fossil alpha-particle recoil tracks: a new method of age determination," *Science*, vol. 155, no. 3766, pp. 1103–1106, 1967.

[131] R. Jonckheere and K. Gögen, "A Monte-Carlo calculation of the size distribution of latent alpha-recoil tracks," *Nuclear Instruments and Methods in Physics Research B*, vol. 183, no. 3-4, pp. 347–357, 2001.

[132] K. Stübner and R. C. Jonckheere, "A Monte-Carlo calculation of the size distribution of latent alpha-recoil tracks in phlogopite: implications for the recoil-track dating method," *Radiation Measurements*, vol. 41, no. 1, pp. 55–64, 2006.

[133] K. Stübner, R. C. Jonckheere, and L. Ratschbacher, "Alpha-recoil track densities in mica and radiometric age determination," *Radiation Measurements*, vol. 40, no. 2–6, pp. 503–508, 2005.

[134] M. Lang, U. A. Glasmacher, B. Moine, C. Müller, R. Neumann, and G. A. Wagner, "Artificial ion tracks in volcanic dark mica simulating natural radiation damage: a scanning force microscopy study," *Nuclear Instruments and Methods in Physics Research B*, vol. 191, no. 1–4, pp. 346–351, 2002.

[135] G. Malow, W. Lutze, and R. C. Ewing, "Alteration effects and leach rates of basaltic glasses: implications for the long-term stability of nuclear waste form borosilicate glasses," *Journal of Non-Crystalline Solids*, vol. 67, no. 1–3, pp. 305–321, 1984.

[136] J.-L. Crovisier, T. Advocat, and J.-L. Dussossoy, "Nature and role of natural alteration gels formed on the surface of ancient volcanic glasses (Natural analogs of waste containment glasses)," *Journal of Nuclear Materials*, vol. 321, no. 1, pp. 91–109, 2003.

[137] J. C. Dran, M. Maurette, and J. C. Petit, "Radioactive waste storage materials: their α-recoil aging," *Science*, vol. 209, no. 4464, pp. 1518–1520, 1980.

[138] J.-C. Dran and J.-C. Petit, "Etching of fission tracks in silicate glasses by means of deionized water," *International Journal of Radiation Applications and Instrumentation. Part D: Nuclear Tracks and Radiation Measurements*, vol. 12, no. 1–6, pp. 851–854, 1986.

[139] J. E. French, "Imogolite: an ideal mineralogical substrate for the prebiotic assembly of long polynucleotides," *Mineralogical Magazine*, vol. 77, no. 5, p. 1109, 2013, F—Goldschmidt Abstracts 2013.

[140] D. Macdougall, "Fission track dating of volcanic glass shards in marine sediments," *Earth and Planetary Science Letters*, vol. 10, no. 4, pp. 403–406, 1971.

[141] R. L. Fleischer, "Alpha-recoil damage: relation to isotopic disequilibrium and leaching of radionuclides," *Geochimica et Cosmochimica Acta*, vol. 52, no. 6, pp. 1459–1466, 1988.

[142] J.-C. Petit, C. Brousse, J.-C. Dran, and G. D. Mea, "Use of fission tracks for deciphering the dissolution mechanism of silicate glasses," *Nuclear Tracks*, vol. 12, no. 1–6, pp. 847–850, 1985.

[143] A. S. Sandhu, J. A. Westgate, and S. J. Preece, "Thermal stability of fission tracks in natural and synthetic glasses: an assessment of compositional effects," *Radiation Measurements*, vol. 31, no. 1, pp. 665–668, 1999.

[144] A. S. Sandhu and K. Singh, "Etching and annealing studies of nuclear tracks in glasses," *Radiation Physics and Chemistry*, vol. 61, no. 3–6, pp. 579–581, 2001.

[145] R. L. Fleischer, "Etching of recoil tracks in solids," *Geochimica et Cosmochimica Acta*, vol. 67, no. 24, pp. 4769–4774, 2003.

[146] K. P. Jochum, A. W. Hofmann, E. Ito, H. M. Seufert, and W. M. White, "K, U and Th in mid-ocean ridge basalt glasses and heat production, K/U and K/Rb in the mantle," *Nature*, vol. 306, no. 5942, pp. 431–436, 1983.

[147] F. Aumento, "The Mid-Atlantic Ridge near 45° N.V. Fission track and ferromanganese chronology," *Canadian Journal of Earth Science*, vol. 6, pp. 1431–1440, 1969.

[148] A. J. W. Gleadow and J. F. Lovering, "The effect of weathering on fission track dating," *Earth and Planetary Science Letters*, vol. 22, no. 2, pp. 163–168, 1974.

[149] C. Weise, K. G. van den Boogaart, R. Jonckheere, and L. Ratschbacher, "Annealing kinetics of Kr-tracks in monazite: implications for fission-track modelling," *Chemical Geology*, vol. 260, no. 1-2, pp. 129–137, 2009.

[150] J.-L. Tien and C.-H. Chen, "Naturally etched tracks in apatites and the correction of fission track dating," *Radiation Measurements*, vol. 31, no. 1, pp. 669–672, 1999.

[151] C. C. Allen, J. L. Gooding, M. Jercinovic, and K. Keil, "Altered basaltic glass: a terrestrial analog to the soil of Mars," *Icarus*, vol. 45, no. 2, pp. 347–369, 1981.

[152] S. Erard and W. Calvin, "New composite spectra of Mars, 0.4–5.7 μm," *Icarus*, vol. 130, no. 2, pp. 449–460, 1997.

[153] J. L. Bandfield, V. E. Hamilton, and P. R. Christensen, "A global view of martian surface compositions from MGS-TES," *Science*, vol. 287, no. 5458, pp. 1626–1630, 2000.

[154] R. Rieder, R. Gellert, R. C. Anderson et al., "Chemistry of rocks and soils at Meridiani Planum from the Alpha Particle X-ray Spectrometer," *Science*, vol. 306, no. 5702, pp. 1746–1749, 2004.

[155] S. W. Squyres, R. E. Arvidson, J. F. Bell III et al., "Ancient impact and aqueous processes at Endeavour crater, Mars," *Science*, vol. 336, no. 6081, pp. 570–576, 2012.

[156] J. Orenberg and J. Handy, "Reflectance spectroscopy of palagonite and iron-rich montmorillonite clay mixtures: implications for the surface composition of Mars," *Icarus*, vol. 96, no. 2, pp. 219–225, 1992.

[157] B. Horgan and J. F. Bell III, "Widespread weathered glass on the surface of Mars," *Geology*, vol. 40, no. 5, pp. 391–394, 2012.

[158] S. W. Squyres, "Urey prize lecture: water on Mars," *Icarus*, vol. 79, no. 2, pp. 229–288, 1989.

[159] J. W. Head III, H. Hiesinger, M. A. Ivanov, M. A. Kreslavsky, S. Pratt, and B. J. Thomson, "Possible ancient oceans on Mars: evidence from Mars orbiter laser altimeter data," *Science*, vol. 286, no. 5447, pp. 2134–2137, 1999.

[160] Shipboard Scientific Parties, "Site 418," in *Initial Reports of the Deep Sea Drilling Project*, LI, LII, LIII, (Part 1), pp. 351–626, US Government Printing Office, Washington, DC, USA, 1980.

[161] M. F. J. Flower, W. Ohnmacht, P. T. Robinson, G. Marriner, and H.-U. Schmincke, "Lithologic and chemical stratigraphy at deep sea drilling sites 417 and 418," in *Initial Reports of the Deep Sea Drilling Project*, LI, LII, LIII, (Part 2), pp. 939–956, US Government Printing Office, Washington, DC, USA, 1980.

[162] G. R. Byerly and J. M. Sinton, "Compositional trends in natural basaltic glasses from deep sea drilling project holes 417D and 418A," in *Initial Reports of the Deep Sea Drilling Project*, LI, LII, LIII, (Part 2), pp. 957–971, US Government Printing Office, Washington, DC, USA, 1980.

[163] S. H. Richardson, S. R. Hart, and H. Staudigel, "Vein mineral ages of old oceanic crust," *Journal of Geophysical Research*, vol. 85, no. 12, pp. 7195–7200, 1980.

[164] J. C. Alt, "Subseafloor processes in mid-ocean ridge hydrothermal systems," in *Seafloor Hydrothermal Systems: Physical, Chemical, Biological, and Geological Interactions*, vol. 91 of *Geophysical Monograph*, pp. 85–114, American Geophysical Union, 1995.

[165] M. Ozima, I. Kaneoka, and M. Yanagisawa, "^{40}Ar–^{39}Ar geochronological studies of drilled basalts from Leg 51 and Leg 52," in *Initial Reports of the Deep Sea Drilling Project*, LI, LII, LIII, (Part 2), pp. 1127–1128, US Government Printing Office, Washington, DC, USA, 1980.

[166] S. Gartner, "Calcareous nannofossils, deep sea drilling project holes 417A, 417D, and 418A," in *Initial Reports of the Deep Sea Drilling Project*, LI, LII, LIII, Part 2, pp. 815–821, US Government Publishing Office, Washington, DC, USA, 1980.

[167] R. D. Müller, M. Sdrolias, C. Gaina, and W. R. Roest, "Age, spreading rates, and spreading asymmetry of the world's ocean crust," *Geochemistry, Geophysics, Geosystems*, vol. 9, no. 4, Article ID Q04006, 19 pages, 2008.

[168] F. J. Vine and D. H. Matthews, "Magnetic anomalies over oceanic ridges," *Nature*, vol. 199, no. 4897, pp. 947–949, 1963.

[169] F. J. Vine, "Spreading of the ocean floor: new evidence," *Science*, vol. 154, no. 3755, pp. 1405–1415, 1966.

[170] C. G. A. Harrison, "Marine magnetic anomalies—the origin of the stripes," *Annual Review of Earth And Planetary Sciences*, vol. 15, pp. 505–543, 1987.

[171] P. D. Rabinowitz, H. Hoskins, and S. M. Asquith, "Geophysical site survey results near deep sea drilling project sites 417 and 418 in the central Atlantic Ocean," in *Initial Reports of the Deep Sea Drilling Project*, LI, LII, LIII, (Part 1), pp. 629–669, US Government Printing Office, Washington, DC, USA, 1980.

[172] Shipboard Scientific Parties, "Introduction and explanatory notes," in *Initial Reports of the Deep Sea Drilling Project*, LI, LII, LIII, (Part 1), pp. 5–22, US Government Printing Office, Washington, DC, USA, 1980.

[173] R. D. Müller, W. R. Roest, J.-Y. Royer, L. M. Gahagan, and J. G. Sclater, "Digital isochrons of the world's ocean floor," *Journal of Geophysical Research B: Solid Earth*, vol. 102, no. 2, Article ID 96JB01781, pp. 3211–3214, 1997.

[174] P. Keary and F. J. Vine, *Global Tectonics*, Blackwell Science, London, UK; Cambridge University Press, 2nd edition, 1996.

[175] D. E. Bird, S. A. Hall, K. Burke, J. F. Casey, and D. S. Sawyer, "Early Central Atlantic Ocean seafloor spreading history," *Geosphere*, vol. 3, no. 5, pp. 282–298, 2007.

[176] J. E. T. Channell, E. Erba, M. Nakanishi, and K. Tamaki, "Late Jurassic–Early Cretaceous timescales and ocean magnetic anomaly block models," in *Geochronology Timescales and Global Stratigraphic Correlation*, W. A. Berggren, Ed., vol. 54, pp. 51–63, SEPM (Society for Sedimentary Geology) Special Publication, 1995.

[177] K. D. Klitgord and H. Schouten, "Plate kinematics of the central Atlantic," in *The Geology of North America, Volume M. The Western North Atlantic Region*, P. R. Vogt and B. E. Tucholke, Eds., pp. 351–378, Geological Society of America, Boulder, Colo, USA, 1986.

[178] W. Roest, "Seafloor spreading pattern of the North Atlantic between 10° and 40° N," *Geologica Ultraiectina*, vol. 48, pp. 1–121, 1987.

[179] A. Schettino and E. Turco, "Breakup of Pangaea and plate kinematics of the central Atlantic and Atlas regions," *Geophysical Journal International*, vol. 178, no. 2, pp. 1078–1097, 2009.

[180] A. J. Biggin, D. J. J. van Hinsbergen, C. G. Langereis, G. B. Straathof, and M. H. L. Deenen, "Geomagnetic secular variation in the Cretaceous Normal Superchron and in the Jurassic," *Physics of the Earth and Planetary Interiors*, vol. 169, no. 1–4, pp. 3–19, 2008.

[181] E. J. W. Jones, S. C. Cande, and F. Spathopoulos, "Evolution of a major oceanographic pathway: the equatorial Atlantic," in *The Tectonics, Sedimentation and Palaeoceanography of the North Atlantic Region*, R. A. Scrutton, M. S. Stoker, G. B. Shimmield, and A. W. Tudhope, Eds., Geological Society Special Publication no. 90, pp. 199–213, The Geological Society, London, UK, 1995.

[182] N. A. Stroncik and H.-U. Schmincke, "Palagonite—a review," *International Journal of Earth Sciences*, vol. 91, no. 4, pp. 680–697, 2002.

[183] I. H. Thorseth, H. Furnes, and O. Tumyr, "A textural and chemical study of Icelandic palagonite of varied composition and its bearing on the mechanism of the glass-palagonite transformation," *Geochimica et Cosmochimica Acta*, vol. 55, no. 3, pp. 731–749, 1991.

[184] R. A. F. Cas and J. V. Wright, *Volcanic Successions: Modern and Ancient*, Allen and Unwin Publishers Ltd, London, UK, 1987.

[185] A. Allaby and M. Allaby, "Axiolitic structure," in *A Dictionary of Earth Sciences*, M. Allaby, Ed., Oxford University Press, 3rd edition, 2008.

[186] C. M. Scarfe, "Secondary minerals in some basaltic rocks from Deep Sea Drilling Project Legs 52 and 53, hole 418A," in *Initial Reports of the Deep Sea Drilling Project*, LI, LII, LIII, (Part 2), pp. 1243–1251, US Government Printing Office, Washington, DC, USA, 1980.

[187] Shipboard Scientific Party, "3. Site 1188," in *Proceedings of the Ocean Drilling Program, Initial Reports*, R. A. Binns, F. J. A. S. Barriga, D. J. Miller et al., Eds., vol. 193, pp. 1–305, Ocean Drilling Program, College Station, Tex, USA, 2002.

[188] S. Umino, "5. Data report: textural variation of units 1256C-18 and 1256D-1 lava pond, with special reference to recrystallization of the base of unit 1256C-18," in *Proceedings of the Ocean Drilling Program, Scientific Results*, D. A. H. Teagle, D. S. Wilson, G. D. Acton, and D. A. Vanko, Eds., vol. 206, pp. 1–32, College Station, Tex, USA, 2007.

[189] J. Gottsmann, A. J. L. Harris, and D. B. Dingwell, "Thermal history of Hawaiian pāhoehoe lava crusts at the glass transition: implications for flow rheology and emplacement," *Earth and Planetary Science Letters*, vol. 228, no. 3-4, pp. 343–353, 2004.

[190] A. Steiner, "Clay minerals in hydrothermally altered rocks at Wairakei, New Zealand," *Clays and Clay Minerals*, vol. 16, no. 3, pp. 193–213, 1968.

[191] I. Techer, T. Advocat, J. Lancelot, and J.-M. Liotard, "Dissolution kinetics of basaltic glasses: control by solution chemistry and protective effect of the alteration film," *Chemical Geology*, vol. 176, no. 1–4, pp. 235–263, 2001.

[192] D. K. Smith and J. R. Cann, "Mid-Atlantic ridge volcanic processes: how erupting lava forms Earth's anatomy," *Oceanus Magazine*, vol. 41, no. 1, pp. 11–14, 1998.

[193] G. R. Lumpkin, "Alpha-decay damage and aqueous durability of actinide host phases in natural systems," *Journal of Nuclear Materials*, vol. 289, no. 1-2, pp. 136–166, 2001.

[194] J. E. French, *U-Pb dating of Paleoproterozoic mafic dyke swarms of the South Indian shield: implications for paleocontinental reconstructions and identifying ancient mantle plume events [Ph.D. thesis]*, University of Alberta, Alberta, Canada, 2007.

[195] J. E. French, "Dendritic zircon formation by deterministic volume-filling fractal growth: implications for the mechanisms of branch formation in dendrites," *American Mineralogist*, vol. 95, no. 5-6, pp. 706–716, 2010.

[196] J. P. Gustafsson, "The surface chemistry of imogolite," *Clays and Clay Minerals*, vol. 49, no. 1, pp. 73–80, 2001.

[197] J. M. Guilbert and C. F. Park Jr., *The Geology of Ore Deposits*, W.H. Freeman and Company, New York, NY, USA, 1986.

[198] A. S. Sandhu, J. A. Westgate, and B. V. Alloway, "Optimizing the isothermal plateau fission-track dating method for volcanic glass shards," *Nuclear Tracks and Radiation Measurements*, vol. 21, no. 4, pp. 479–488, 1993.

[199] J. A. Westgate and N. D. Naeser, "Tephrochronology and fission-track dating," in *Dating Methods for Quaternary Deposits*, N. W. Rutter and N. R. Catto, Eds., pp. 15–28, Geological Association of Canada, St. John's, Canada, 1995.

[200] A. H. Jaffey, K. F. Flynn, L. E. Glendenin, W. C. Bentley, and A. M. Essling, "Precision measurement of half-lives and specific activities of ^{235}U and ^{238}U," *Physical Review C*, vol. 4, no. 5, pp. 1889–1906, 1971.

[201] N. E. Holden and D. C. Hoffman, "Spontaneous fission half-lives for ground-state nuclides," *Pure and Applied Chemistry*, vol. 72, pp. 1525–1562, 2000.

[202] G. Faure, *Principles of Isotope Geology*, John Wiley & Sons, New York, NY, USA, 2nd edition, 1986.

[203] A. Simonetti, M. R. Buzon, and R. A. Creaser, "In-situ elemental and Sr isotope investigation of human tooth enamel by laser ablation-(MC)-ICP-MS: successes and pitfalls," *Archaeometry*, vol. 50, no. 2, pp. 371–385, 2008.

[204] R. H. Steiger and E. Jäger, "Subcommission on geochronology: convention on the use of decay constants in geo- and cosmochronology," *Earth and Planetary Science Letters*, vol. 36, no. 3, pp. 359–362, 1977.

[205] M. Trieloff, M. Falter, and E. K. Jessberger, "The distribution of mantle and atmospheric argon in oceanic basalt glasses," *Geochimica et Cosmochimica Acta*, vol. 67, no. 6, pp. 1229–1245, 2003.

[206] H. Cheng, R. L. Edwards, J. Hoff, C. D. Gallup, D. A. Richards, and Y. Asmerom, "The half-lives of uranium-234 and thorium-230," *Chemical Geology*, vol. 169, no. 1-2, pp. 17–33, 2000.

[207] G. Bigazzi, "Length of fission tracks and age of muscovite samples," *Earth and Planetary Science Letters*, vol. 3, pp. 434–438, 1967.

[208] R. F. Galbraith and G. M. Laslett, "Some calculations relevant to thermal annealing of fission tracks in apatite," *Proceedings of the Royal Society A: Mathematical, Physical and Engineering Sciences*, vol. 419, no. 1857, pp. 305–321, 1988.

[209] J. G. Sclater and L. Wixon, "The relationship between depth and age and heat flow and age in the Western North Atlantic," in *The Geology of North America, Vol. M, The Western North Atlantic Region*, P. R. Vogt and B. E. Tucholke, Eds., pp. 257–270, Geological Society of America, New York, NY, USA, 1986.

[210] R. Tada and R. Siever, "Pressure solution during diagenesis," *Annual Review of Earth and Planetary Sciences*, vol. 17, pp. 89–118, 1989.

[211] H. L. Lescinsky and L. Benninger, "Pseudo-borings and predator traces: artifacts of pressure—dissolution in fossiliferous shales," *Palaios*, vol. 9, no. 6, pp. 599–604, 1994.

[212] R. Yamada, T. Tagami, and S. Nishimura, "Assessment of overetching factor for confined fission-track length measurement in zircon," *Chemical Geology*, vol. 104, no. 1–4, pp. 251–259, 1993.

[213] A. J. W. Gleadow, I. R. Duddy, P. F. Green, and J. F. Lovering, "Confined fission track lengths in apatite: a diagnostic tool for thermal history analysis," *Contributions to Mineralogy and Petrology*, vol. 94, no. 4, pp. 405–415, 1986.

[214] R. C. Jonckheere and G. A. Wagner, "On the occurrence of anomalous fission tracks in apatite and titanite," *American Mineralogist*, vol. 85, no. 11-12, pp. 1744–1753, 2000.

[215] D. Vincent, R. Clocchiatti, and Y. Langevin, "Fission-track dating of glass inclusions in volcanic quartz," *Earth and Planetary Science Letters*, vol. 71, no. 2, pp. 340–348, 1984.

[216] R. L. Fleischer and P. B. Price, "Fission track evidence for the simultaneous origin of tektites and other natural glasses," *Geochimica et Cosmochimica Acta*, vol. 28, no. 6, pp. 755–IN5, 1964.

[217] W. Gentner, B. P. Glass, D. Storzer, and G. A. Wagner, "Fission track ages and ages of deposition of deep-sea microtektites," *Science*, vol. 168, no. 3929, pp. 359–361, 1970.

[218] D. Storzer and G. A. Wagner, "Fission track ages of North American tektites," *Earth and Planetary Science Letters*, vol. 10, no. 4, pp. 435–440, 1971.

[219] G. Wagner and P. Van den Haute, *Fission-Track Dating*, Kluwer Academic Publishers, Dodrecht, The Netherlands, 1992.

[220] D. S. Miller and G. A. Wagner, "Fission-track ages applied to obsidian artifacts from South America using the plateau-annealing and the track-size age-correction techniques," *Nuclear Tracks*, vol. 5, no. 1-2, pp. 147–155, 1981.

[221] T. Tsuruta, "Reduction in etching time for fission tracks in diallyl phthalate resin," *Radiation Measurements*, vol. 34, no. 1–6, pp. 167–170, 2001.

[222] N. Yasuda, M. Yamamoto, N. Miyahara, N. Ishigure, T. Kanai, and K. Ogura, "Measurement of bulk etch rate of CR-39 with atomic force microscopy," *Nuclear Instruments and Methods in Physics Research B*, vol. 142, no. 1-2, pp. 111–116, 1998.

[223] P. B. Price and R. M. Walker, "Observation of fossil particle tracks in natural micas," *Nature*, vol. 196, no. 4856, pp. 732–734, 1962.

[224] P. B. Price and R. M. Walker, "Fossil tracks of charged particles in mica and the age of minerals," *Journal of Geophysical Research*, vol. 68, no. 16, pp. 4847–4862, 1963.

[225] R. Jonckheere and P. Van den Haute, "On the frequency distributions per unit area of the projected and etchable lengths of surface-intersecting fission tracks: influences of track revelation, observation and measurement," *Radiation Measurements*, vol. 30, no. 2, pp. 155–179, 1999.

[226] J. Barbarand, A. Carter, I. Wood, and T. Hurford, "Compositional and structural control of fission-track annealing in apatite," *Chemical Geology*, vol. 198, no. 1-2, pp. 107–137, 2003.

[227] H. Matzke, "Radiation damage in nuclear materials," *Nuclear Instruments and Methods in Physics Research B*, vol. 65, no. 1-4, pp. 30–39, 1992.

[228] R. F. Galbraith, "Some remarks on fission-track observational biases and crystallographic orientation effects," *American Mineralogist*, vol. 87, no. 7, pp. 991–995, 2002.

[229] P. P. Metha and Rama, "Annealing effects in muscovite and their influence on dating by fission track method," *Earth and Planetary Science Letters*, vol. 7, no. 1, pp. 82–86, 1969.

[230] S. Krishnaswami, D. Lal, N. Prabhu, and D. Macdougall, "Characteristics of fission tracks in zircon: applications to geochronology and cosmology," *Earth and Planetary Science Letters*, vol. 22, no. 1, pp. 51–59, 1974.

[231] D. Storzer and G. A. Wagner, "Correction of thermally lowered fission track ages of tektites," *Earth and Planetary Science Letters*, vol. 5, pp. 463–468, 1968.

[232] G. Bigazzi, F. P. Bonadonna, M. A. Laurenzi, and S. Tonarini, "A test sample for fission-track dating of glass shards," *Nuclear Tracks and Radiation Measurements*, vol. 21, no. 4, pp. 489–497, 1993.

[233] F. Renard, D. K. Dysthe, J. G. Feder, P. Meakin, S. J. S. Morris, and B. Jamtveit, "Pattern formation during healing of fluid-filled cracks: an analog experiment," *Geofluids*, vol. 9, no. 4, pp. 365–372, 2009.

[234] J. C. Dran, J. P. Duraud, Y. Langevin, M. Maurette, and J. C. Petit, "Contribution of radiation damage studies to the understanding of the Oklo phenomenon," Natural Fission Reactors IAEA-TC-119/12, 1978.

[235] D. W. Davis and T. E. Krogh, "Preferential dissolution of ^{234}U and radiogenic Pb from α-recoil-damaged lattice sites in zircon: implications for thermal histories and Pb isotopic fractionation in the near surface environment," *Chemical Geology*, vol. 172, no. 1-2, pp. 41–58, 2001.

[236] H. Matzke, "Actinide behavior and radiation damage produced by α-decay in materials to solidify nuclear waste," *Inorganica Chimica Acta*, vol. 94, no. 1–3, pp. 142–143, 1984.

[237] W. Primak, "Radiation effects in silicate glasses pertinent to their application as a radioactive waste storage medium," *Nuclear Technology*, vol. 60, no. 2, pp. 199–205, 1983.

[238] W. J. Weber and F. P. Roberts, "A review of radiation effects in solid nuclear waste forms," *Nuclear Technology*, vol. 60, no. 2, pp. 178–198, 1983.

[239] P. Cattermole, "Volcanic flow development at Alba Patera, Mars," *Icarus*, vol. 83, no. 2, pp. 453–493, 1990.

[240] R. E. Ernst, E. B. Grosfils, and D. Mège, "Giant dike swarms: Earth, Venus, and Mars," *Annual Review of Earth and Planetary Sciences*, vol. 29, pp. 489–534, 2001.

[241] R. V. Morris, J. L. Gooding, H. V. Lauer Jr., and R. B. Singer, "Iron mineralogy of a Hawaiian palagonitic soil with Mars-like spectral and magnetic properties," in *Proceedings of the Abstracts: 21st Lunar and Planetary Science Conference*, pp. 811–812, Houston, Tex, USA, March 1990.

[242] N. G. Barlow, "Crater size-frequency distributions and a revised Martian relative chronology," *Icarus*, vol. 75, no. 2, pp. 285–305, 1988.

[243] P. Toulmin III, A. K. Baird, B. C. Clark et al., "Geochemical and mineralogical interpretation of the Viking inorganic chemical results," *Journal of Geophysical Research*, vol. 82, no. 28, pp. 4625–4634, 1977.

[244] W. W. Barker, S. A. Welch, and J. F. Banfield, "Biogeochemical weathering of silicate minerals," *Reviews in Mineralogy and Geochemistry*, vol. 35, no. 1, pp. 391–428, 1997.

7

Object Detection in Ground-Penetrating Radar Images using a Deep Convolutional Neural Network and Image Set Preparation by Migration

Kazuya Ishitsuka ⓘ,[1] Shinichiro Iso ⓘ,[2,3] Kyosuke Onishi ⓘ,[4] and Toshifumi Matsuoka ⓘ[3]

[1]*Division of Sustainable Resources Engineering, Hokkaido University, Sapporo 065-0068, Japan*
[2]*School of Creative Science and Engineering, Waseda University, Tokyo 169-8555, Japan*
[3]*Fukada Geological Institute, Tokyo 113-0021, Japan*
[4]*Geology and Geotechnical Engineering Research Group, Public Works Research Institute, Tsukuba 305-8516, Japan*

Correspondence should be addressed to Kazuya Ishitsuka; ishitsuka@eng.hokudai.ac.jp

Academic Editor: Yun-tai Chen

Ground-penetrating radar allows the acquisition of many images for investigation of the pavement interior and shallow geological structures. Accordingly, an efficient methodology of detecting objects, such as pipes, reinforcing steel bars, and internal voids, in ground-penetrating radar images is an emerging technology. In this paper, we propose using a deep convolutional neural network to detect characteristic hyperbolic signatures from embedded objects. As a first step, we developed a migration-based method to collect many training data and created 53510 categorized images. We then examined the accuracy of the deep convolutional neural network in detecting the signatures. The accuracy of the classification was 0.945 (94.5%)–0.979 (97.9%) when using several thousands of training images and was much better than the accuracy of the conventional neural network approach. Our results demonstrate the effectiveness of the deep convolutional neural network in detecting characteristic events in ground-penetrating radar images.

1. Introduction

Ground-penetrating radar (GPR) is an effective technology for the nondestructive investigation of the shallow subsurface based on the transmitting and receiving of an electromagnetic wave. The radar system transmits a short pulse toward the target and records the signal that is backscattered by discontinuities in the dielectric permittivity. Because the dielectric permittivity relates to embedded subsurface features, GPR systems have been widely used for road inspection, mine detection, and geological/archaeological studies. Recent GPR systems with air-coupled antennas have been mounted onto vehicles for recording signals in a broader and faster way (e.g., [1, 2]). However, with the increasing data amount, the detection of target signatures becomes a challenging and time-consuming task.

For the automatic detection of characteristic features in GPR images, a neural network (NN)-based methodology

was developed [3, 4]. Frequency features of GPR images were used for detection using an NN [3] while spatial signatures of GPR images were used after binarization using a threshold [4]. The successful application of an NN has also been demonstrated (e.g., [5]). Despite these successful applications, the NN-based method is often unable to recognize complex features. Moreover, accurate recognition requires an ideal noise condition and appropriate preprocessing (e.g., clutter removal), whose assumptions sometimes fail.

Recently, the deep convolutional neural network (CNN) has been successfully used in pattern recognition (e.g., [6, 7]) and classification (e.g., [8, 9]). Compared with the conventional NN, the architecture of the deep CNN consists of several alternations of convolution and pooling layers, which is an analog of the receptive field. The architecture has a great advantage in terms of representing images with multiple levels of abstraction. The deep CNN improves the accuracy of the target detection of GPR images and the versatility of the approach.

The present paper proposes the use of the deep CNN to increase the accuracy and versatility of target detection from GPR images. For versatility, we used the original pattern of GPR images as input images without conducting a prior feature extraction. Many training images are needed for the successful training of the deep CNN and improved accuracy. We therefore developed an algorithm based on the migration procedure for the first-step extraction of training images with target signatures.

2. Deep CNN

We first introduce an algorithm to prepare many training images from a two-dimensional section of GPR images in Section 2.1. We then describe the architecture and roles of each component of the deep CNN in Section 2.2. Sections 2.3 and 2.4 give the specifics of CNN architecture and a description of the accuracy index used in this study.

2.1. Semi-automatic Collection of Training Images. Normalization is a necessary preprocessing step for standardization in collecting training images and conducting CNN learning. This study applied local contrast normalization to input images. Resulting images thus have zero mean and a standard deviation of 1.

To extract hyperbolic signatures originating from embedded objects, we propose the following algorithm based on the difference between the envelope of the original signal (without migration) and the envelope of the signal after migration procedure. A signal reflected from an object has a characteristic hyperbolic curve in the GPR data section, because the distance between radar and target decreases as the radar approaches the target and the distance increases as the radar moves away from the target (Figure 1(a)). The migration procedure focuses the hyperbolic signal and reconstructs the true shape of an object at the true location. The shape of the signal therefore varies with migration.

The processing conducted to extract reflection events in this study is (1) frequency–wavenumber migration, (2) application of the Hilbert transform to signals with and without migration, (3) calculation of the envelope of signals with and without migration (Figures 1(b) and 1(c)), (4) calculation of the difference between the two envelopes, (5) extraction of changed areas by binarization of the difference map (Figure 1(d)), and (6) division of the original data into small batches and classification of them into two classes. The first class contains hyperbolic signatures (event images) while the second class does not contain these characteristics (nonevent images). The classification involves checking if the corresponding area of the binarized maps contains reflection events.

As a migration algorithm, we adopted a frequency–wavenumber migration algorithm [10] because it is an efficient algorithm with which to process a large amount of data. The migration algorithm calculates migrated signals through Fourier transformation and Stolt interpolation [11]. The envelope of the signal A(t) is defined by the magnitude of the complex signal via the Hilbert transformation $A(t) = \sqrt{f^2(t) + g^2(t)}$, where the complex signal can be defined as $A(t)\exp\{i\bullet\theta(t)\} = f(t)+i\bullet g(t)$ [12]. After the above processing, images classified into the wrong class were removed manually, because the above algorithm sometimes misclassified noisy images as images with the characteristic reflection events. Although fully automatic image set preparation is preferable, there is currently no perfect algorithm for GPR image classification. Therefore, the manual check is necessary as a final part of the image set preparation.

2.2. Architecture of the Deep CNN. The deep CNN is a feedforward NN that comprises a stack of alternating convolution layers and pooling layers and then fully connected layers and softmax layers [13] (Figure 2). Each convolutional layer receives inputs from a set of units located in a small neighborhood of the previous layer, referred to as the local receptive field. In each local receptive field, the dot product of the weights and input is calculated, and a bias term then added (Eq. (1)). This operation is called filtering in the convolution layer. The filter then moves along the input vertically and horizontally, repeating the same computation for each local receptive field. The feature represented by the filtering can be emphasized through the layer.

$$u_{i,j,k} = \sum_{p=0}^{M-1}\sum_{q=0}^{N-1} x_{i+p,j+q}w_{p,q,k} + b_k \qquad (1)$$

Here $x_{i,j}$ and $u_{i,j}$ denote ith and jth pixels of input and output images. $w_{p,q,k}$ indicates the kth filter having a rectangular size of $M \times N$, which moves along the input space to convolute with the input image $x_{p,q}$. p and q denote a pixel within the rectangular area of the filter. b_k indicates the bias term corresponding to the kth filter. The rectified linear units (ReLU) activation function is applied after the convolutional layer (Eq. (2)).

$$y_{i,j,k} = \max\{0, u_{i,j,k}\} \qquad (2)$$

Here $y_{i,j,k}$ is pixel (i,j) of the output corresponding to the kth filter. The advantage of the ReLU function is its nonsaturating nonlinearity. In other words, the gradient of the ReLU function is not zero even for an extreme input value. It was also found that the computation is much faster than the computation of the conventional sigmoid function [14].

Each convolution layer possessing the ReLU function is followed by a max pooling layer having a certain pooling size. A max pooling layer extracts the maximum value within the pooling window from the output feature map generated by the previous convolutional layer (Eq. (3)). The pooling operation reduces arising information redundancy because the convolution windows of neighboring locations overlap and enhances the universality of slight position changes of an object.

$$z_{i,j,k} = \max_{(p,q)\in Q}\{y_{p,q,k}\} \qquad (3)$$

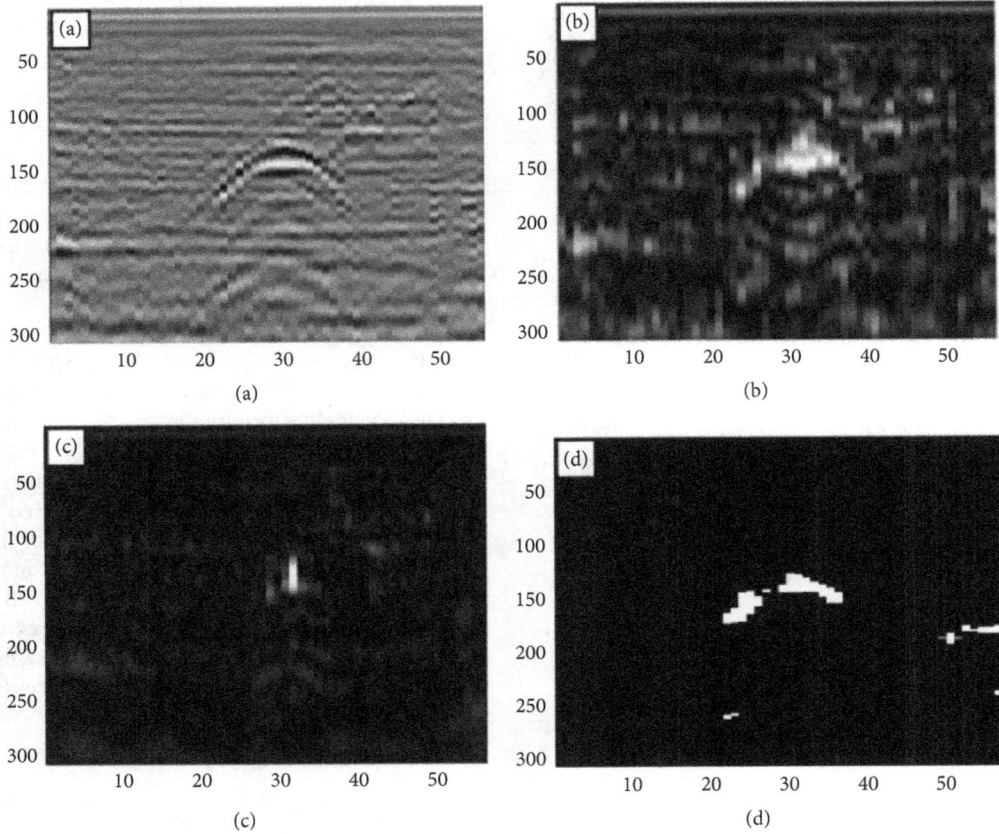

FIGURE 1: (a) Part of an original GPR image that contains the characteristic reflection event from a buried object. (b) Envelope of the GPR image of (a). (c) Envelope of the GPR image after migration. (d) Binarized image of the difference in the envelope without and with the migration procedure.

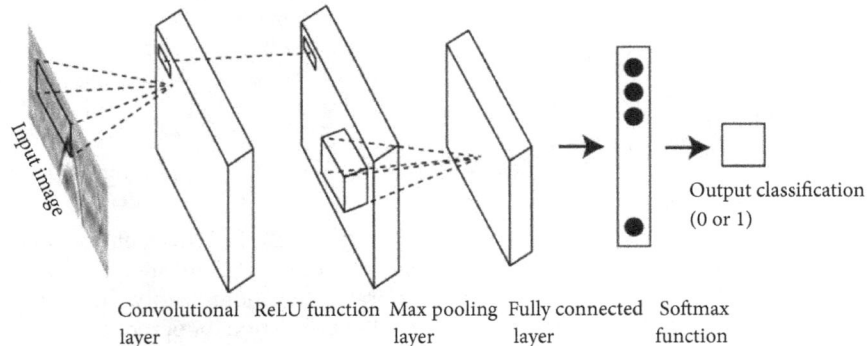

FIGURE 2: Conceptual diagram of image classification using the deep CNN.

Here Q is the pooling size and $z_{i,j,k}$ is pixel (i, j) of the output of the pooling layer corresponding to the kth filter. A larger pooling size generally results in worse performance because the operation throws away too much information.

Fully connected layers correspond to a standard multi-layer NN. The layer consists of a stack of several perceptrons that represents a nonlinear relationship between the weighted sums of inputs and outputs. Each perceptron has the weights and biases as unknown parameters, which are optimized in the learning process.

The softmax function is applied to the output layer. The softmax function calculates posterior probabilities (p_l) over each class (Eq. (4)). In this study, a class is defined as whether the input figure contains the target signatures or not.

$$p_l = \frac{\exp\left(a_l\right)}{\sum_{k=1}^{K} \exp\left(a_k\right)} \quad (4)$$

Here a_l indicates the weighted sum of inputs to the l_{th} unit in the output layer.

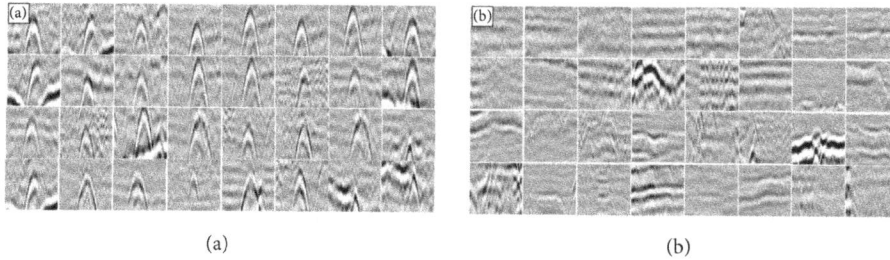

(a)   (b)

FIGURE 3: Examples of extracted GPR images with characteristic reflection events: (a) training images categorized into reflection events, and (b) training images categorized into nonreflection events.

The CNN contains unknown parameters: the filtering coefficients and biases in convolution layers and the weights and biases in fully connected layers. The learning process optimizes these parameters via a backpropagation algorithm. The backpropagation algorithm iteratively estimates parameters using the gradient of an objective function. In the case of classification, cross entropy is usually adopted as the objective function (Eq. (5)).

$$C = -\sum_{l=1}^{n} d_l \log_e p_l \qquad (5)$$

Here d_l is 1 if the classified result is correct and zero if not.

2.3. Configuration.
The input layer has dimensions of 65×65 corresponding to the size of the extracted original image (e.g., Figure 3). Figure 3 shows typical images of two classes (i.e., reflection event/no reflection event). For the convolutional layer, we used 20 convolution filters having a window size of 5×5 as a default setting, resulting in 20 feature maps with a size of 61×61. In this case, the numbers of weights and biases are respectively $5 \times 5 \times 20$ and 20. The accuracy of the CNN depends on the window size; this is evaluated in Section 4. No spatial zero padding is used in the convolution filters, and the convolution stride is fixed to 1 pixel. A max pooling layer with a pooling size of 2×2 has a stride of 2 pixels. The size of the resulting 20 feature maps is 30×30. In this study, we used a simple configuration: one convolution layer and one max pooling layer. This is because the accuracy did not significantly increase in our data when the number of these layer increases. A fully connected layer is specified as having an output size of 2. Therefore, the numbers of weights and biases are respectively $30 \times 30 \times 2$ and 10. The softmax function divides an input image into two classes; i.e., it categorizes on the basis of whether an input image contains target signatures (reflection events) or not.

2.4. Accuracy Evaluation of the Deep CNN.
We quantified the accuracy of the deep CNN using the total accuracy (Ac) as follows:

$$Ac = \frac{c_{est}}{c_{val}} \qquad (6)$$

Here c_{est} is the number of correctly categorized images in the validation data and c_{val} is the number of validation images.

The accuracy is ranged from 0 to 1, and a higher value of the index indicates higher accuracy of overall classification. In addition, we calculated the true positive ratio (TPR) and the true negative ratio (TNR) to investigate how the deep CNN can detect reflection events correctly (Eq. (7) and (8)).

$$TPR = \frac{c_{est,e}}{c_{val,e}} \qquad (7)$$

$$TNR = \frac{c_{est,n}}{c_{val,n}} \qquad (8)$$

Here $c_{est,e}$ and $c_{est,n}$ are the number of correctly categorized images as reflection events and nonreflection event in the validation data, while $c_{val,e}$ and $c_{val,n}$ are the number of validation images with reflection events and without reflection events. The TPR is the accuracy in detecting reflection images, and the TNR is the accuracy in detecting nonreflection images.

3. GPR Images

We used image data acquired by a vehicle-mounted GPR system. The system acquired GPR trace data with 1-cm spacing along a road, recording 305 samples during 29.79 ns in the depth direction. In this configuration, GPR data were acquired along four lines (namely 8009-1, 8009-2, 8009-3 and 8009-4). The corresponding numbers of traces were 548,700, 556,980, 552,580 and 569,940.

To prepare training data, we first extracted square areas within a window size of 65×65. The size corresponds to 65 cm in the horizontal direction and 6.35 ns in the vertical direction. Then, according to the method described in Section 2.1, the square areas were automatically classified according to whether there were characteristic reflection events. Images with noisy areas were sometimes also classified as images having a reflection event, and we checked the horizontal and vertical continuity of the difference in envelopes with and without migration and discarded the misclassified images that did not show spatial continuity. Subsequently, the extracted areas were manually checked, and the images of reflection were reclassified from target objects and images without such events (e.g., inclined geological boundaries). About 39% of images were reclassified. Examples of classified images with characteristic reflection events are shown in Figure 3. The number of extracted images was 21,879 (1875 for reflection events / 20,004 for nonreflection events) on line

TABLE 1: Accuracy of the cross-validation test using the deep CNN. One line was used for validation images while other lines were used for training images.

The line for validation images	Accuracy	True Positive Ratio (TPR)	True Negative Ratio (TNR)
8009-1	0.945	0.794	0.959
8009-2	0.978	0.719	0.989
8009-3	0.958	0.702	0.991
8009-4	0.979	0.741	0.980

TABLE 2: Accuracy of the deep CNN depending on the window size of the convolutional layer.

The window size	Accuracy	True Positive Ratio (TPR)	True Negative Ratio (TNR)
3×3	0.928	0.789	0.941
5×5	0.945	0.794	0.959
11×11	0.945	0.828	0.956
15×15	0.949	0.799	0.963
20×20	0.947	0.801	0.961

8009-1, 8702 (501/8201) on line 8009-2, 13,293 (1532/11,761) on line 8009-3, and 9636 (263/9423) on line 8009-4.

4. Results of Classification

For accuracy evaluation, we used one of the four lines as validation data and the other three lines as training data. In other words, we calculated four accuracies corresponding to the four lines as validation images. During the training of the deep CNN, we monitored the learning curve to avoid overfitting. Among the four obtained accuracies, the accuracy was a maximum (0.979) when 8009-4 was used as the validation images and a minimum (0.945) when 8009-1 was used as the validation images (Table 1). The TPR had a maximum of 0.794 and a minimum of 0.702, while the TNR had a maximum of 0.959 and a minimum of 0.991 (Table 1). The accuracy of reflection image detection (TPR) was inferior to that of nonreflection image detection (TNR), probably owing to the variation in reflection features and fewer training images.

Visualization of the output images of the convolutional and pooling layers clarified the characteristics of the deep CNN. Figures 4(a) and 4(d) shows one of the input images with and without reflection events. The optimal filtering of the convolutional layers produced the images shown in Figures 4(b) and 4(e). Figures 4(b) and 4(e) shows that the filtered images had emphasized features in a certain preferential direction. After application of ReLU functions and the max pooling layer, the characteristic parts were exhibited by emphasizing areas with larger positive amplitude (Figures 4(c) and 4(f)).

The window size may affect accuracy because the convolutional layer extracts the features of input data. We therefore examined accuracy depending on the architecture of the deep CNN. Specifically, we tested the deep CNN with the convolution layer having a window size of 3, 5, 11, 15, and 20. Lines 8009-2, 8009-3, and 8009-4 were used as training data while line 8009-1 was used as validation data. The result of the examination is presented in Table 2. The table shows that the accuracy for a window size of 15 was 0.949, which

was superior to the accuracy for other window sizes. The total accuracy decreased with the window size decreasing from 15 and also with the window size increasing above 15 (Table 2). The correlation of the total accuracy with the window size was due to the TNR value; i.e., TNR was a maximum when a window size of 15 was used, and the accuracy monotonically decreased with the window size decreasing from 15 (Table 2). Since the size of training images is 65×65, the window size of 15 is 23 % of the total image size. Meanwhile, there was no apparent correlation of TPR with the window size (Table 2).

To show the advantage of the deep CNN, we compared the results of the deep CNN with those of the conventional feedforward NN. The feedforward NN consists of layers of perceptrons, with perceptrons between layers being fully connected [15]. Because the performance of the NN depends on the number of layers and perceptrons, we examined the accuracy for several architectures: 10 perceptrons per layer (one hidden layer), 20 perceptrons per layer (one hidden layer), 5 perceptrons per layer (two hidden layers), and 10 perceptrons per layer (two hidden layers). For the NN with two hidden layers, we used the same number of perceptrons per layer for simplicity. In the examination, data on the 8009-1 line were used as validation data and the other data (of lines 8009-2, -3, and -4) were used as training data to optimize the neural network.

The results of the conventional NN approach are shown in Table 3. The accuracy of the overall classification was highest (0.707) for 10 perceptrons per layer and one hidden layer and lowest (0.513) for 10 perceptrons per layer and two hidden layers (Table 3). In the classification of data without a characteristic reflection event, TNR ranged 0.518–0.752 (Table 3). Meanwhile, the TPR ranged 0.237–0.466 (Table 3). The total accuracy of the deep CNN exceeded 0.9, showing that the deep CNN performed well. In addition, the results imply that the classification accuracy of the NN highly depends on the architecture of the NN in contrast with the case for the deep CNN.

It is widely known that the deep CNN requires many training images. We thus examined the accuracy depending on the number of training GPR images. As stated above, there

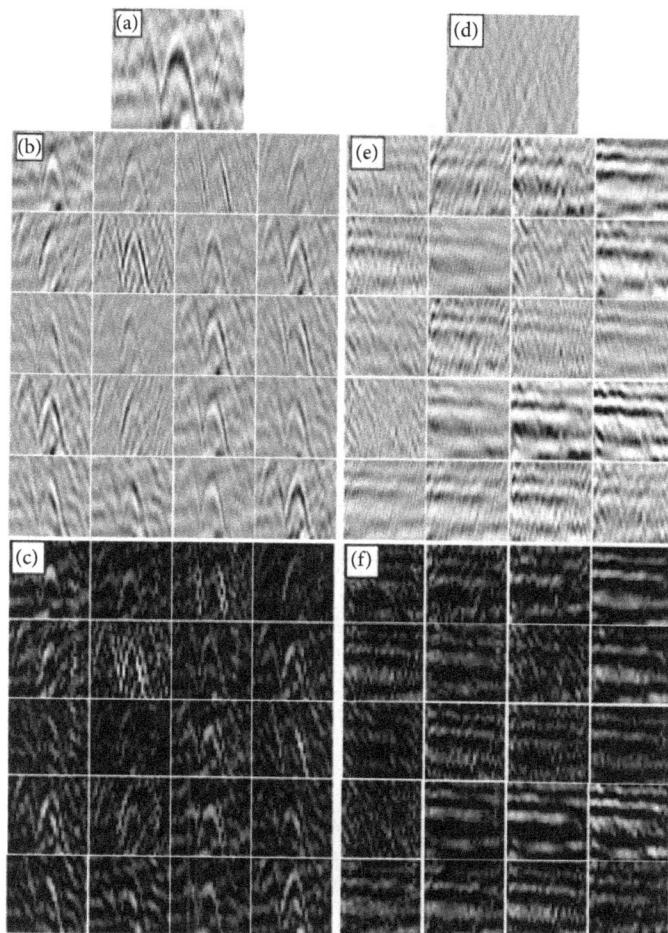

FIGURE 4: Visualization of images for the deep CNN: (a) and (d) inputs of reflection and nonreflection images, (b) and (e) examples of the images following the convolutional layer, and (c) and (f) examples of the images following the max pooling layer.

TABLE 3: Accuracy using the NN depending on the number of perceptrons and layers.

The number of perceptrons per a layer	Accuracy	True Positive Ratio (TPR)	True Negative Ratio (TNR)
10	0.707	0.237	0.752
20	0.594	0.393	0.613
5×5	0.589	0.387	0.609
10×10	0.513	0.466	0.518

were 2296 training images for reflection events and 29,385 for nonreflection events when lines 8009-2, 3 and 4 were used as training images. We then examined the accuracy when using 90% (2066 images for reflection and 26,447 for nonreflection), 70% (1607 and 20,570), 50% (1148 and 14,692), 30% (668 and 8816) and 10% (230 and 2939) of training data. The 8009-4 line was used as validation data. Table 4 shows that the total accuracy decreased with a decreasing number of training images: the accuracy was 0.945 when 90% and 70% of training data were used, and the values respectively decreased to 0.937 and 0.941 when using 30% and 10% of images (Table 4). An examination of the TPR and TNR shows that the decrease in the total accuracy is attributed to TPR.

In particular, the decrease in accuracy was notable when the number of images fell below about 1000 (30%–50%) (Table 4). Meanwhile, TNR was almost constant over the percentages of images tested in this study. Although the total accuracy was lower for smaller number of training images used, the total accuracy remained high (>0.9).

The visualization and examination of correctly and incorrectly classified images by the deep CNN further clarifies the characteristics of the deep CNN approach. The correctly classified images apparently contain or do not contain reflection events (Figures 5(a) and 5(c)). On the other hand, incorrectly classified images display an incomplete signature of reflection events, which is difficult to be categorized even by visual

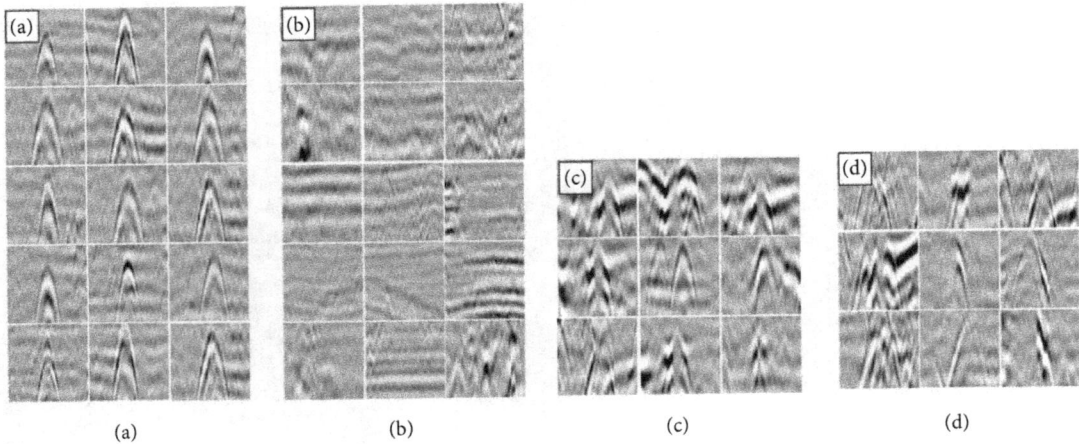

FIGURE 5: Classified images using the deep CNN: (a) images originally categorized into images with reflection events and correctly classified by the deep CNN, (b) images originally categorized into images without reflection events and correctly classified by the deep CNN, (c) images originally categorized into images with reflection events but recognized as nonreflection images by the deep CNN, and (d) images originally categorized into images without a reflection event but recognized as reflection images by the deep CNN.

TABLE 4: Accuracy depending on the number (the percentage) of training images in the deep CNN. The percentage is that of all training data.

The percentage	Accuracy	True Positive Ratio (TPR)	True Negative Ratio (TNR)
90 %	0.945	0.741	0.964
70 %	0.945	0.765	0.962
50 %	0.944	0.740	0.963
30 %	0.937	0.725	0.957
10 %	0.941	0.612	0.972

inspection (Figures 5(b) and 5(d)). The facts demonstrate that the algorithm has successfully classified the apparent reflection events or nonreflection event images. Further strategies for the recognition of the incomplete reflection signatures would increase the applicability of the deep CNN approach.

5. Conclusions

We examined the accuracy of the deep CNN in detecting a characteristic reflection pattern in GPR images. To prepare training images, we used the difference in envelopes obtained without and with a migration procedure. We found the classification accuracy of the deep CNN ranged 0.945–0.979. The accuracy was slightly improved by a few percentage points by tuning the window size of the convolutional layer. Comparison with the conventional NN showed the high accuracy of the deep CNN. Our results demonstrate that a large number of training data and an effective methodology improve the effectiveness of object detection in GPR images.

Acknowledgments

The authors acknowledge Canaan Geo Research for providing GPR data used in this study. This study was supported in part by general account budgets of the universities and the institutions the authors are affiliated with.

References

[1] T. Saarenketo and T. Scullion, "Road evaluation with ground penetrating radar," Journal of Applied Geophysics, vol. 43, no. 2-4, pp. 119–138, 2000.

[2] A. Benedetto and S. Pensa, "Indirect diagnosis of pavement structural damages using surface GPR reflection techniques," Journal of Applied Geophysics, vol. 62, no. 2, pp. 107–123, 2007.

[3] W. Al-Nuaimy, Y. Huang, M. Nakhkash, M. T. C. Fang, V. T. Nguyen, and A. Eriksen, "Automatic detection of buried utilities and solid objects with GPR using neural networks and pattern recognition," Journal of Applied Geophysics, vol. 43, no. 2–4, pp. 157–165, 2000.

[4] P. Gamba and S. Lossani, "Neural detection of pipe signatures in ground penetrating radar images," IEEE Transactions on Geoscience and Remote Sensing, vol. 38, no. 2, pp. 790–797, 2000.

[5] M. R. Shaw, S. G. Millard, T. C. K. Molyneaux, M. J. Taylor, and J. H. Bungey, "Location of steel reinforcement in concrete using ground penetrating radar and neural networks," NDT & E International, vol. 38, no. 3, pp. 203–212, 2005.

[6] R. Girshick, J. Donahue, T. Darrell, and J. Malik, "Rich feature hierarchies for accurate object detection and semantic segmentation," in Proceedings of the 2014 IEEE Conference on Computer Vision and Pattern Recognition, pp. 580–587, 2014.

[7] J. Ding, B. Chen, H. Liu, and M. Huang, "Convolutional neural network with data augmentation for SAR target recognition," IEEE Geoscience and Remote Sensing Letters, vol. 13, no. 3, pp. 364–368, 2016.

[8] A. Krizhevsky, I. Sutskever, and G. E. Hinton, "ImageNet classification with deep convolutional neural networks," *Proc. Advances in Neural Information Processing Systems*, vol. 25, pp. 1090–1098, 2012.

[9] S. Chen, H. Wang, F. Xu, and Y.-Q. Jin, "Target classification using the deep convolutional networks for SAR images," *IEEE Transactions on Geoscience and Remote Sensing*, vol. 54, no. 8, pp. 4806–4817, 2016.

[10] R. H. Stolt, "Migration by fourier transform," *Geophysics*, vol. 43, no. 1, pp. 23–48, 1978.

[11] N. Smitha, D. R. Ullas Bharadwaj, S. Abilash, S. N. Sridhara, and V. Singh, "Kirchhoff and F-K migration to focus ground penetrating radar images," *International Journal of Geo-Engineering*, vol. 7, no. 1, 2016.

[12] M. T. Tarner, F. Koehler, and R. E. Sheriff, "Complex seismic traces analysis," *Geophysics*, vol. 44, pp. 1485–1501, 1979.

[13] Y. LeCun, Y. Bengio, and G. Hinton, "Deep learning," *Nature*, vol. 521, pp. 436–444, 2015.

[14] X. Glorot, A. Bordes, and Y. Bengio, "Deep sparse rectifier neural networks," in *Proceedings of the 14th International Conference on Artificial Intelligence and Statistics*, pp. 315–323, 2011.

[15] M. Van Der Baan and C. Jutten, "Neural networks in geophysical applications," *Geophysics*, vol. 65, no. 4, pp. 1032–1047, 2000.

8

Hydromagnetic Stability of Metallic Nanofluids (Cu-Water and Ag-Water) using Darcy-Brinkman Model

J. Ahuja,[1] U. Gupta,[2] and R. K. Wanchoo[2]

[1]*Energy Research Centre, Panjab University, Chandigarh 160014, India*
[2]*Dr. S. S. Bhatnagar University Institute of Chemical Engineering & Technology, Panjab University, Chandigarh 160014, India*

Correspondence should be addressed to U. Gupta; dr_urvashi_gupta@yahoo.com

Academic Editor: Yun-tai Chen

Thermal convection of a nanofluid layer in the presence of imposed vertical magnetic field saturated by a porous medium is investigated for both-free, rigid-free, and both-rigid boundaries using Darcy-Brinkman model. The effects of Brownian motion and thermophoretic forces due to the presence of nanoparticles and Lorentz's force term due to the presence of magnetic field have been considered in the momentum equations along with Maxwell's equations. Keeping in mind applications of flow through porous medium in geophysics, especially in the study of Earth's core, and the presence of nanoparticles therein, the hydromagnetic stability of a nanofluid layer in porous medium is considered in the present formulation. An analytical investigation is made by applying normal mode technique and Galerkin type weighted residuals method and the stability of Cu-water and Ag-water nanofluids is compared. Mode of heat transfer is through stationary convection without the occurrence of oscillatory motions. Stability of the system gets improved appreciably by raising the Chandrasekhar number as well as Darcy number whereas increase in porosity hastens the onset of instability. Further, stability of the system gets enhanced as we proceed from both-free boundaries to rigid-free and to both-rigid boundaries.

1. Introduction

The concept of nanofluids has improved the heat transfer mechanism by replacing the suspension of micrometer sized particles with nanometer sized particles in conventional fluids. These nanometer sized particles are called nanoparticles which may be metals, metal oxides, carbides, nitrides, or semiconductors. The host liquids may be water, ethylene glycol, propylene glycol, and so forth. The magnificent idea of introducing nanofluids first came into the mind of Choi [1] who claimed the enhanced heat transfer with the addition of nanoparticles. Due to the ultra fine size of nanoparticles, nanofluids have overcome the limitations of micrometer and millimeter sized particles such as settling down in fluid, erosion, and clogging in channel or low thermal conductivity of fluids. Eastman et al. [2] found that dispersion of ultrafine particles in regular fluids improves the physical properties of that fluid. The enhanced physical properties of

nanofluids can be utilized in a vast variety of applications [3, 4]. A well comprehensive model for the enhanced thermal conductivity of nanofluids has been given by Wang et al. [5].

The problem of thermal convection for regular fluids has been discussed in length in a treatise by Chandrasekhar [6]. The problem of thermal convection for nanofluids has been initiated by Kim et al. [7]. Tzou [8] studied the problem analytically and used eigenfunction expansion method to solve the conservation equations given by Buongiorno and he found that critical Rayleigh number is reduced with the addition of nanoparticles. Nield and Kuznetsov [9, 10] investigated the onset of thermal convection in a nanofluid layer for porous/nonporous medium. Kuznetsov and Nield [11] further extended the problem using Darcy-Brinkman model. Effect of rotation on thermal convection has been accounted for by Bhadauria and Agarwal [12] and Chand and Rana [13] in porous/nonporous medium and it was established that addition of Coriolis force term in momentum equation increases the stability of the system. Gupta et al. [14] and Yadav et al. [15] were the first authors for studying convection problem in the presence of imposed magnetic field for bottom heavy and top heavy distribution of nanoparticles, respectively. It was found that magnetic field postpones the onset of thermal convection and the mode of heat transfer is through oscillatory motions for bottom heavy distribution whereas it is through stationary convection for top heavy arrangement of nanoparticles.

The onset of thermal instability of a fluid layer in porous medium has its major application in geophysics particularly in underground reservoirs and in enhanced oil recovery in addition to the usual industrial applications. Nanofluids in porous medium emerge out to have usage in porous foam and microchannel heat sinks which are used for electronic cooling. Applying magnetic field on horizontal layer of nanofluid exhibits some notable features which make it essential to investigate the effects of magnetic field in porous medium. The present paper formulates to present this effect using Darcy-Brinkman model for three types of boundaries: both-free, rigid-free, and both-rigid. The thermal instability problem is analyzed within the framework of normal mode technique and one term weighted residuals method. Lorentz force term is added in the momentum equation in addition to the body and buoyancy forces and an extra viscous term is added in the Darcy equation for consideration of Darcy-Brinkman model and it gives rise to two additional parameters: Darcy number and Chandrasekhar number. Stability of the system gets improved appreciably by raising the applied magnetic field/hence Chandrasekhar number as well as Darcy number whereas increase in porosity hastens the onset of instability. Oscillatory motions are not possible and stability of the system increases appreciably as we proceed from both-free boundaries to rigid-free and to both-rigid boundaries.

2. Formulation of Problem and Conservation Equations in Porous Medium

A horizontal layer of nanofluid of infinite length and thickness d in a homogenous porous medium is considered which is assumed to be at rest initially. Disturbance is caused by heating from beneath the layer so that T_0, T_1 are the temperatures and ϕ_0, ϕ_1 ($\phi_1 > \phi_0$) are the nanoparticle's volume fractions at lower and upper boundaries, respectively (as shown in Figure 1). Porous medium considered here has porosity ε and medium permeability k_1. Gravity force $\mathbf{g} = (0, 0, -g)$ is acting vertically downwards and magnetic field $\mathbf{H} = (0, 0, H)$ is acting in vertically upwards direction. The Darcy velocity is denoted by "$\mathbf{q}_D$" which is related to "$\mathbf{q}$" the nanofluid velocity as $\mathbf{q}_D = \varepsilon\mathbf{q}$.

The conservation equations for mass, nanoparticles, momentum, and thermal energy in porous medium using Brinkman's model in the presence of magnetic field are [6, 9]

$$\nabla \cdot \mathbf{q}_D = 0, \tag{1}$$

$$\frac{\partial \phi}{\partial t} + \frac{\mathbf{q}_D}{\varepsilon} \cdot \nabla\phi = \nabla \cdot \left[D_B\nabla\phi + D_T\frac{\nabla T}{T} \right], \tag{2}$$

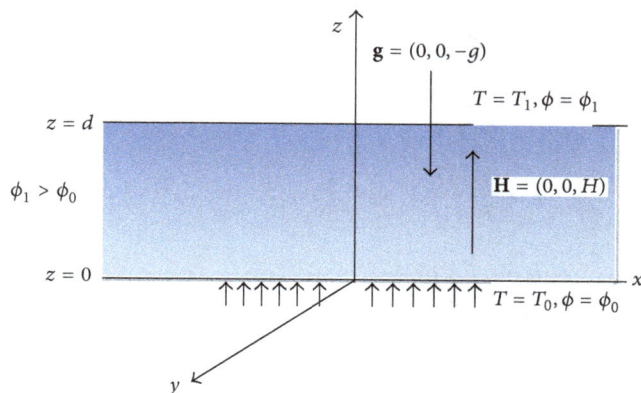

FIGURE 1: Physical system and geometry.

$$\frac{\rho_f}{\varepsilon}\frac{\partial \mathbf{q}_D}{\partial t}$$
$$= -\nabla p + \tilde{\mu}\nabla^2\mathbf{q}_D - \frac{\mu}{k_1}\mathbf{q}_D + \rho\mathbf{g} + \frac{\mu_e}{4\pi}(\nabla \times \mathbf{h}) \times \mathbf{H}, \tag{3}$$

$$(\rho c)_m\frac{\partial T}{\partial t} + (\rho c)_f\,\mathbf{q}_D \cdot \nabla T$$
$$= \left(k_m\nabla^2 T\right) + \varepsilon\,(\rho c)_p\left[D_B\nabla\phi \cdot \nabla T + D_T\frac{\nabla T \cdot \nabla T}{T}\right], \tag{4}$$

$$\frac{d\mathbf{h}}{dt} = (\mathbf{H} \cdot \nabla)\frac{\mathbf{q}_D}{\varepsilon} + \eta\nabla^2\mathbf{h}, \tag{5}$$

$$\nabla \cdot \mathbf{h} = 0, \tag{6}$$

where η is the magnetic diffusivity and ρ is the nanofluid's density which is given by

$$\rho = \phi\rho_p + (1 - \phi)\rho_f$$
$$\cong \phi\rho_p + (1 - \phi)\{\rho(1 - \beta(T - T_0))\}, \tag{7}$$

where $\mathbf{q}_D$, D_B, D_T, ρ_f, ρ_p, $(\rho c)_f$, $(\rho c)_m$, $(\rho c)_p$, ϕ, μ, $\tilde{\mu}$, μ_e, (h_x, h_y, h_z), p, and k_m denote, respectively, the Darcy velocity, the Brownian diffusion coefficient, the thermophoretic diffusion coefficient, the density of the fluid, the density of nanoparticles, the heat capacity of the fluid, the heat capacity of the medium, the heat capacity of the nanoparticles, the nanoparticles volume fraction, the viscosity of the fluid, the effective viscosity, the magnetic permeability, the components of magnetic field, the pressure, and the effective thermal conductivity of the porous medium.

Now the above equations are nondimensionalized using

$$(\tilde{x}, \tilde{y}, \tilde{z}) = \frac{(x, y, z)}{d},$$

$$(\tilde{u}, \tilde{v}, \tilde{w}) = \frac{(u, v, w)\,d}{\alpha_m},$$

$$\tilde{p} = \frac{pk_1}{\mu\alpha_m},$$

$$\tilde{t} = \frac{t\alpha_m}{\sigma d^2},$$

$$\tilde{\phi} = \frac{\phi - \phi_0}{\phi_1 - \phi_0},$$

$$\tilde{T} = \frac{T - T_1}{T_0 - T_1},$$

$$\tilde{h} = \frac{\eta}{H\alpha_m}h,$$

$$\text{where } \alpha_m = \frac{k_m}{(\rho c)_f}, \ \sigma = \frac{(\rho c)_m}{(\rho c)_f}.$$

$$\tag{8}$$

Then the nondimensional forms of (1)–(6) after omitting the symbol ~ are

$$\nabla \cdot \mathbf{q} = 0, \tag{9}$$

$$\frac{\text{Dr}}{\text{Pr}_1}\frac{\partial \mathbf{q}}{\partial t}$$

$$= -\nabla p + \text{Dr}\nabla^2 \mathbf{q} - \mathbf{q} - \text{Rm}\,\hat{e}_z - \text{Rn}\,\phi\hat{e}_z + \text{Ra}\,T\hat{e}_z \tag{10}$$

$$+ Q\left[\left(\frac{\partial h_x}{\partial z} - \frac{\partial h_z}{\partial x}\right)\hat{e}_x - \left(\frac{\partial h_z}{\partial y} - \frac{\partial h_y}{\partial z}\right)\hat{e}_y\right],$$

$$\frac{\partial T}{\partial t} + \mathbf{q}\cdot\nabla T = \nabla^2 T + \frac{N_b}{\text{Le}}\nabla\phi\cdot\nabla T + \frac{N_a N_b}{\text{Le}}\nabla T\cdot\nabla T, \tag{11}$$

$$\frac{1}{\sigma}\frac{\partial\phi}{\partial t} + \frac{1}{\varepsilon}\mathbf{q}\cdot\nabla\phi = \frac{1}{\text{Le}}\nabla^2\phi + \frac{N_a}{\text{Le}}\nabla^2 T, \tag{12}$$

$$\nabla \cdot \mathbf{h} = 0, \tag{13}$$

$$\frac{\varepsilon}{\sigma}\frac{\text{Pr}_2}{\text{Pr}_1}\frac{d\mathbf{h}}{dt} = \left(\frac{\partial u}{\partial z}\hat{e}_x + \frac{\partial v}{\partial z}\hat{e}_y + \frac{\partial w}{\partial z}\hat{e}_z\right) + \varepsilon\nabla^2\mathbf{h}, \tag{14}$$

where

$$\text{Pr}_1 = \frac{\mu}{\rho\alpha_m};$$

$$\text{Pr}_2 = \frac{\mu}{\rho\eta};$$

$$\text{Le} = \frac{\alpha_m}{D_B};$$

$$\text{Ra} = \frac{(\rho g\beta dk_1(T_h - T_c))}{\mu\alpha_m};$$

$$\text{Dr} = \frac{\tilde{\mu}k_1}{\mu d^2};$$

$$\text{Rm} = \left[\phi_0\rho_p + \frac{\rho(1-\phi_0)}{\mu\alpha_m}\right]gdk_1;$$

$$\text{Rn} = \left(\frac{(\rho_p - \rho)(\phi_1 - \phi_0)}{\mu\alpha_m}\right)gdk_1;$$

$$N_a = \frac{(D_T(T_h - T_c))}{(D_B T_c(\phi_1 - \phi_0))};$$

$$N_b = \left[\frac{\varepsilon(\rho c)_p}{(\rho c)_f}\right](\phi_1 - \phi_0),$$

$$Q = \left(\frac{\mu_e H^2 k_1}{4\pi\eta\mu}\right),$$

$$\tag{15}$$

the Prandtl number, the magnetic Prandtl number, the Lewis number, the thermal Rayleigh number, the Darcy number, the basic density Rayleigh number, the concentration Rayleigh number, the modified diffusivity ratio, the modified particle-density increment, and the Chandrasekhar number are the various nondimensional parameters, respectively. It is worthwhile to mention that Boussinesq approximation has been used to linearize the system of equations.

3. Primary Flow and Disturbance Equations

The primary flow is described by the state of the system which is at rest whereas temperature, pressure, and volume fraction of nanoparticles are varying in the vertical direction; that is,

$$\mathbf{q} = 0,$$

$$T = T_p(z),$$

$$p = p_p(z), \tag{16}$$

$$\phi = \phi_p(z),$$

$$\phi_p = -N_a T_p + (1 + N_a)(1 - z),$$

and by putting above equations in (11) we get

$$\frac{d^2 T_p}{dz^2} - \frac{(1+N_a)N_b}{\text{Le}}\frac{dT_p}{dz} = 0. \tag{17}$$

Equation (17) along with the boundary conditions gives the solution as

$$T_p = \frac{1 - e^{-(1+N_a)N_b(1-z)/\text{Le}}}{1 - e^{-(1+N_a)N_b/\text{Le}}}. \tag{18}$$

Using the parametric values of nanofluid parameters (Le ranges from 10^2 to 10^3, N_a is less than 10) and neglecting the terms of second and higher order in the expansion of exponential function, we get the best approximate solution as

$$T_p = 1 - z,$$

$$\phi_p = z. \tag{19}$$

Disturbance on the primary flow is caused by heating from beneath the nanofluid layer and these disturbances are assumed to be small so that

$$\mathbf{q}\,(u, v, w) = 0 + \mathbf{q}'\,\left(u', v', w'\right),$$

$$T = T_p + T',$$

$$p = p_p + p', \tag{20}$$

$$\mathbf{h}\,\left(h_x, h_y, h_z\right) = \mathbf{H} + \mathbf{h}'\,\left(h'_x, h'_y, h'_z\right),$$

$$\phi = \phi_p + \phi',$$

with $T_p = 1 - z$, $\phi_p = z$. By applying these perturbations to (9)–(14), the resulting equations are linearized by using the concept of linear theory. The obtained system of perturbation equations will be analyzed within the framework of normal modes and single term Galerkin method.

4. Normal Mode Analysis

Now the disturbances are examined by using the technique of superposition of basic modes which are described by the pattern

$$(w, T, \phi, h_z, \xi, \zeta)$$

$$= (W\,(z), \Theta\,(z), \Phi\,(z), K\,(z), X\,(z), Z\,(z)) \tag{21}$$

$$\cdot \exp\left(ik_x x + ik_y y + st\right),$$

where k_x and k_y represent the wave numbers in horizontal and vertical directions, respectively, and s represents the growth rate parameter. The system of perturbation equations reduces to

$$\frac{\mathrm{Dr}}{\mathrm{Pr}_1} sZ = \mathrm{Dr}\left(D^2 - \alpha^2\right) - Z + QDX,$$

$$\frac{\mathrm{Dr}}{\mathrm{Pr}_1} s\left(D^2 - \alpha^2\right)W = \mathrm{Dr}\left(D^2 - \alpha^2\right)^2 W$$

$$- \left(D^2 - \alpha^2\right)W + \mathrm{Rn}\,\alpha^2\Phi - \mathrm{Ra}\,\alpha^2\Theta$$

$$+ QD\left(D^2 - \alpha^2\right)K,$$

$$\left\{\frac{\varepsilon\,\mathrm{Pr}_2}{\sigma\,\mathrm{Pr}_1}s - \varepsilon\left(D^2 - \alpha^2\right)\right\}K = DW, \tag{22}$$

$$\left\{\frac{\varepsilon\,\mathrm{Pr}_2}{\sigma\,\mathrm{Pr}_1}s - \varepsilon\left(D^2 - \alpha^2\right)\right\}X = DZ,$$

$$s\Theta - W - \left(D^2 - \frac{N_b}{\mathrm{Le}}D - 2\frac{N_a N_b}{\mathrm{Le}}D - \alpha^2\right)\Theta$$

$$+ \frac{N_b}{\mathrm{Le}}D\Phi = 0,$$

$$\frac{1}{\sigma}s\Phi - \frac{W}{\varepsilon} - \frac{N_a}{\mathrm{Le}}\left(D^2 - \alpha^2\right)\Theta - \frac{1}{\mathrm{Le}}\left(D^2 - \alpha^2\right)\Phi = 0.$$

After the process of elimination, the above set of equations reduce to

$$\left[\frac{\mathrm{Dr}}{\mathrm{Pr}_1}s\left(D^2 - \alpha^2\right) - \mathrm{Dr}\left(D^2 - \alpha^2\right)^2 + \left(D^2 - \alpha^2\right)\right]$$

$$\cdot\left[\varepsilon\left(D^2 - \alpha^2\right) - \frac{\varepsilon\,\mathrm{Pr}_2}{\sigma\,\mathrm{Pr}_1}s\right]W$$

$$+ \left[\varepsilon\left(D^2 - \alpha^2\right) - \frac{\varepsilon\,\mathrm{Pr}_2}{\sigma\,\mathrm{Pr}_1}s\right]\left[\mathrm{Ra}\,\alpha^2\Theta - \mathrm{Rn}\,\alpha^2\Phi\right]$$

$$+ Q\left(D^2 - \alpha^2\right)D^2 W = 0, \tag{23}$$

$$s\Theta - W = \left(D^2 - \alpha^2\right)\Theta + \frac{N_b}{\mathrm{Le}}\left(D\Theta - D\Phi\right) - 2\frac{N_a N_b}{\mathrm{Le}}$$

$$\cdot D\Theta,$$

$$\frac{1}{\sigma}s\Phi + \frac{W}{\varepsilon} = \frac{N_a}{\mathrm{Le}}\left(D^2 - \alpha^2\right)\Theta + \frac{1}{\mathrm{Le}}\left(D^2 - \alpha^2\right)\Phi$$

$$= 0,$$

where $D \equiv d/dz$ and $\alpha = (k_x^2 + k_y^2)^{1/2}$ is the resultant wave number in nondimensional form. Now, single term approximation of Galerkin weighted residuals approach is adopted to solve (23). In this approach, choice of trial functions W_p, Θ_p, and Φ_p depends on the relevant boundary conditions and we write

$$W = \sum_{p=1}^{N} A_p W_p,$$

$$\Theta = \sum_{p=1}^{N} B_p \Theta_p, \tag{24}$$

$$\Phi = \sum_{p=1}^{N} C_p \Phi_p.$$

For one term approximation we put $p = 1$. Let us substitute (24) into (23) and make use of orthogonality to the trial functions; we obtain a system of three equations in three unknowns A_1, B_1, C_1. Elimination of these unknowns from the obtained set of equations gives the eigenvalue equation.

5. Results and Discussions

5.1. Free-Free Boundaries and Special Cases. The boundary conditions on both-free boundary surfaces are

$$W = 0,$$

$$D^2 W = 0,$$

$$\Phi = 0, \tag{25}$$

$$\Theta = 0,$$

at the lower and upper boundaries;

$$K = 0,$$

on the boundaries [6]. Trial functions satisfying the above boundary conditions can be chosen as $W = A_1 \sin \pi z$, $\Theta = B_1 \sin \pi z$, and $\Phi = C_1 \sin \pi z$. Let us substitute this solution in (23) and follow the process of integration and elimination of A_1, B_1, C_1 from the obtained set of equations; the eigenvalue equation becomes

$$\left[-\frac{\mathrm{Dr}}{\mathrm{Pr}_1} sJ - \mathrm{Dr}J^2 - J \right] \left[\varepsilon J + \frac{\varepsilon \, \mathrm{Pr}_2}{\sigma \, \mathrm{Pr}_1} s \right] (s + J) \left(\frac{s}{\sigma} + \frac{J}{\mathrm{Le}} \right)$$

$$+ \left[\varepsilon J + \frac{\varepsilon \, \mathrm{Pr}_2}{\sigma \, \mathrm{Pr}_1} s \right]$$

$$\cdot \left[\mathrm{Ra} \, \alpha^2 \left(\frac{s}{\sigma} + \frac{J}{\mathrm{Le}} \right) + \mathrm{Rn} \, \alpha^2 \left(\frac{s+J}{\varepsilon} + \frac{JN_a}{\mathrm{Le}} \right) \Phi \right] \tag{26}$$

$$+ Q\pi^2 J (s + J) \left(\frac{s}{\sigma} + \frac{J}{\mathrm{Le}} \right) = 0.$$

At the state of marginal stability, when the amplitudes of small disturbances grow or damp aperiodically then the transition from stability to instability takes place via a stationary pattern of motions which is described by $s = i\omega = 0$. Then the eigenvalue equation (26) reduces to

$$\mathrm{Ra}^{\mathrm{stat}} = \frac{1}{\alpha^2} \left[\mathrm{Dr} \left(\pi^2 + \alpha^2 \right)^3 + \left(\pi^2 + \alpha^2 \right)^2 \right.$$

$$\left. + \frac{Q\pi^2}{\varepsilon} \left(\pi^2 + \alpha^2 \right) \right] - \mathrm{Rn} \left[\frac{\mathrm{Le}}{\varepsilon} + N_a \right], \tag{27}$$

$$\mathrm{Ra}^{\mathrm{stat}} + \mathrm{Rn} \left(\frac{\mathrm{Le}}{\varepsilon} + N_a \right) = \pi^2 \frac{(1 + x)}{x} \left[\mathrm{Dr} \, \pi^2 (1 + x)^2 \right.$$

$$\left. + \pi^2 (1 + x) + \frac{Q}{\varepsilon} \right], \tag{28}$$

$$\mathrm{Ro} = \pi^2 \frac{(1 + x)}{x} \left[\mathrm{Dr} \, \pi^2 (1 + x)^2 + \pi^2 (1 + x) + \frac{Q}{\varepsilon} \right], \tag{29}$$

where $\mathrm{Ro} = \mathrm{Ra}^{\mathrm{stat}} + \mathrm{Rn} \left(\frac{\mathrm{Le}}{\varepsilon} + N_a \right)$, $\alpha^2 = \pi^2 x$.

Case 1. When $(\mathrm{Rn} = 0, Q = 0)$, (28) turns out to be

$$\mathrm{Ra}^{\mathrm{stat}} = \pi^2 \frac{(1 + x)}{x} \left[\mathrm{Dr} \, \pi^2 (1 + x)^2 + \pi^2 (1 + x) \right], \tag{30}$$

which confirms that our result coincides with the Brinkman model of Kuznetsov and Nield [11]. The values of thermal Rayleigh number and wave number at which instability sets in (critical values) are found by taking $(d\mathrm{Ra}/dx)_{x=x_c} = 0$.

Case 2. When $\mathrm{Dr} = 0, Q \neq 0$, (29) reduces to

$$\mathrm{Ro} = \pi^2 \frac{(1 + x)}{x} \left[\pi^2 (1 + x) + \frac{Q}{\varepsilon} \right]. \tag{31}$$

As a function of x, (31) gets its lowest value when

$$x^2 = \frac{Q}{\varepsilon} + 1. \tag{32}$$

For $(Q = 0)$,

$$x_c = 1 \, (\alpha_c = \pi), \tag{33}$$

$$\mathrm{Ra}_c = 4\pi^2 - \mathrm{Rn} \left(\frac{\mathrm{Le}}{\varepsilon} + N_a \right),$$

$$\mathrm{Ro}_c = \mathrm{Ra} + \mathrm{Rn} \left(\frac{\mathrm{Le}}{\varepsilon} + N_a \right) = 4\pi^2, \tag{34}$$

which is in confirmation with the outcome of Horton-Rogers Lapwood model of Nield and Kuznetsov [9]. By omitting the nanoparticle's term and magnetic field term we get the critical Rayleigh number as $\mathrm{Ra}_c = 4\pi^2$.

Case 3. When $\mathrm{Dr} \to \infty, Q \neq 0$, (28) reduces to

$$\frac{1}{\mathrm{Dr}} \left(\mathrm{Ra}^{\mathrm{stat}} + \mathrm{Rn} \left(\frac{\mathrm{Le}}{\varepsilon} + N_a \right) \right)$$

$$= \pi^2 \frac{(1 + x)}{x} \left[\pi^2 (1 + x)^2 + \frac{1}{\mathrm{Dr}} \pi^2 (1 + x) + \frac{1}{\mathrm{Dr}} \frac{Q}{\varepsilon} \right]. \tag{35}$$

The right-hand side attains its minimum value at $\alpha = \pi/\sqrt{2}$ and its minimum value is given by

$$\frac{1}{\mathrm{Dr}} \left(\mathrm{Ra}^{\mathrm{stat}} + \mathrm{Rn} \left(\frac{\mathrm{Le}}{\varepsilon} + N_a \right) \right) = \frac{27}{4} \pi^4, \tag{36}$$

$$\mathrm{Ra}^{\mathrm{stat}} = \frac{27}{4} \pi^4 \mathrm{Dr} - \mathrm{Rn} \left(\frac{\mathrm{Le}}{\varepsilon} + N_a \right). \tag{37}$$

From (34) it is clear that in the absence of nanoparticles and magnetic field critical value of thermal Rayleigh number is $\mathrm{Ra}_c = 4\pi^2$ for the case when Darcy number vanishes ($\mathrm{Dr} = 0$). The value of critical Rayleigh number for the case when Dr tends to infinity is $\mathrm{Ra}_c = 657.5$ (from (37)) which coincides with its value for the regular fluid. Thus when Dr is large, the nanofluid behaves like a regular fluid. It is clear from (27) that the suspension of nanoparticles in conventional fluids lowers the critical value of Rayleigh number as all the parameters $\mathrm{Rn}, \mathrm{Le}, N_a$ are positive for the present configuration of nanoparticles and the expression $\mathrm{Rn}(\mathrm{Le}/\varepsilon + N_a)$ appears with negative sign. Thus the system with the distribution of nanoparticles at the top of the fluid layer is less stable as compared to regular fluid and bottom heavy distribution of nanoparticles.

5.2. Rigid-Free Boundaries. Let us now consider the lower boundary of the fluid to be a rigid surface while the upper boundary surface is free. Then the relevant conditions on the boundary surfaces for W are

$$W = 0,$$

$$DW = 0,$$

at lower boundary,

$$W = 0,$$ (38)

$$D^2W = 0,$$

at upper boundary.

The trial function appropriate to these boundaries is

$$W_1 = z^2 (1 - z)(3 - 2z),$$

$$\Theta_1 = z(1 - z),$$ (39)

$$\Phi_1 = z(1 - z).$$

By making use of single term Galerkin method and orthogonality, the eigenvalue equation in the present case becomes

$$\frac{420}{390}\varepsilon\left[\left(\frac{Dr}{Pr_1}s + 1\right)\left(\frac{36}{5} + \frac{24}{35}\alpha^2 + \frac{19}{630}\alpha^4\right)\right.$$

$$\left.+ Dr\left(\frac{108}{5}\alpha^2 + \frac{36}{35}\alpha^4\frac{19}{630}\alpha^6\right)\right]\left[\frac{s}{\sigma} + \frac{10 + \alpha^2}{Le}\right]$$

$$\cdot\left(10 + \alpha^2 + s\right) + \frac{420}{390}\frac{\varepsilon}{\sigma}\frac{Pr_2}{Pr_1}$$

$$\cdot s\left[\left(\frac{Dr}{Pr_1}s + 1\right)\left(\frac{12}{35} + \frac{19}{630}\alpha^2\right)\right.$$

$$\left.+ Dr\left(\frac{36}{5} + \frac{24}{35}\alpha^2 + \frac{19}{630}\alpha^4\right)\right]\left[\frac{s}{\sigma} + \frac{10 + \alpha^2}{Le}\right](10$$
(40)
$$+ \alpha^2 + s) + Q\frac{420}{390}\left[\frac{36}{5} + \frac{12}{35}\alpha^2\right]\left[\frac{s}{\sigma} + \frac{10 + \alpha^2}{Le}\right]$$

$$\cdot\left(10 + \alpha^2 + s\right) - \varepsilon\left[\left(\frac{3}{10} + \frac{13}{420}\alpha^2\right)\right.$$

$$\left.+ \frac{\varepsilon}{\sigma}\frac{Pr_2}{Pr_1}s\frac{13}{420}\right]Ra\,\alpha^2\left[\frac{s}{\sigma} + \frac{10 + \alpha^2}{Le}\right]$$

$$-\varepsilon\left[\left(\frac{3}{10} + \frac{13}{420}\alpha^2\right) + \frac{\varepsilon}{\sigma}\frac{Pr_2}{Pr_1}s\frac{13}{420}\right]Rn\,\alpha^2\left[10\right.$$

$$\left.+ \alpha^2\left(\frac{1}{\varepsilon} + \frac{N_a}{Le}\right) + \frac{s}{\varepsilon}\right] = 0.$$

For stationary convection, the eigenvalue equation (40) takes the form

$$Ra^{stat} + Rn\left[\frac{Le}{\varepsilon} + N_a\right]$$

$$= \frac{28}{13\alpha^2}\left[\frac{1}{3}\left\{\left(4536 + 432\alpha^2 + 19\alpha^4\right)\right.\right.$$
(41)
$$\left.\left.+ Dr\left(13608\alpha^2 + 648\alpha^4 + 19\alpha^6\right)\right\} + \frac{72Q}{\varepsilon}\left(21\right.\right.$$

$$\left.\left.+ \alpha^2\right)\right]\left(\frac{10 + \alpha^2}{126 + 13\alpha^2}\right).$$

When $Dr = 0$ and $Q = 0$, (41) attains its minimum value as $Ro_c = 56.97$ at $\alpha_c = 3.96$. In the presence of magnetic field, that is, for $Dr = 0$, $Q = 50$, and $\varepsilon = 0.4$, the critical value increases appreciably as is given by $Ro_c = 1882.11$ at $\alpha_c = 13.22$. Thus magnetic field has a strong stabilizing effect for the case of rigid-free boundaries. When Dr is large, comparable to unity, that is, $(Dr = 1, Q = 50, \varepsilon = 0.4)$, critical Rayleigh number is given by $Ro_c = 998.823$ at $\alpha_c = 1.581$, while, in the presence of magnetic field $(Dr = 1, Q = 50, \varepsilon = 0.4)$, it increases significantly; that is, $Ro_c = 5052.82$ at $\alpha_c = 4.493$. This shows that magnetic field and Darcy number both inhibit the onset of convection and contribute largely towards the stability of the system.

5.3. *Rigid-Rigid Boundaries.* Let us now consider that both boundaries of the fluid layer are rigid. Therefore, vanishing of normal and horizontal components of velocity to the rigid surface (no slip condition) leads to the following conditions on boundaries:

$$W = 0,$$

$$DW = 0,$$

$$\Theta = 0$$

$$\Phi = 0,$$ (42)

at lower and upper boundary;

$$K = 0,$$

$$DK = 0,$$

for perfectly conducting boundaries. The suitable trial functions satisfying these boundary conditions are

$$W_1 = z^2 (1 - z)^2,$$

$$\Phi_1 = z(1 - z),$$ (43)

$$\Theta_1 = z(1 - z).$$

Using the approximation of one term Galerkin method and orthogonality, the eigenvalue equation becomes

$$\frac{14}{3}\varepsilon\left[\left(\frac{Dr}{Pr_1}s+1\right)\left(\frac{24}{30}+\frac{4}{105}\alpha^2+\frac{1}{630}\alpha^4\right)\right.$$

$$+Dr\left(\frac{72}{30}\alpha^2+\frac{6}{105}\alpha^4+\frac{1}{630}\alpha^6\right)\right]\left[\frac{s}{\sigma}+\frac{10+\alpha^2}{Le}\right]$$

$$\cdot\left(10+\alpha^2+s\right)+\frac{14}{3}\frac{\varepsilon}{\sigma}\frac{Pr_2}{Pr_1}$$

$$\cdot s\left[\left(\frac{Dr}{Pr_1}s+1\right)\left(\frac{2}{105}+\frac{1}{630}\alpha^2\right)\right.$$

$$+Dr\left(\frac{24}{30}+\frac{4}{105}\alpha^2+\frac{1}{630}\alpha^4\right)\right]\left[\frac{s}{\sigma}+\frac{10+\alpha^2}{Le}\right]$$

$$\cdot\left(10+\alpha^2+s\right)+Q\frac{14}{3}\left[\frac{24}{30}+\frac{2}{105}\alpha^2\right]\left[\frac{s}{\sigma}\right] \tag{44}$$

$$+\frac{10+\alpha^2}{Le}\right]\left(10+\alpha^2+s\right)-\varepsilon\left[\left(\frac{2}{30}+\frac{1}{140}\alpha^2\right)\right.$$

$$+\frac{\varepsilon}{\sigma}\frac{Pr_2}{Pr_1}\frac{s}{140}\right]Ra\,\alpha^2\left[\frac{s}{\sigma}+\frac{10+\alpha^2}{Le}\right]$$

$$-\varepsilon\left[\left(\frac{2}{30}+\frac{1}{140}\alpha^2\right)+\frac{\varepsilon}{\sigma}\frac{Pr_2}{Pr_1}\frac{s}{140}\right]Rn\,\alpha^2\left[10\right.$$

$$+\alpha^2\left(\frac{1}{\varepsilon}+\frac{N_a}{Le}\right)+\frac{s}{\varepsilon}\right]=0.$$

Let us discuss the case of stationary convection by putting $s = 0$ in (44). The expression for Rayleigh number for both-rigid boundaries becomes

$$Ra^{stat}+Rn\left[\frac{Le}{\varepsilon}+N_a\right]$$

$$=\frac{28}{3\alpha^2}\left[\frac{1}{3}\left\{\left(504+24\alpha^2+\alpha^4\right)\right.\right.$$

$$+Dr\left(1512+36\alpha^2+\alpha^4\right)\alpha^2\right\}+\frac{4Q}{\varepsilon}\left(42+\alpha^2\right)\right] \tag{45}$$

$$\cdot\left(\frac{10+\alpha^2}{28+3\alpha^2}\right).$$

When $Dr = 0$ and $Q = 0$, (45) attains its minimum value as $Ro_c = 72.94$ at $\alpha_c = 4.791$. In the presence of magnetic field, that is, for $Dr = 0, Q = 50, \varepsilon = 0.4$, the critical value increases appreciably as is given by $Ro_c = 2108.45$ at $\alpha_c = 15.95$. Thus magnetic field predominantly stabilizes the nanofluid layer system for the case of rigid-rigid boundaries. When Dr is large, comparable to unity, $(Dr = 1, Q = 50, \varepsilon = 0.4)$, critical Rayleigh number is given by $Ro_c = 1986.99$ at $\alpha_c = 1.92$ while, in the presence of magnetic field $(Dr = 1, Q = 50, \varepsilon = 0.4)$, it increases significantly, that is, $Ro_c = 7514.95$ at $\alpha_c = 5.19$. Thus, both Chandrasekhar number and Darcy number contribute significantly towards the stability of the system and this stability increases as we move from both-free boundaries to rigid-free boundaries and then to both-rigid boundaries.

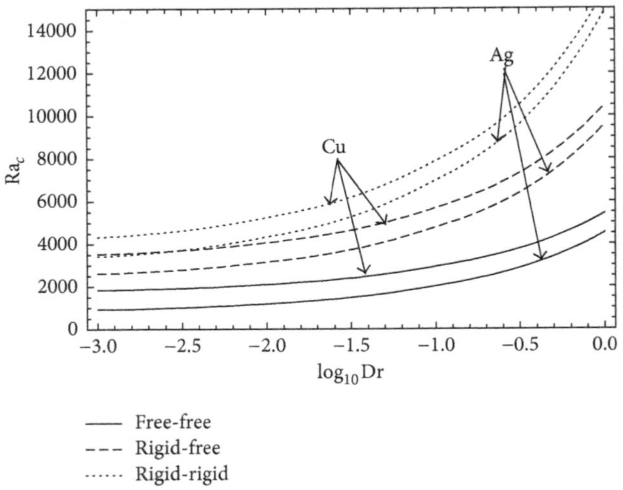

— Free-free
--- Rigid-free
..... Rigid-rigid

FIGURE 2: Plot of Ra_c with Darcy number for fixed $Q = 250$ and $\varepsilon = 0.4$.

6. Numerical Results and Discussion

Let us now consider the numerical/graphical investigation of the problem by considering numerical values of various parameters under consideration $(Q = 250, \varepsilon = 0.4, Dr = 0.01)$ for rigid-rigid, rigid-free, and free-free boundaries. To carry out computations, (27), (41), and (45) are used for both-free, rigid-free, and both-rigid boundaries, respectively. The values of nanofluid parameters for $\Delta\phi = 0.001$ are $Rn = 0.392, N_a = 0.5$, and $Le = 5000$ for Cu-water nanofluid and $Rn = 0.465, N_a = 0.5$, and $Le = 5000$ for Ag-water nanofluid. Figures 2–7 show the graphical results for stationary convection for Cu-water nanofluid and Ag-water nanofluid in which concentration of nanoparticles at the upper boundary is more than that at the lower boundary.

Figures 2-3 are plot of Ra_c and α_c with the variation in Darcy number for Cu-water and Ag-water nanofluids for three different boundaries. Clearly, Ra_c increases and α_c decreases with the increase in Dr for all the three boundaries. Further, the curves for rigid-free boundaries lie between the curves for both-rigid and both-free boundaries which confirm the earlier result that rigid-rigid boundaries exhibit higher stability than rigid-free and both-free boundaries. Also, the curves showing the effect of Darcy number for Ag-water nanofluid lie below the curves for Cu-water nanofluid which means that Cu-water nanofluid exhibits higher stability as compared to Ag-water nanofluid in the present configuration.

Figures 4-5 correspond to the variation of Ra_c and α_c with the variation in Chandrasekhar number for fixed values of $Dr = 0.01, \varepsilon = 0.4$ and Figures 6-7 show the variation of porosity for fixed $Dr = 0.01, Q = 250$ for both types of nanofluids. The values of Ra_c as well as α_c increase appreciably with the increase in Chandrasekhar number and decrease significantly with the rise in porosity. Thus the applied magnetic field and porosity have opposing effects of stabilizing/destabilizing the fluid layer system. Further, (27), (41), and (45) show that thermal Rayleigh numbers

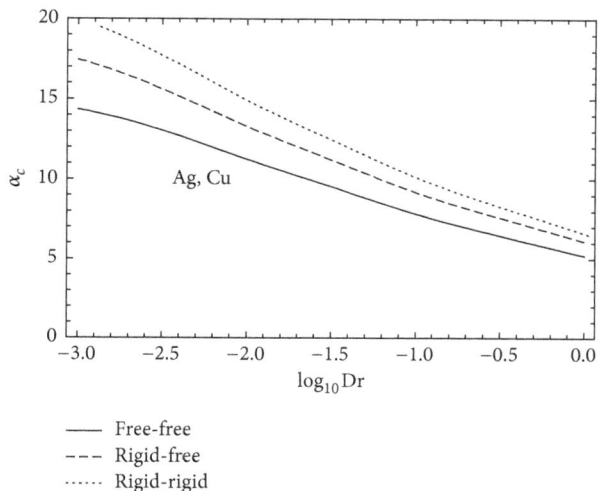

FIGURE 3: Plot of α_c with Darcy number for fixed $Q = 250$ and $\varepsilon = 0.4$.

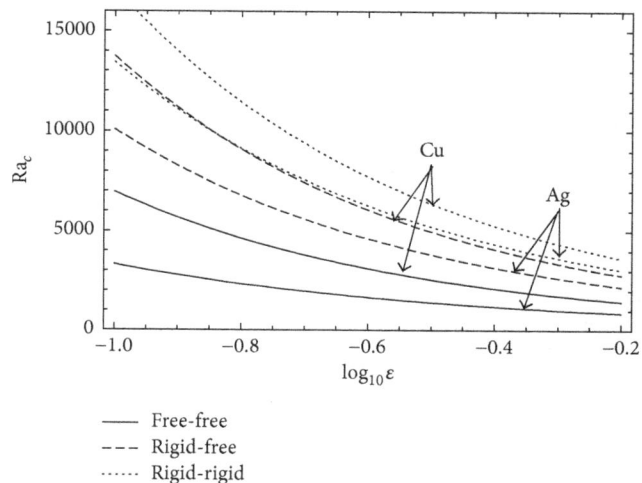

FIGURE 4: Plot of Ra_c with Chandrasekhar number for fixed $Dr = 0.01$ and $\varepsilon = 0.4$.

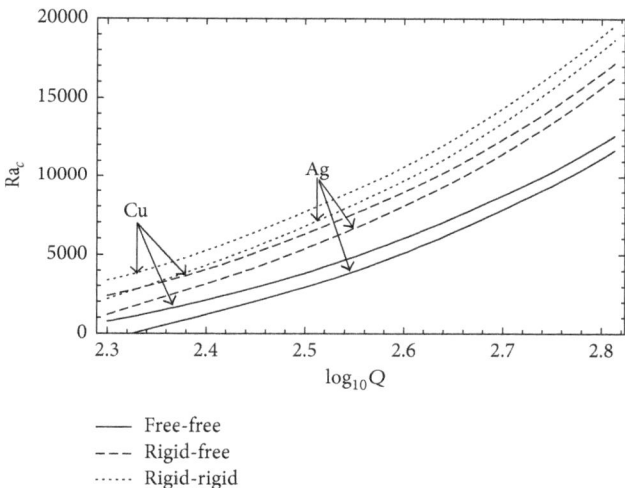

FIGURE 5: Plot of α_c with Chandrasekhar number for fixed $Dr = 0.01$ and $\varepsilon = 0.4$.

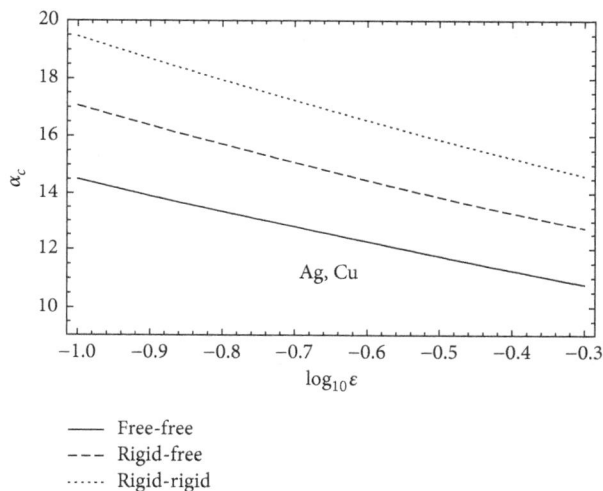

FIGURE 6: Plot of Ra_c with porosity for fixed $Dr = 0.01$ and $Q = 250$.

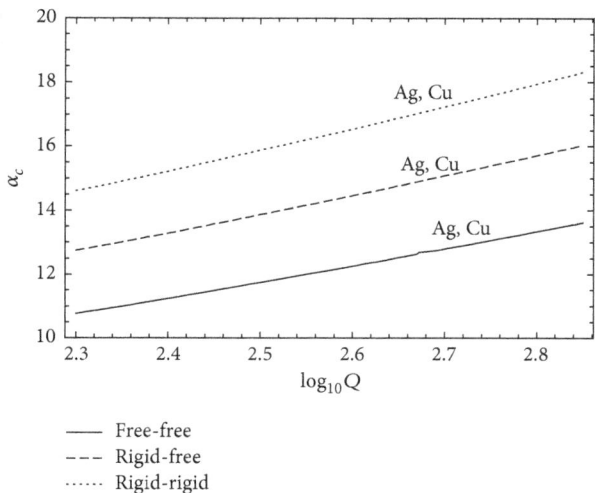

FIGURE 7: Plot of α_c with porosity for fixed $Dr = 0.01$ and $Q = 250$.

for stationary convection are independent of both Prandtl number and magnetic Prandtl number. Further increase in concentration of nanoparticles makes the system unstable by decreasing the critical value of thermal Rayleigh number. It is due to the fact that the parameters Rn, Le, N_a are all positive and the expression $Rn(Le/\varepsilon + N_a)$ appears with negative sign. However this can well be compensated by increasing the magnetic field and Darcy number. Further the stability of Cu-water nanofluid is higher than that of Ag-water nanofluid in the presence of magnetic field in porous medium.

7. Conclusions

Keeping in mind the application of flow through porous medium in the presence of magnetic field in geophysics particularly in underground reservoirs, enhanced oil recovery, soil sciences and in hydrology; we have investigated the impact of vertical magnetic field on a nanofluid layer using Darcy-Brinkman model for three different boundaries.

The presence of nanoparticles is an essential feature of these processes which introduces two additional effects: Brownian motion and thermophoretic forces. The analysis is carried out within the framework of linear stability theory, normal mode analysis, and single term Galerkin approximation. It is found that the instability sets in through the mode of stationary convection instead of oscillatory motions. The condition for the occurrence of oscillatory motions is that the two buoyancy forces (density gradient of nanoparticles and density variation due to heating from the bottom) must act in opposite directions. The value of critical Rayleigh number is decreased to an appreciable extent due to the presence of nanoparticles on the top whereas it increases for the bottom heavy configuration of nanoparticles. This is due to the fact that the top heavy arrangement works in tandem together with the known fact of thermal conductivity enhancement. While for bottom heavy arrangement of nanoparticles, buoyancy forces play antagonistic roles which results in generating oscillatory motions. It is this aspect that triggers the reversing trend in the two modes of configuration. From (28), it is clear that the suspension of nanoparticles in conventional fluids lowers the critical value of Rayleigh number as all the parameters Rn, Le, N_a are positive for the present configuration of nanoparticles and the expression Rn(Le/ε + N_a) appears with negative sign. Thus the system with the distribution of nanoparticles at the top of the fluid layer is less stable as compared to regular fluid and bottom heavy distribution of nanoparticles. It is figured out that the mode of stationary convection is independent of both; the Prandtl number and magnetic Prandtl number. It has been found that the critical values of wave number and Rayleigh number (α_c, Ra$_c$) exhibit a significant rise/fall with the rise in magnetic field parameter Q/porosity ε. Thus magnetic field is found to delay the onset of convection while porosity advances the same. With the increase in Darcy number, Ro$_c$ increases while α_c decreases. It seems that the heat transfer characteristic of the nanofluid will get enhanced with the increase in Darcy number and magnetic field. Further, as one shifts from both-free boundaries to both-rigid boundaries, α_c shows an appreciable increase along with moderate increase in Ro$_c$. Thus, the system with both-rigid boundaries is found to have more stability as compared to rigid-free boundaries which in turn are more stable than free-free boundaries. Also, Cu-water nanofluid exhibits higher stability than Ag-water nanofluid in the present configuration of the system.

Competing Interests

The authors declare that they have no competing interests.

Acknowledgments

The financial assistance in the form of Research Project (Ref. no. 25(0247)/15/EMR-II) of Council of Scientific and Industrial Research, New Delhi, India, and Senior Research Fellowship (Serial no. 2061040991, Ref. no. 20-06-2010(i) EU-IV) of University Grants Commission, New Delhi, India, is gratefully acknowledged by the authors.

References

[1] S. Choi, "Enhancing thermal conductivity of fluids with nanoparticles," in *Development and Applications of Non-Newtonian Flows*, D. A. Siginer and H. P. Wang, Eds., vol. 66 of *FED-231/MD*, pp. 99–105, ASME, New York, NY, USA, 1995.

[2] J. A. Eastman, S. U. S. Choi, S. Li, W. Yu, and L. J. Thompson, "Anomalously increased effective thermal conductivities of ethylene glycol-based nanofluids containing copper nanoparticles," *Applied Physics Letters*, vol. 78, article 718, 2001.

[3] S. Ozturk, Y. A. Hassan, and V. M. Ugaz, "Interfacial complexation explains anomalous diffusion in nanofluids," *Nano Letters*, vol. 10, no. 2, pp. 665–671, 2010.

[4] J. A. Lewis, "Colloidal processing of ceramics," *Journal of the American Ceramic Society*, vol. 83, no. 10, pp. 2341–2359, 2000.

[5] W. Wang, L. Lin, Z. X. Feng, and S. Y. Wang, "A comprehensive model for the enhanced thermal conductivity of nanofluids," *Journal of Advanced Research in Physics*, vol. 3, no. 2, Article ID 021209, 2012.

[6] S. Chandrasekhar, *Hydrodynamic and Hydromagnetic Stability*, Dover, New York, NY, USA, 1981.

[7] J. Kim, Y. T. Kang, and C.-K. Choi, "Analysis of convective instability and heat transfer characteristics of nanofluids," *Physics of Fluids*, vol. 16, no. 7, pp. 2395–2401, 2004.

[8] D. Y. Tzou, "Instability of nanofluids in natural convection," *ASME Journal of Heat Transfer*, vol. 130, no. 7, Article ID 072401, 9 pages, 2008.

[9] D. A. Nield and A. V. Kuznetsov, "Thermal instability in a porous medium layer saturated by a nanofluid," *International Journal of Heat and Mass Transfer*, vol. 52, no. 25-26, pp. 5796–5801, 2009.

[10] D. A. Nield and A. V. Kuznetsov, "The onset of convection in a horizontal nanofluid layer of finite depth," *European Journal of Mechanics—B/Fluids*, vol. 29, no. 3, pp. 217–223, 2010.

[11] A. V. Kuznetsov and D. A. Nield, "Thermal instability in a porous medium layer saturated by a nanofluid: Brinkman model," *Transport in Porous Media*, vol. 81, no. 3, pp. 409–422, 2010.

[12] B. S. Bhadauria and S. Agarwal, "Natural convection in a nanofluid saturated rotating porous layer: a nonlinear study," *Transport in Porous Media*, vol. 87, no. 2, pp. 585–602, 2011.

[13] R. Chand and G. C. Rana, "On the onset of thermal convection in rotating nanofluid layer saturating a Darcy-Brinkman porous medium," *International Journal of Heat and Mass Transfer*, vol. 55, no. 21-22, pp. 5417–5424, 2012.

[14] U. Gupta, J. Ahuja, and R. K. Wanchoo, "Magneto convection in a nanofluid layer," *International Journal of Heat and Mass Transfer*, vol. 64, pp. 1163–1171, 2013.

[15] D. Yadav, R. Bhargava, and G. S. Agrawal, "Thermal instability in a nanofluid layer with a vertical magnetic field," *Journal of Engineering Mathematics*, vol. 80, no. 1, pp. 147–164, 2013.

9

Geomorphology Characterization of Ica Basin and its Influence on the Dynamic Response of Soils for Urban Seismic Hazards in Ica, Peru

Isabel Bernal ⓘ, Hernando Tavera, Wilfredo Sulla, Luz Arredondo, and Javier Oyola

Instituto Geofísico del Perú (IGP), Lima, Peru

Correspondence should be addressed to Isabel Bernal; ybernal@igp.gob.pe

Academic Editor: Rudolf A. Treumann

We evaluated the influence of the geomorphology of Peru's Ica Basin on the dynamic response of soils of the city of Ica. We applied five geophysical methods: spectral ratio (H/V), frequency-wavenumber (F-K), multichannel analysis of surface waves (MASW), multichannel analysis of microtremor (MAM), and Gravimetric Analysis. Our results indicate that the soils respond to two frequency ranges: $F0$ (0.4–0.8 Hz) and $F1$ (1.0–3.0 Hz). The F-K, which considers circular arrays, shows two tendencies with a jump between 1.0 and 2.0 Hz. MASW and MAM contribute to frequencies greater than 2.0 Hz. The inversion curve indicates the presence of three layers of 4, 16, and 60 m with velocities of 180, 250, and 400 m/s. The Bouguer anomalies vary between −17.72 and −24.32 mGal and with the spectral analysis we identified two deposits, of 60 m and 150 m of thickness. Likewise, the relationship between the velocities of 400 and 900 m/s, with the frequency = 1.5 Hz, allows us to determine the thickness for the layers of 60 (slightly alluvial to moderately compact) and 150 m (soil-rock interface). These results suggest that the morphology of the Ica Basin plays an important role in the dynamic behavior of the soils to low frequency.

1. Introduction

Ica Basin (IB) is a depression located in western central Peru between the Coastal and Western Andean mountains (Figure 1). In lower basin is located the urban area of Ica city. The main geodynamic events affecting Ica are earthquakes, inundations, debris flows, rock falls, and sandy eolic deposit.

Ica has been severely damaged by earthquakes, such as the quakes in 1942 (7.8 Mw) and 1996 (7.6 Mw) and most recently the 8.0 Mw Pisco earthquake of 2007. The Pisco earthquake generated maximum intensities of VII-VIII on the Modified Mercalli Intensity Scale within a 250 km radius, including the cities Pisco, Ica, and Chincha. This was one of the largest earthquake of the last 300 years [1] and showed particular characteristics such as its duration (120 s) and a complex rupture process that induced a local tsunami. The most significant structural damage was observed in adobe and "quincha" houses, which resulted in more than 590 fatalities and 320 injuries [2]. The structural damage observed in more than 12 villages around Ica, Lima, and Huancavelica was mainly

associated with local site effects (i.e., soil liquefaction along the coastline and in weakly consolidated soils), the age of structures, and landslides on the roads [3]. The study area is located on thick alluvial deposits composed of pebbles and small blocks embedded in a silty sand matrix [4]. This soil type and quality will contribute to generating damage on the surface when earthquake occurs; therefore, it is necessary to determine the sedimentary basin's geometry to better understand its dynamic behavior [5]. Understanding this depends on the soil and basin's physical and geomechanical properties (e.g., stratigraphy, lithology, layer thickness, and basal rock), because they control propagation velocity of shear waves (Vs).

In this study, we evaluate the influence of the geomorphology characterization of the Ica Basin and the dynamic response of soils for urban seismic hazards in Ica. To estimate these characteristics, we applied five geophysical methods: spectral ratio (H/V), frequency-wavenumber (F-K), multichannel analysis of surface waves (MASW), multichannel analysis of microtremor (MAM), and gravimetric method. We then used these results to know the seismic and

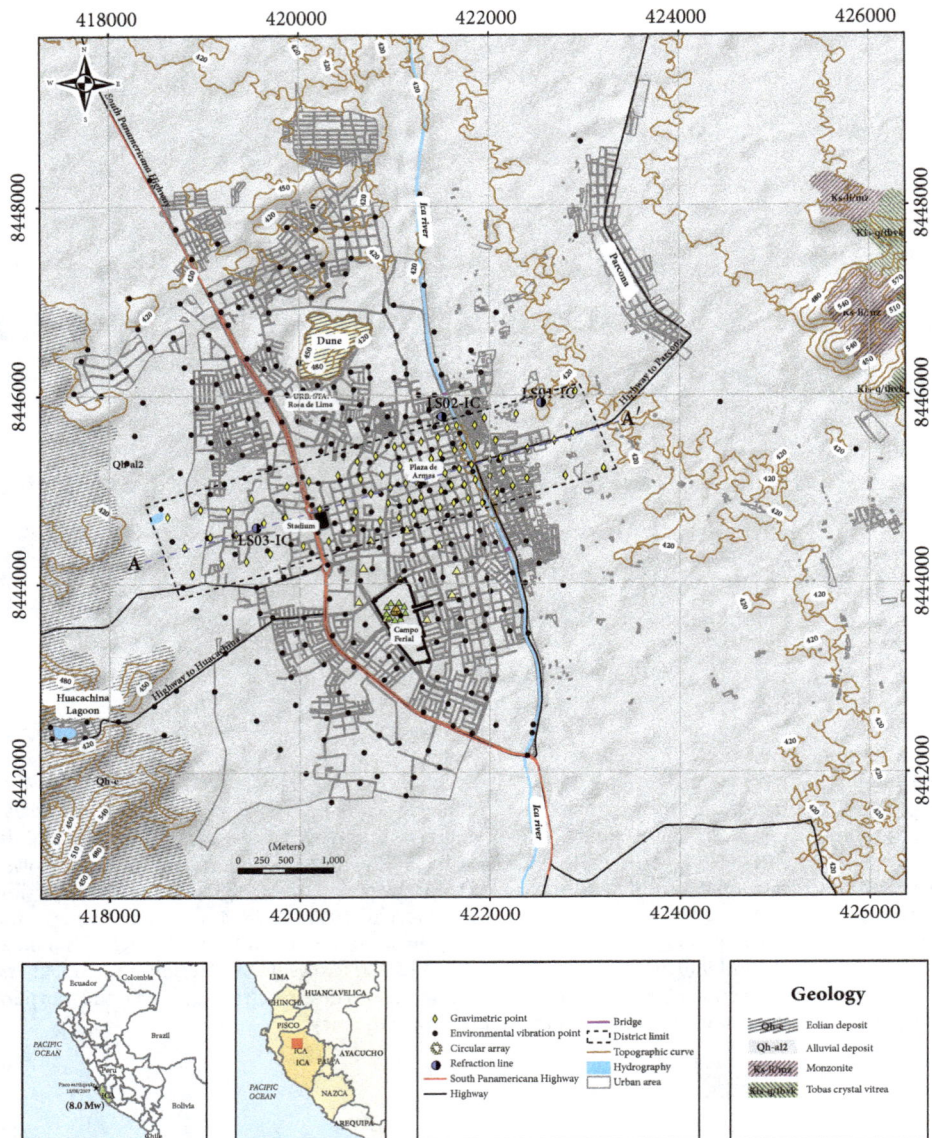

FIGURE 1: Geological setting of Ica, Peru. Black dots correspond to the locations of environmental vibration record, yellow diamonds correspond to gravimetric measurements, and yellow and green triangles correspond to circular seismic arrays. The LS01–LS03 line represents the linear seismic array. The A-A$'$ labels indicate the orientation of the gravimetric profile.

geophysical properties of soils of Ica city and generate a two-dimensional (2D) model of the Ica Basin.

1.1. Geological Framework.

The city of Ica represents 36% of the total surface of the Ica Department and is located in the lower part of the Ica River Basin (Figure 1). According to Gomez et al. [6], the most representative geomorphological features in the area are the dunes, which are formed by coastal winds near the shoreline and the plain or alluvial valley the city sits on. The rocky basement of this region is characterized by a Precambrian coastal basal complex composed of metamorphic rocks and in surface by quaternary deposits. The soils in this area consist of sands and silty-sands with some fine contents. From a geotechnical perspective, Ica's urban area is characterized by soils with low bearing capacity (1.0–2.0 kg/cm^2), although some areas toward the southwest and southeast show very low (<1.0 kg/cm^2) and medium (2.0–3.0 kg/cm^2) bearing capacities, respectively.

2. Methods

2.1. Description of the Methods.

In order to know the influence of the geomorphology of the Ica Basin on the dynamic response of soils for urban in Ica, we applied five geophysical methods: spectral ratio (H/V), frequency-wavenumber (F-K), multichannel analysis of surface waves (MASW), multichannel analysis of microtremor (MAM), and gravimetric method.

2.2. Spectral Ratio (H/V) Method.

2.2. Spectral Ratio (H/V) Method. This method allows calculating the empirical soil transfer function (FTE) from the spectral ratio of the horizontal and vertical component of an environmental vibration record (natural noise and/or noise generated by human activity) considering that the vertical component is not affected by the sedimentary deposits [7–10]. These spectra allow us to know the dynamic parameters of the soil such as the fundamental frequency, the dominant period, and the maximum relative amplifications of the soil. Nakamura [11] reaffirms that the spectral quotient is a reliable estimate of the site transfer function for *S* waves, allowing identification of the fundamental frequency of resonance of sedimentary deposits [12, 13].

2.3. Frequency-Wavenumber (F-K) Method. This method allows obtaining the velocity profile of the shear waves (*Vs*) and thickness of sedimentary deposits. This method considers that the array of sensors is traversed by a flat-wave front [14, 15] of known frequency, velocity, and direction of propagation, given in a two-dimensional space defined by the wavenumber in the direction of *Kx, Ky* [16], Socco et al. 2010. Finally, the transformation frequency number of wave frequency *F-K* [14, 17, 18] allows obtaining the dispersion curve to determine the phase velocity of Rayleigh waves according to their vibration mode [19, 20]. The fundamental vibration mode is characterized by attenuation in amplitude as the depth increases and the superior modes (first-mode, second-mode, etc.) by presenting varying amplitudes at different depth levels [21–24]. Likewise, the nature of the higher modes results from the constructive interference of wave reflection in the Earth's crust [25–27], Foti et al. 2014.

For *F-k*, the most sensitive parameters are associated with the reliability range of each seismic array (Figure 2(a)) because they depend on distance (*D*), wavelength (*λ*), and number of waves (*K*), where *K*min and *K*max, given in a two-dimensional space *Kx* and *Ky* (Figures 2(b) and 2(c)), define the greatest and least contribution of energy to propagating waves. In Figure 2(d), the discontinuous curves sectorize the dispersion curves and delimit the highest resolution zones for the dispersion curve, identifying low energy zones (lower frequency values) and aliasing zones with several energy peaks (greater frequency values). The first is associated with the boundary imposed by the width of the central lobe of the array's response function, while aliasing is associated with the minimum spacing between geophones.

For the inversion of the dispersion curve, the neighbourhood algorithm [28] is considered, which makes use of Voronoi's cell decomposition of the spatial parameters, based on an approximation of the "misfit" function, which is progressively refined during the inversion process. The misfit is proportional to the error in the adjustment of the empirical dispersion curve with the theoretical curve obtained with the proposed velocity profile. This parameter must tend toward low values. For this approach, more than 500 speed models are generated to consider a misfit less than 0.2. The misfit function is defined by the following equation [29]:

$$\text{misfit} = \sqrt{\sum_{i=1}^{n_F} \frac{(x_{di} - x_{ci})^2}{\sigma_i^2 n_F}}, \quad (1)$$

where x_{di} is the velocity of the frequency curve f_i, x_{ci} is the velocity of the calculated curve at the frequency curve f_i, σ_i^2 is the uncertainty of the frequency sample, and n_F is the sample frequency number. Finally, the dispersion curve with its different modes, through a nonlinear process, is inverted in order to look for a theoretical profile that fits this experimental dispersion curve.

In order to validate the results, the velocity models (*Vs*) obtained through this process were inverted to obtain a theoretical transfer function (FTT) by applying the Thomson-Haskell method for horizontal stratified media subject to SH wave action [20, 30], to finally overlay the FTT with the empirical transfer function (FTE).

2.4. MASW and MAM Methods. Both methods make it possible to determine the one-dimensional seismic profile of waves (*Vs*) by means of surface wave measurement tests, the resolution of which differs at surface and deep levels, respectively. Multichannel arrays of sensors located at predetermined distances along an axis along the ground surface are considered. MASW considers waves generated by an impulsive energy source at predetermined points and MAM considers the recording of environmental vibrations. From these methods we obtain dispersion curves of Rayleigh waves (phase velocity of the superficial waves versus frequency) and their inversion allows us to determine the profile of *S* wave velocity (*Vs*) [31, 32], Socco et al., 2010.

2.5. Gravimetric Method. This method allows the depth of the soil-rock interface to be determined from the variation of gravity acceleration on the ground. The method detects variations in densities in geological units present in the subsoil (density > 2 gm/cm3 is associated with rocks and lower with sediments).

The gravimetric data were corrected by free-air using regional (Shuttle Radar Topographic Mission, SRTM) and local elevation models (50 × 50 meters' resolution grid). The Oasis Montaj software from Geosoft and an average rock density of 2.5 g/cm3 [33] were used to correct Bouguer. For topographic correction, the methodology proposed by Kane [34] and Nagy [35] is considered, in order to obtain a grid of topographic correction, which through a sampling operation assigns the correction value to each gravimetric point. Finally, the Bouguer anomaly values are triangular interpolated.

In order to estimate the depth of the anomalies, the spectral analysis method proposed by Spector (1968) and Grant (1970) is used, which allows the grid of Bouguer's anomaly to be transformed into the space domain and the frequency domain. The values corresponding to each slope of the spectrum, divided by 4π, allow knowing the average depth of the center of mass of each anomaly. The first line slope is associated with the depth of the masses generating the regional anomaly, the second with the depth of the intermediate sources, and the third with the more superficial sources.

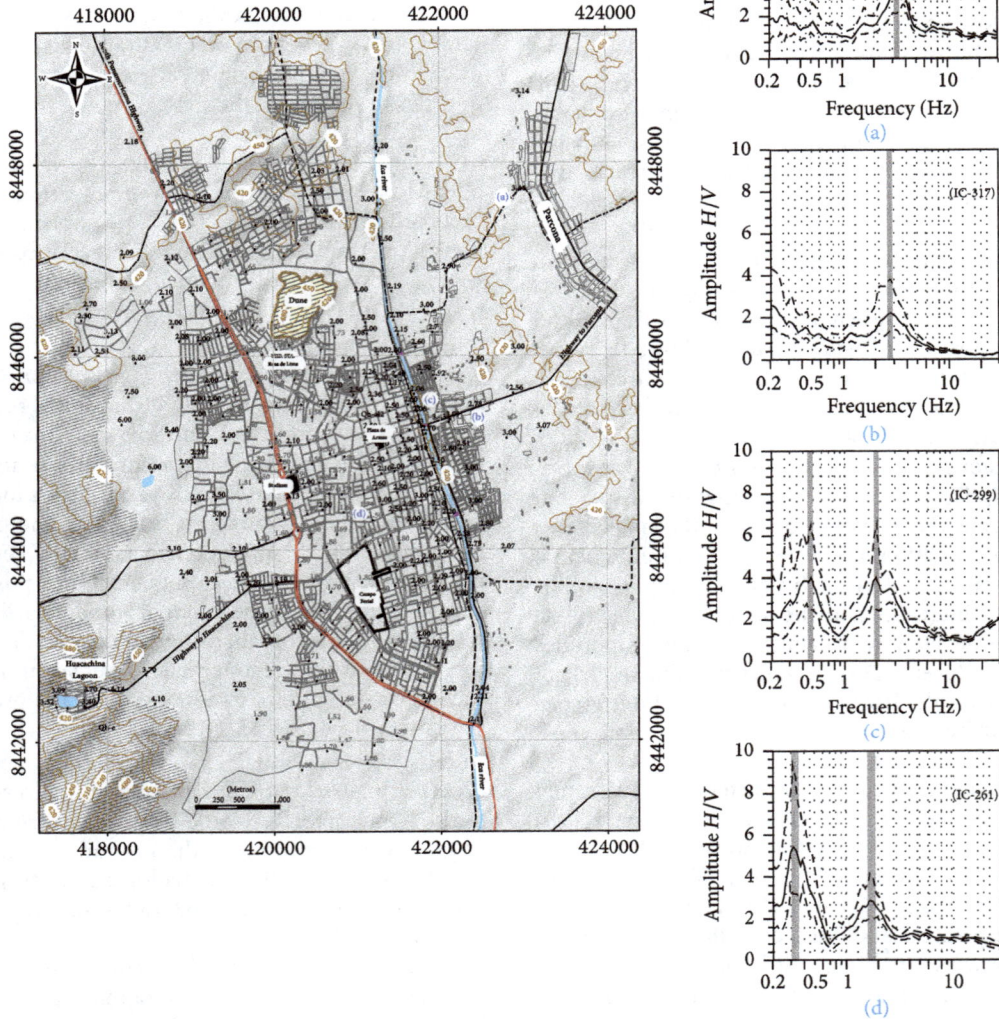

FIGURE 2: Spatial distribution of the frequency $F1$ ($F > 1.0$ Hz) and examples of spectral ratios obtained in several locations.

3. Data Acquisition

Figure 1 shows the locations of the individual measurement sites discussed in this study. To apply the H/V technique, we used microtremor data collected from 300 measurement points using a Lennartz LE-3D/5s seismometer and a CityShark digitizer, with a duration of 15 minutes per measuring point. To select which points to record, we considered the study area's geological and geomorphological characteristics, as well as the distribution of urban areas and accessibility. To apply the F-K method, we used microtremors data obtain by mean circular arrays of seismometers with 10, 30, 100, and 400 m radius, acquiring between approximately 30 minutes and 4 hours of data on each array, depending on its diameter. We considered the center of the arrays the "Campo

Ferial of Ica." For these arrays we used 10 Guralp 3-channel seismometers, each with a 24-bit Reftek digitizer.

The MASW and MAM methods use linear arrays of geophones (sensors), located at predefined distances along an axis on the surface. The MASW method considers the waves generated by an impulsive energy source at predefined sites. In the MAM method, use environmental vibrations. Both methods allow us to obtain the dispersion curve of the surface waves (phase velocity versus frequency) and its inversion allows determining the S-waves velocity profile (Vs) [31, 32], Socco et al., 2010. For both methods, we used an ES-300 instrument equipped with 24 sensors, with a sensitivity of 4.5 Hz. We assembled three arrays 144 and 240 m long, in the center and the boundary of the Ica Basin. On the other hand, for gravimetric method, we performed 80 gravimetric

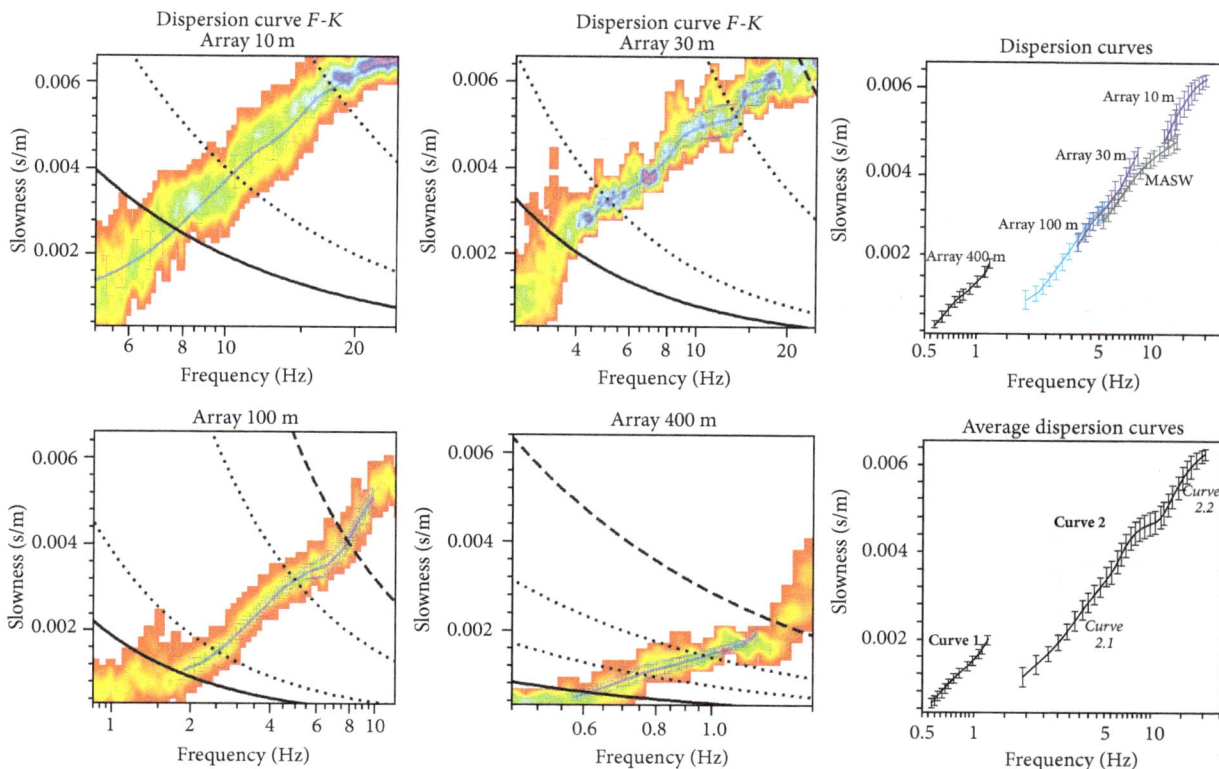

FIGURE 3: Dispersion curves (slowness versus frequency) obtained using the *F-K* method, which considers circular arrays with radii of 10, 30, 100, and 400 m. Dashed lines indicate reliability ranges and the bars in the curves are associated with the dispersion of slowness for each frequency. The right panels show the dispersion curves for different seismic arrays and the solid black line represents the average.

measurements distributed in five parallel lines (SW-NE), separated by an average distance of 300 m (Figure 1). The distance between each measurement is about 200 m. For the measurements we used a Lacoste & Romberg gravimeter with an accuracy of ±0.01 mGal. We applied an absolute gravity correction, using as reference base a point near the Rio Grande Tunnel (978215.134 mGal) south of Ica (Figure 1). The position of each site (coordinates and ellipsoidal elevation) used the WGS84 system and was determined using a Nikon DTM-322 Total Station.

4. Results

The predominant frequency (Fr), shear wave velocity (Vs) of the different soil layers, and the depth of the soil-rock interface are three important parameters in the characterization of physical and dynamic properties of soils to know the urban seismic hazards in Ica.

4.1. Predominant Frequency (Fr).
The frequency analysis (Fr) shows that soil of Ica responds in two frequency ranges (Figure 2), $F0$ ($F < 1.0$ Hz) and $F1$ ($F > 1.0$ Hz), with amplifications varying from factors of 2 to 6 depending on the location. For $F0$ we observed Fr between 0.4 and 0.8 Hz, with relative amplifications up to a factor of 5. For $F1$, Frs lower than 2.0 Hz are distributed in the center of the Ica city and along the "Panamericana Sur" road. Toward the eastern and western borders of Ica, $F1$ showed higher frequencies with relative amplifications of up to a factor of 6. Likewise, near

"Santa Rosa de Lima Urbanization" (to the north), the Ica River (to the east), and the Huacachina Lagoon (to the southwestern), we observe Frs for $F0$ of 0.35, 0.40, and 0.48 and 1.8, 2.6, and 3.0 for $F1$, respectively. It is evident that the central part of the basin shows low $F0$ and $F1$ values, increasing gradually toward the borders of the basin.

Figure 2 shows four representative spectral ratios curves labeled (a), (b), (c), and (d). (d), located in the central area, is characterized by the predominance of $F0$ (0.4 Hz) over $F1$ (2.0 Hz). In (c), which is close to the Ica River, $F0$ and $F1$ are similar, whereas, in (a) and (b), to the east of the Ica River, we observe predominant $F1$ values between 2.0 and 4.0 Hz. These results show that in Ica's urban area there are two Fr ranges, $F0$ and $F1$. Although $F0$ tends to disappear, $F1$ shows higher values as the distance from/to the east from the basin's center increases. With these results we can infer that the dynamic behavior of the soils in Ica changes because the soil-rock interface presents an irregular geomorphology.

4.2. Shear Wave Velocity (Vs)

4.2.1. 1D Profile Using the F-K Method.
Figure 3 shows the tendencies of the dispersion curves obtained by different seismic arrays. The reliability ranges (dashed lines) delimit the areas of maximum resolution for the dispersion curve. In this case, for a radius of 10 m the frequency range is 10 to 15 Hz, for 30 m it is 5.0 to 8.0 Hz, for 100 m it is 2.5 to 4.5 Hz, and for 400 m it is 0.8 to 1.5 Hz. We observe that the energy between the curves varies strongly between 1 and 2 Hz,

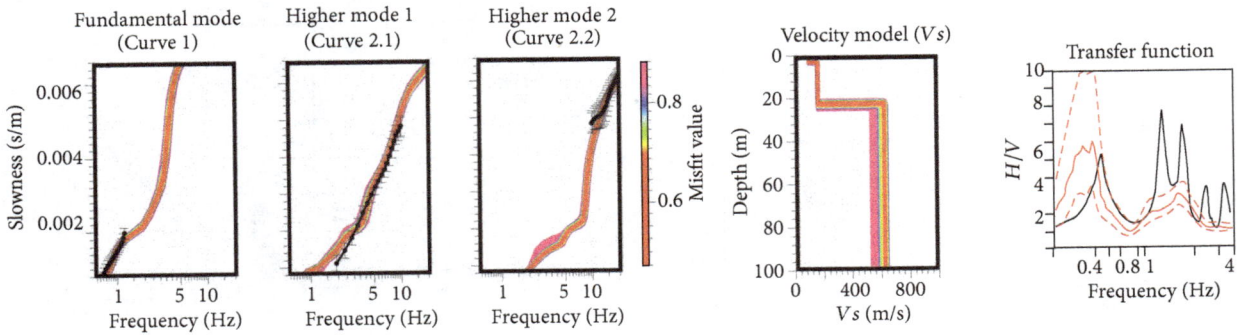

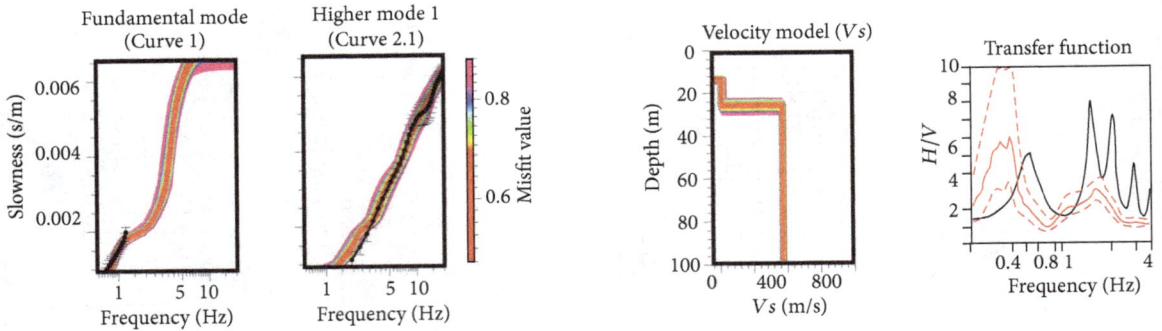

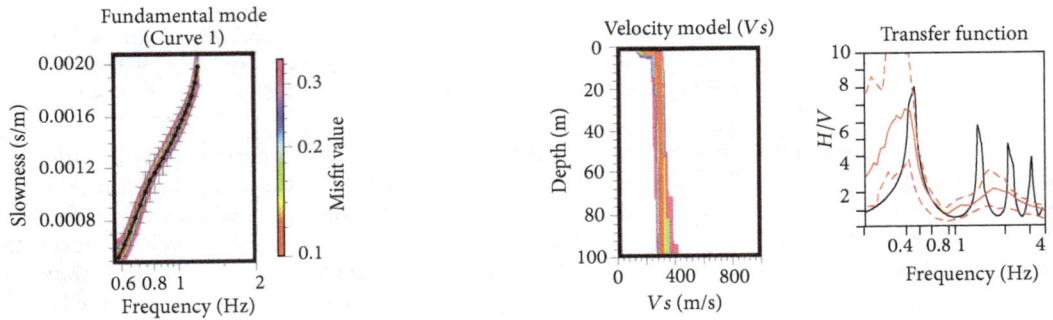

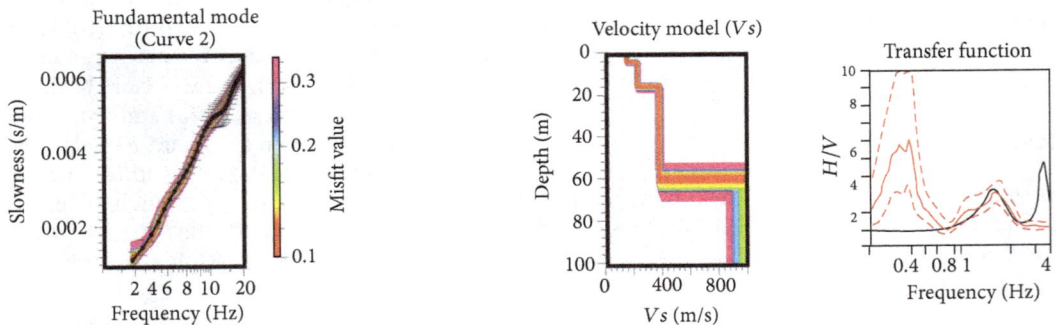

FIGURE 4: Results for the four tests, from left to right: the inversion of the dispersion curves, the velocity profiles of shear waves (Vs), and the correspondence of the theoretical transfer function (FTT; black line) obtained from the velocity profiles inversion and the empirical transfer function (FTE; red line) obtained from the spectral ratios for site IC-33.

with a jump that defines two tendencies. The average of the curves is between 2 and 20 Hz, with a moderate deflection at 8 Hz. These tendencies are associated with two frequency ranges corresponding to different vibration modes of the Rayleigh waves. Our results show that the velocities for *Curve 1* vary between 600 and 2000 m/s for frequencies between 0.6 and 1.0 Hz, whereas for *Curve 2* they vary between 170 and 800 m/s for frequencies between 2.0 and 15 Hz. Because of the complexity of the dispersion curve, we conducted testing to obtain phase velocities combining the fundamental and higher modes of the dispersion curve, according to Figure 4. We then inverted the data subsets to reconstruct

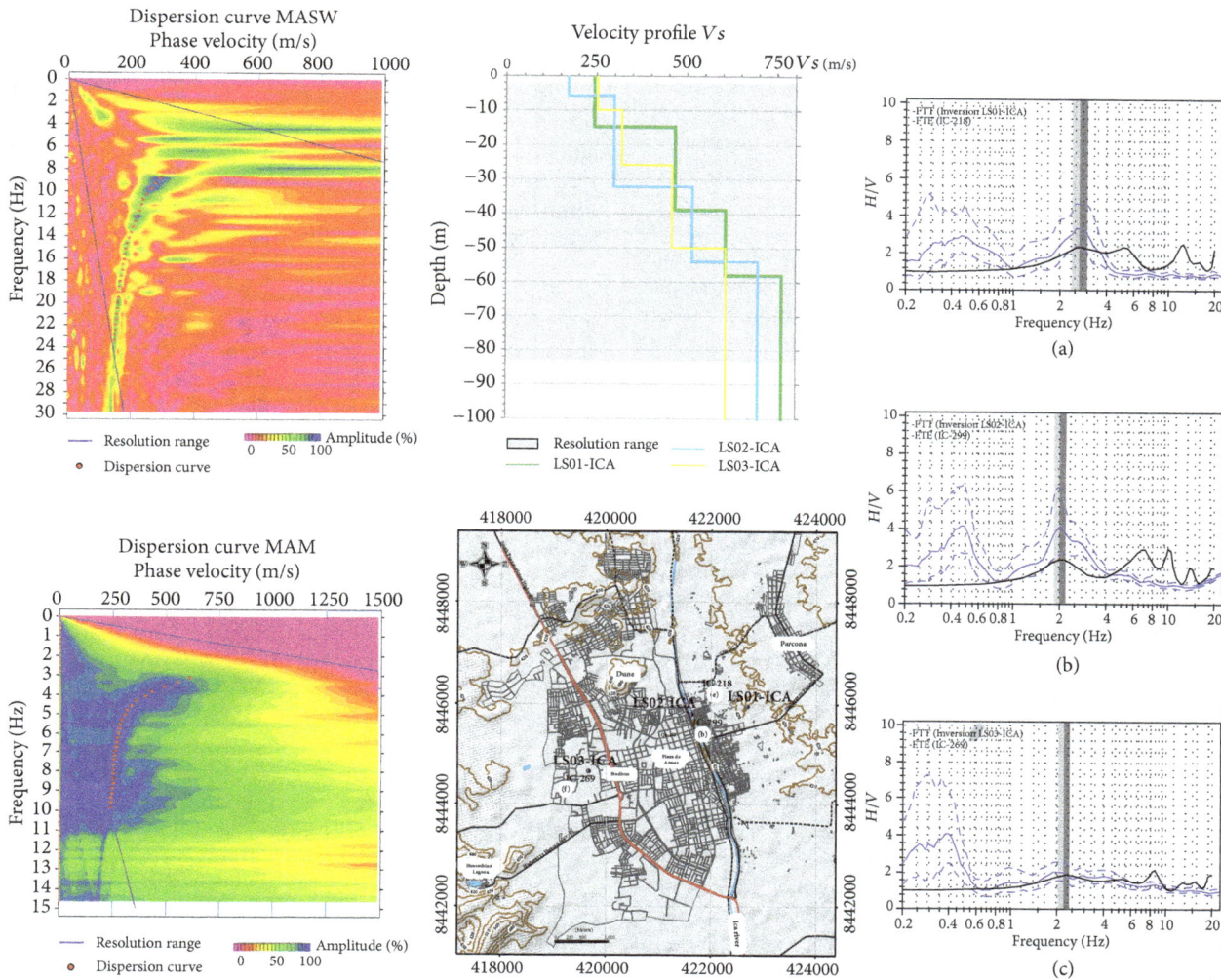

FIGURE 5: The left panels show the dispersion curves obtained with multichannel analysis of surface waves (MASW; top) and multichannel analysis of microtremors (MAM; bottom) methods. Here the averages and their inversions allow us to determine the velocity profile. The central plots show the velocity models (top) for the three linear arrays (bottom). The validation of the results is shown in the plots to the right. The inversion allowed us to obtain a theoretical transfer function (FTT; black line), which is superposed to the empirical transfer function (FTE; blue line) and shows a good correlation (degree of correspondence) with the fundamental frequency defined by the vertical gray line.

velocity profiles. We evaluated the effects of frequency range and combining of mode by comparing the inverted models obtained from the empirical transfer function (FTE) datasets (H/V: IC-33) with the theoretical transfer function (FTT) datasets theoretically obtained from the $F-K$ method.

The results from four different tests are described thus: *Test 1:* Curve 1 corresponds to a fundamental mode of the Rayleigh waves and Curve 2 is subdivided into the first and second superior modes of the Rayleigh waves (2.1 and 2.2, resp.). *Test 2:* Curve 1 corresponds to a fundamental mode of the Rayleigh waves and Curve 2 corresponds to a first superior mode. *Test 3:* Curve 1 corresponds to a fundamental mode. *Test 4:* Curve 2 corresponds to a fundamental mode. For the first three tests, we obtain velocities lower than 180 m/s at 25 m and 300 m/s at 38 and 136 m, respectively. These results are not consistent with the geology, geomorphology, and stratigraphy of the study area; therefore, we consider these scenarios not representative of the area.

Unlike these tests, Test 4 considers three shallow interfaces located at 4, 16, and 60 m. The first low-velocity layer would correspond to alluvial material and sandy soils, followed by the two layers showing velocities of 250 and 400 m/s, composed of moderately consolidated alluvial material to weakly compacted materials. Here as the depth increases, Vs increases above 900 m/s. In this last case, FTT and FTE coincide with the fundamental frequency of 1.8 Hz; thus, this result is consistent with the areas geology, geomorphology, and stratigraphy.

4.2.2. 1D Profile Using MASW and MAM. We performed MASW and MAM surveys on the borders and in the central part of the basin. The combination of MASW and MAM techniques allowed us to obtain velocity profiles at depths up to 60 and 100 m. The obtained results are consistent with a model consisting of three layers (Figure 5); the first with Vs between 170 and 180 m/s is composed of loose alluvial material (sandy soils); the second with Vs between 220 and 300 m/s is composed of moderately consolidated alluvial material; and the third with Vs between 400 and 460 m/s is composed of weakly compacted materials. It is important to note that as the depth of the layer increases Vs reaches values above 600 m/s. These Vs values correspond to layers 40 m thick to the east of the city and 60 m thick to the west. At greater depths, Vs increases to 800 to 900 m/s. In Figure 6 we present the results of the inversion, and we observe that for the fundamental frequency there is a correspondence between FTT and FTE represented by the gray line in each plot.

4.3. Depth of Soil-Rock Interface, Using Gravimetric Analysis.

In Figure 7(a) we show the corrected Bouguer anomaly, which we obtained using the spectral analysis method proposed by Spector (1968) and Grant (1970). In Figure 7(b), we identify three gravimetric sources, associated with a residual or shallow anomaly (sources 1 and 2) and a regional or deeper (source 3) anomaly. In Figure 7(c), we present the gravimetric profile including our interpretation. To the west, we observe that the sedimentary layer is 150 m thick, decreasing to 60 m as the topographic elevation increases. These results provide evidence that the geomorphology of the Ica Basin's soil-rock interface is quite irregular because thicker sediment layers are on the basin's western border.

5. Discussions

The spectral ratio curves are useful for determining the soil responses; its resolution is related to the impedance contrast of the materials, allowing defining different frequencies and/or frequency ranges at specific sites [36]. In some cases, a single peak indicates a homogeneous soil, and in other cases, more peaks are consistent with heterogeneous soils. However these peaks may not be directly associated with soil stratigraphy, but rather with nonlinear effects that sometimes lead to inadequate interpretation [37]. Hence, it is important to carefully analyze each peak frequency.

The soils in Ica respond to two frequency ranges ($F0$: 0.4–0.8 Hz and $F1$: 1.0–3.0 Hz). Following the methodology of Semblat et al. (2002), the maximum relative amplifications are analyzed in terms of amplitude, frequency, and location to evaluate their correspondence with geomorphology. In the central part of Ica, $F0$ shows an amplification factor of four, which decreases rapidly toward the west and gradually toward the east (Parcona village). $F1$ shows amplification factors between two and three in the sites on the right margin of the Ica River and amplification factors of five on the left margin. We observe that these values increase rapidly toward Parcona village, which is 30 m higher with respect to the elevation of the river. In general, these results show a correspondence of $F0$ with regional sources that are modulated by the Ica Basin's geomorphology and a correspondence of $F1$ directly with the stratigraphy of the sediments deposited on the basin. The results using seismic methods allowed us to determine that the Ica Basin's shallow stratigraphic limit fluctuates between 50 and 60 m depth with Vs between 600 and 900 m/s. The gravimetric profiles also show that sediment

thickness is variable along the profile, with layers of ~150 m to the west and 60 m to the east.

To determine the depth of the more representative interfaces, we applied the relation To = $4H/Vs$ [38], considering Vs values of 400 and 900 m/s, with an average frequency of 1.5 Hz. Using these parameters, we found two interfaces, one at a depth of 66 and one at 150 m. The first interface appears to correlate with moderately consolidated to slightly compacted alluvial materials, and the second appears to correlate with the soil-rock interface. These results agree with those obtained from gravity measurements. On the other hand, in Figure 7 we show the polynomial fit used to determine layer thickness, which we then used to construct the 2D model for the city of Ica. The results show that the basin consists of an irregular concave surface with depths of 60 and 150 m in the center of Ica, increasing rapidly to the west and decreasing gradually to the east.

Our results allow us to conclude that $F1$ corresponds to the fundamental frequency of Ica subsoil and $F0$ is harmonic with a regional origin [20] modulated by the Ica Basin. Finally, the frequency variations at depth that are associated with the physical characteristics of the soil, local topography, and geomorphology (dunes, small hills, and plateaus) allow us to characterize and infer the geometry and alluvial contents of Ica. The depth and irregularities of the basin generate seismic waves associated with resonance effects within layers of heterogeneous composition.

6. Conclusions

In this study we determined that the soils in the city of Ica respond to two frequency ranges: $F0$ (0.4–0.8 Hz) and $F1$ (1.0–3.0 Hz). $F0$ is associated with a regional source modulated by basin geomorphology and $F1$ is associated with a local source that corresponds to the dynamic response of the sediment layer.

Gravimetric and seismic analysis results show that the depth of the rock-soil interface under Ica varies from 150 m to the west to 60 m to the east. The 2D model suggests that the Ica Basin's geomorphology consists of a concave structure with depths that range from 120 to 150 m in the center and decrease rapidly to the west and gradually to the east.

The correlation of the results obtained using the seismic, geophysical, and geotechnical methods suggests that the Ica Basin's irregular geomorphology plays an important role in the dynamic response of Ica's soils to low frequencies, which produces a variation in the frequencies and relative amplification in soils despite a relatively flat surface topography. The structural damage associated with the 2007 Pisco earthquake was larger in the west, which can be explained by the thicker layer there and the variable dynamic behavior.

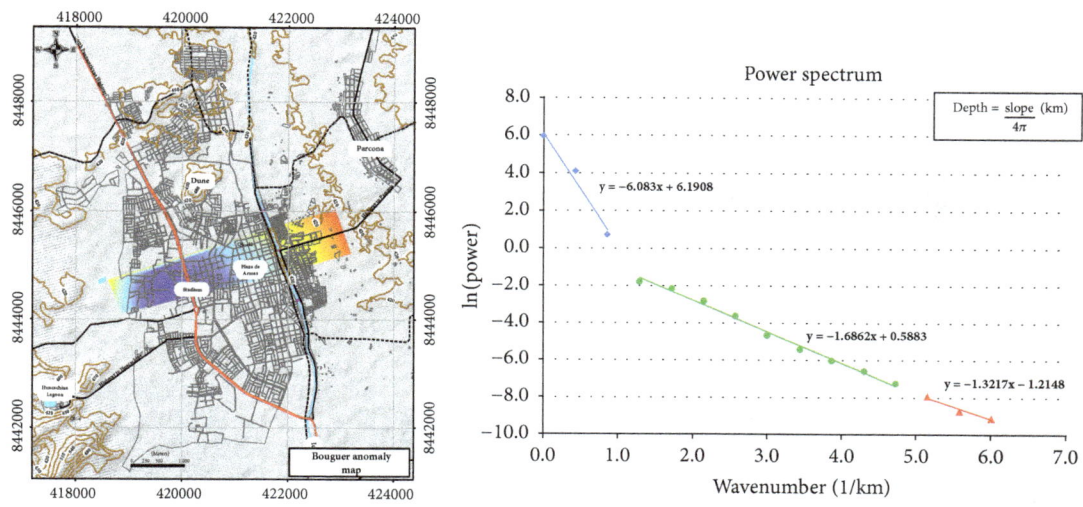

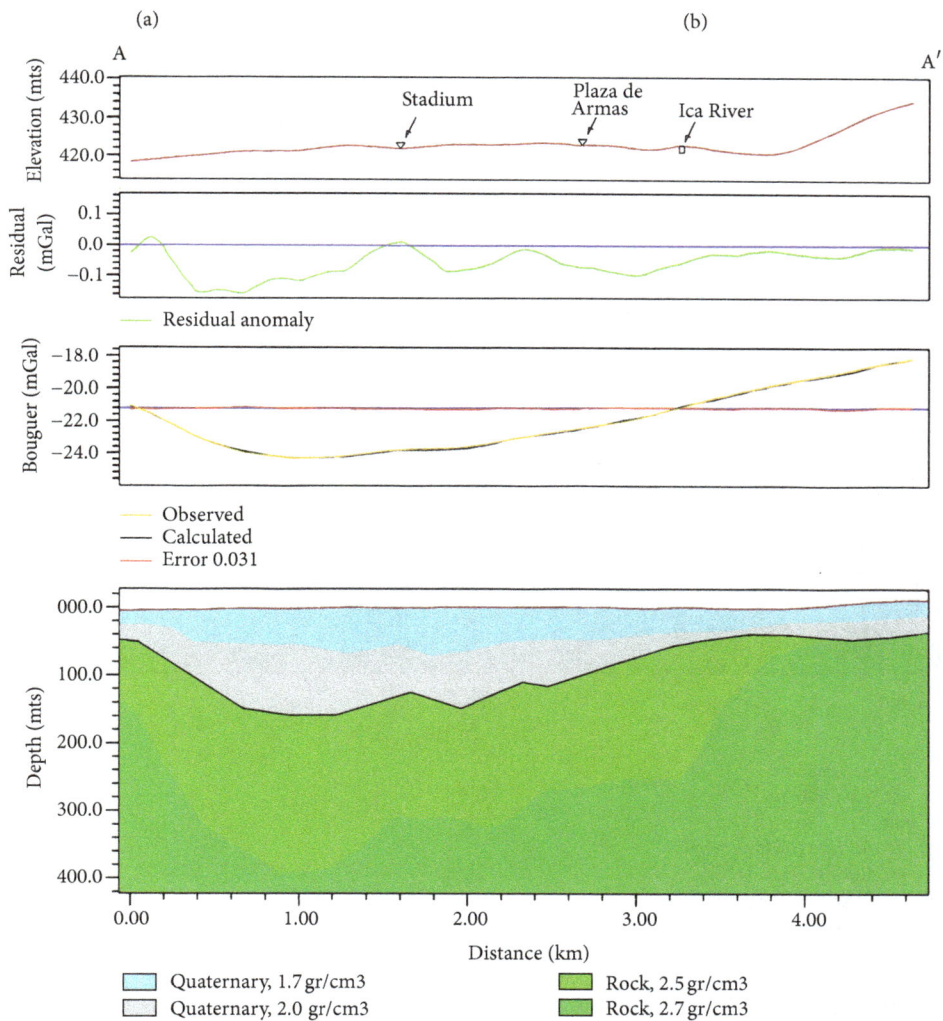

FIGURE 6: (a) Bouguer gravity anomaly map; (b) spectral analysis derived from the Bouguer gravity anomaly; and (c) gravimetric profile A-A′ (see Figure 1) showing sedimentary deposits and the rocky basement.

(a)

(b)

(c)

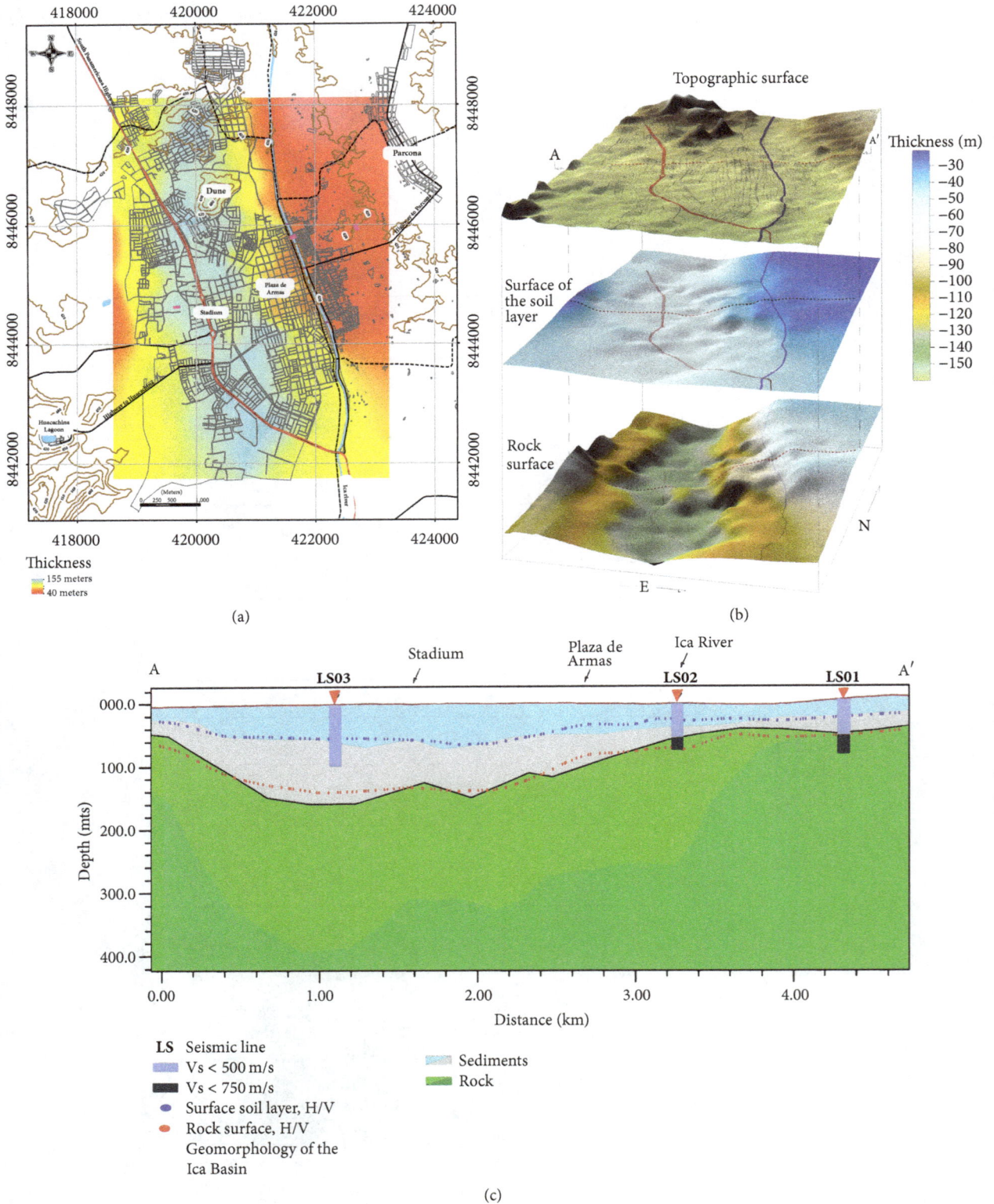

FIGURE 7: (a) Contour depth map derived from the predominant frequencies; (b) 3D schema of the Ica Basin; and (c) cross section of the Ica Basin (A-A′), superposed on the results obtained using seismic and gravimetric methods.

Acknowledgments

This research was supported by Geophysical Institute of Peru (IGP). The authors would like to thank Betrand Guillier and Marc Wathelet of the "Institut de Recherche pour le Developpement (IRD)," for training on the use of tool of Geopsy and their valuable contributions to the manuscript. Also thanks are due to Hugo Perfettini (IRD) for the support, advice, and contribution to their professional life.

References

[1] H. Tavera and I. Bernal, "The pisco (Peru) earthquake of 15 August 2007," *Seismological Research Letters*, vol. 79, no. 4, pp. 510–515, 2008.

[2] INDECI, *Reporte del sismo del*, Instituto Nacional de Defensa Civil, 2007.

[3] H. Tavera, I. Bernal, F. Stresser, M. Arango-Gavina, J. Alarcon, and J. Bommer, "Ground Motions observed during the 15 August 2007 Pisco, Peru, event," *Bulletin of Earthquake Eingineering*, 2008.

[4] W. León and V. Torres, "Mapa geológico actualizado del cuadrángulo de Ica (29 - i)," *INGEMMET, DGR*, 2001.

[5] S. Bonnefoy-Claudet, S. Baize, L. F. Bonilla et al., "Site effect evaluation in the basin of Santiago de Chile using ambient noise measurements," *Geophysical Journal International*, vol. 176, no. 3, pp. 925–937, 2009.

[6] J. C. Gomez, S. Ortiz, and C. Chiroque, *Informe técnico, Geología y geotecnia de la ciudad de Ica*, 2013.

[7] K. Kanai and T. Tanaka, "On microtremors. VIII," *Bulletin of the Earthquake Research Institute Tokyo University*, vol. 39, pp. 97–114, 1961.

[8] M. Nogoshi and T. Igarashi, "On the amplitude characteristics of microtremor (part 2)," *Journal of the Seismological Society of Japan*, vol. 24, pp. 26–40, 1971.

[9] Y. Nakamura, "Method for dynamic characteristics estimation of subsurface using microtremor on the ground surface," *Quarterly Report of RTRI (Railway Technical Research Institute) (Japan)*, vol. 30, no. 1, pp. 25–33, 1989.

[10] H. Okada, "The microtremors survey method, Geophysical Monograph Series 12," *Society of Exploration Geophysics of Japan*, p. 155, 2004.

[11] Y. Nakamura, "Clear Identification of Fundamental Idea of Nakamura's Technique and its applications in the hill zone of México City," *Bulletin of the Seismological Society of America*, vol. 82, pp. 24–43, 2000.

[12] J. Lermo and F. J. Chavez-García, "Site effect evaluation using spectral ratios with only one station," *Bulletin of the Seismological Society of America*, vol. 83, pp. 1574–1594, 1993.

[13] F. Leyton, S. Sepulveda, M. Astroza et al., "Zonificación sísmica de la cuenca de Santiago," in *Congreso Chileno de Sismología e Ingeniería Antisφsmica, X Jornadas*, pp. 22–27, Chile, 2010.

[14] R. Lacoss, E. Kelly, and M. Toksöz, "Estimation of seismic noise structure using arrays," *Geophysics*, vol. 34, no. 1, pp. 21–38, 1969.

[15] T. Kvaerna and F. Ringdahl, "Stability of various fk estimation techniques, Norsar semiannual technical summary," Tech. Rep., 1, –86, 1986.

[16] S. Foti, R. Lancellota, L. V. Socco, and L. Sambuelli, "Application of FK analysis of surface waves for geotechnical characterization," in *Proceedings of the Proc. 4th International Conference on Recent Advances in Geotechnical Earthquake Engineering and Soil Dynamics, Paper No. 1.14*, San Diego, California, 2001.

[17] K. Tokimatsu, Y. Miyadera, and S. Kuwayama, "Determination of Shear Wave Velocity Structures from Spectrum Analyses of Short-Period Microtremors," in *Proceedings of the 10th World Conf. on Earthquake Eng*, vol. 1, pp. 253–258, 1992.

[18] D. J. Zywicki, *Advanced Signal Processing Methods Applied to Engineering Analysis of Seismic Surface Waves*, 1999.

[19] K. Aki and P. G. Richards, *Quantitative Seismology*, University Science Books, 2002.

[20] K. Aki and P. Richards, *Quantitative Seismology: Theory and Methods*, New York, 1980.

[21] M. Wathelet, *Array recordings of ambient vibrations: surface-wave inversion*, Liège University, Belgium, 2005a.

[22] M. Wathelet, *Geopsy geophysical signal database for noise array processing*, Software, LGIT, Grenoble, France, 2005b.

[23] M. Wathelet, "An improved neighborhood algorithm: Parameter conditions and dynamic scaling," *Geophysical Research Letters*, vol. 35, no. 9, Article ID L09301, 2008.

[24] D. Fäh, G. Stamm, and H. Havenith, "Analysis of three-component ambient vibration array measurements," *Geophysical Journal International*, vol. 172, pp. 199–213, 2008.

[25] K. Aki, "Space and time spectra of stationary stochastic waves, with special reference to microtremors," *Bull, Earthquake Res. Inst. Tokyo Univ*, vol. 35, pp. 415–456 (1 plate), 1957.

[26] J. Xia, R. D. Miller, and C. B. Park, "Estimation of near-surface shear-wave velocity by inversion of Rayleigh waves," *Geophysics*, vol. 64, no. 3, pp. 691–700, 1999.

[27] S. Foti, L. Sambuelli, L. V. Socco, and C. Strobbia, "Experiments of joint acquisition of seismic refraction and surface wave data," *Near Surface Geophysics*, vol. 1, pp. 119–129, 2003.

[28] M. Wathelet, D. Jongmans, M. Ohrnberger, and S. Bonnefoy-Claudet, "Array performance for ambient vibrations on a shallow structure and consequences over vs inversion," *Journal of Seismology*, vol. 12, pp. 1–19, 2008.

[29] M. Wathelet, D. Jongmans, and M. Ohrnberger, "Surface-wave inversion using a direct search algorithm and its application to ambient vibration measurements," *Near Surf Geophys*, vol. 2, pp. 211–221, 2004.

[30] F. J. Sánchez-Sesma, V. J. Palencia, and F. Luzón, "Estimation of local site effects during earthquakes: an overview," *ISET Journal of Earthquake Technology*, vol. 39, no. 3, pp. 167–193, 2002.

[31] C. B. Park, R. D. Miller, and J. Xia, "Multichannel analysis of surface waves," *Geophysics*, vol. 64, no. 3, pp. 800–808, 1999.

[32] C. B. Park and R. D. Miller, "Roadside passive multichannel analysis of surface waves (MASW)," *Journal of Environmental & Engineering Geophysics*, vol. 13, no. 1, pp. 1–11, 2008.

[33] W. Hinze, "New standards for reducing gravity data, The North American gravity database," *Geophysics*, vol. 70, no. 4, pp. 25–32, 2005.

[34] M. F. Kane, "A Comprehensive System of Terrain Corrections Using a Digital Computer," *Geophysics*, vol. 27, no. 4, pp. 455–462, 1962.

[35] D. Nagy, "The gravitational attraction of a right rectangular prism," *Geophysics*, vol. 31, no. 2, pp. 362–371, 1966.

[36] M. Pilz, S. Parolai, M. Picozzi et al., "Shear wave velocity model of the Santiago de Chile basin derived from ambient noise measurements: A comparison of proxies for seismic site conditions and amplification," *Geophysical Journal International*, vol. 182, no. 1, pp. 355–367, 2010.

10

Influence of Error in Estimating Anisotropy Parameters on VTI Depth Imaging

S. Y. Moussavi Alashloo, D. P. Ghosh, Y. Bashir, and W. Y. Wan Ismail

Center of Seismic Imaging, Universiti Teknologi PETRONAS, 32610 Seri Iskandar, Malaysia

Correspondence should be addressed to S. Y. Moussavi Alashloo; y.alashloo@gmail.com

Academic Editor: Alexey Stovas

Thin layers in sedimentary rocks lead to seismic anisotropy which makes the wave velocity dependent on the propagation angle. This aspect causes errors in seismic imaging such as mispositioning of migrated events if anisotropy is not accounted for. One of the challenging issues in seismic imaging is the estimation of anisotropy parameters which usually has error due to dependency on several elements such as sparse data acquisition and erroneous data with low signal-to-noise ratio. In this study, an isotropic and anelliptic VTI fast marching eikonal solvers are employed to obtain seismic travel times required for Kirchhoff depth migration algorithm. The algorithm solely uses compressional wave. Another objective is to study the influence of anisotropic errors on the imaging. Comparing the isotropic and VTI travel times demonstrates a considerable lateral difference of wavefronts. After Kirchhoff imaging with true anisotropy, as a reference, and with a model including error, results show that the VTI algorithm with error in anisotropic models produces images with minor mispositioning which is considerable for isotropic one specifically in deeper parts. Furthermore, over- or underestimating anisotropy parameters up to 30 percent are acceptable for imaging and beyond that cause considerable mispositioning.

1. Introduction

It is well-known that hydrocarbon reservoirs and overlying strata are commonly anisotropic [1, 2]. In reality, it is rare to have media with elliptical or weak anisotropy properties. However, anellipticity (deviation of wavefield from ellipse) has been commonly observed in the Earth's subsurface, and it is a significant characteristic of elastic wave propagation [3, 4].

Another challenging issue in depth imaging is the computation of the travel time taken by a seismic wave from source to receiver. An efficient method to compute travel times is solving the eikonal equation by employing finite differences [5, 6]. Different techniques have been introduced to solve the eikonal equation, such as embedding methods, single-pass methods, sweeping methods, and iterative methods [7]. The main difference of these techniques is in how they cope with the complication of multivalued solutions and in finding solutions in the vicinity of cusps and discontinuities [8]. Anisotropy was initially added to an eikonal solver algorithm by Dellinger [9]. The embedding and iterative methods are both time consuming, particularly in heterogeneous and anisotropic conditions [7]. Fast sweeping methods are originally proposed for isotropic media [10]; however, a modification is executed to handle the anisotropic condition [11]. Single-pass or fast marching method (FMM) is another tool for computing travel times but is not generally applicable for anisotropic medium [5]. This algorithm has since been modified to work for anisotropy [12, 13].

In this study, a prestack depth migration algorithm is developed based on an anelliptic VTI compressional wave equation. Fomel's anelliptic approximation [14] for both phase and group velocity of P-wave are employed to derive the eikonal equation. The fast marching finite difference approach is used as our eikonal solver since it is fast and stable for travel time computation. In anisotropic study, four anisotropic models are used: a true model which is exactly similar to the model employed for forward modelling, a model with values 30 percent less than the true model, a model with values 40 percent less than the true model, and a model with values 30 percent more than the true model.

The calculated travel times are compared and employed in a standard Kirchhoff migration to obtain the image of the subsurface. The Marmousi model, as a complex model, is used to test the algorithm. Finally, we analyze both isotropic and VTI images qualitatively.

2. Methodology

We develop a new algorithm to incorporate anelliptic VTI travel times into prestack depth imaging. Our workflow for PSDM consists of the following: (1) travel time computation and (2) Kirchhoff depth migration. Step (1) provides travel times for the Kirchhoff migration. The algorithm is discussed in detail below.

2.1. VTI Fast Marching Eikonal Solver. The anisotropic wavefield propagation under a high frequency assumption is defined by the eikonal equation:

$$\left(\frac{\partial t}{\partial x}\right)^2 + \left(\frac{\partial t}{\partial y}\right)^2 + \left(\frac{\partial t}{\partial z}\right)^2 = \frac{1}{v^2(\theta)}, \quad (1)$$

where t is the travel time, x, y, z are the Cartesian coordinates, and $v(\theta)$ is the VTI phase velocity. The FMM solves (1) by considering the fact that the direction of energy propagation follows the group velocity equation. This method is similar to a ray that is perpendicular to wavefronts defined by phase velocity. This ray is called the travel time gradient. A wave equation is needed as a kernel of fast marching algorithm. Fomel [14] enhanced the anelliptic qP wave approximation proposed by Muir and Dellinger [15] through replacing the linear approximation with a nonlinear one. By using the shifted hyperbola approximation, he obtains the following equation for P wave phase velocity:

$$v^2(\theta) \approx \frac{1}{2}e(\theta) + \frac{1}{2}\sqrt{e^2(\theta) + 4(q-1)ac\sin^2\theta\cos^2\theta}, \quad (2)$$

where $a = c_{11}, c = c_{33}$, where $c_{ij}(X)$ are the density-normalized components of the elastic tensor, θ is the phase angle, and q and $e(\theta)$ are the anellipticity coefficient and the elliptical component of the velocity, respectively, defined by

$$q = \frac{1+2\delta}{1+2\varepsilon} = \frac{1}{1+2\eta}, \quad (3)$$

$$e(\theta) = a\sin^2\theta + c\cos^2\theta, \quad (4)$$

where ε and δ are Thomsen's parameters, and $\eta = (\varepsilon - \delta)/(1+2\delta)$. Similarly, for approximating the group velocity, the shifted hyperbola approach is applied on the Muir's

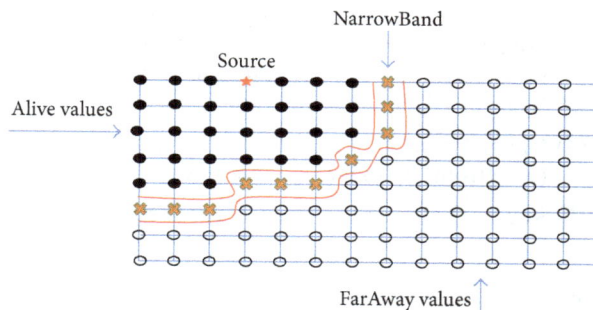

FIGURE 1: Schematic illustration of fast marching method.

approximation to unlinearize the equation. The new group velocity approximation is

$$\frac{1}{V^2(\Theta)}$$

$$\approx \frac{1+2Q}{2(1+Q)}E(\Theta) \quad (5)$$

$$+ \frac{1}{2(1+Q)}\sqrt{E^2(\Theta) + 4(Q^2-1)AC\sin^2\Theta\cos^2\Theta},$$

where $Q = 1/q$, $A = 1/a$, $C = 1/c$, Θ is the group angle, and $E(\Theta)$ is the elliptical part:

$$E(\Theta) = A\sin^2\Theta + C\cos^2\Theta. \quad (6)$$

Approximations in (2) and (5) are used for ray tracing in locally homogeneous cells needed in this algorithm. An attentively selected order of travel time evaluation is the main advantage of the FMM. This method is an upwind method which means if a wave propagates from left to right, a difference scheme should be applied for reaching upwind to the left to collect information to construct the solution downwind to the right.

According to Fomel [16], although the algorithm follows a certain procedure, the grid points are divided into three classes, namely, *Alive*, points which are behind the wavefront and have been already computed; *NarrowBand*, points on the wavefront awaiting assessment; and *FarAway*, which remains untouched ahead of the wavefront (Figure 1). In other words, the source positions at the beginning of the evaluation are considered *Alive*. Given that they are initial points, their travel time is zero. All points that are one grid point away are taken as *NarrowBand*, and their travel times are computed analytically. All other grid points are marked as *FarAway* and have an "infinitely large" travel time value [5, 16]. The algorithm includes the following main steps:

(1) Find the point with the minimum travel time among the *NarrowBand* points.

(2) Tag the point as *Alive* and remove it from *NarrowBand*.

(3) Check the neighbours of the minimum point that are not *Alive*. If any of them are categorized as *FarAway*, update them as *NarrowBand*. It means the wavefront is advanced, and the minimum point is behind it.

(4) Update travel times for points on the NarrowBand (wavefront) by solving (1) numerically.

(5) Repeat the loop until all points are behind the wavefront.

For updating, one to three neighbour points are selected and their travel time values need to be less than the current value. After choosing the points, the quadratic equation,

$$\sum_j \left(\frac{t_i - t_j}{\Delta x_{ij}} \right)^2 = s_i^2, \tag{7}$$

should be solved for t_i which is the updated travel time value. t_j is the travel time at the neighbouring points, s_i is the slowness ($s = 1/v$) at the point i, and Δx_{ij} is the grid size in ij direction.

2.2. Kirchhoff Depth Migration.

To image subsurface structures, which is normally recorded as geometrically unfocused, a migration procedure is needed to eliminate the influences on the wave during propagation in the subsurface. Migration technique moves seismic events to their true position, and it weakens diffractions. Since the prestack data have irregular spatial sampling, Kirchhoff migration is often the choice for prestack imaging [17]. Kirchhoff migration is dependent upon the solution of Kirchhoff integral to the wave equation and upon Green's function theory. Its general form as an integral expression is defined by

$$I(\xi)$$
$$= \int_{\Omega_\xi} W(\xi, m, h) D\left[t = t_D(\xi, m, h), m, h\right] dm \, dh. \tag{8}$$

The image $I(\xi)$, given in a two-dimensional space $\xi = (z_\xi, x_\xi)$, can be determined from the integral on the data values $D(t, m, h)$ at the time $t_D(\xi, m, h)$ and weighted by an appropriate element $W(\xi, m, h)$. The time factor $t_D(\xi, m, h)$ is defined as the total computed time while the waves travel from the source s to the image point $I(\xi)$ and propagate back to the receiver point r at the surface. Overall, for Kirchhoff depth migration, we need two inputs which are seismic data and travel times. An anelliptic VTI fast marching approach is applied on the velocity models to compute travel times, and synthetic data are created via VTI finite difference forward modeling.

3. Numerical Examples

The results of the new VTI PSDM algorithm are demonstrated and discussed in this section. The efficiency of the above-described approach is tested for isotropic and VTI travel time computation on the 2D Marmousi model

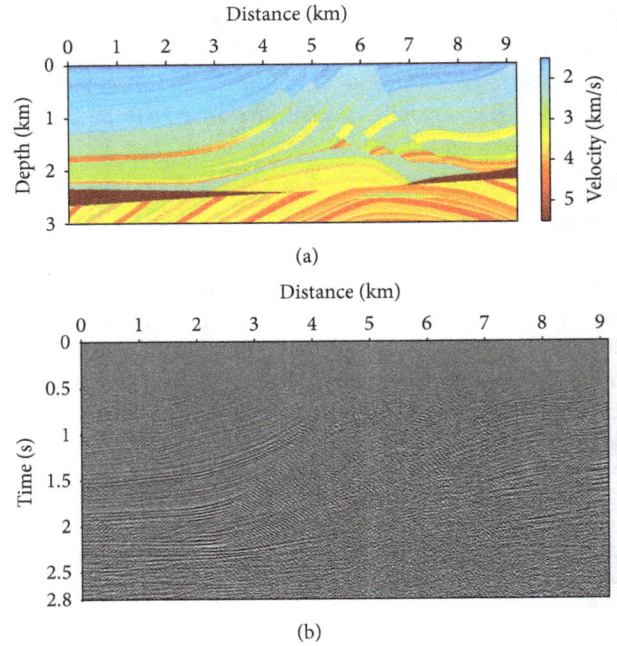

FIGURE 2: (a) Marmousi velocity model and (b) Marmousi synthetic shot gathers.

(Figure 2). The true η model was generated based on the velocity model which varies between 0.03 and 0.1 (Figure 3(a)). Different percentages of error were introduced to the true η model (Figures 3(b)–3(d)). The VTI first arrival travel times are computed and compared with the isotropic travel times for sources at $x = 5$ km (Figure 4). These results indicate that the eikonal solver algorithm is able to cover all the desired area in the model even with high complexity such as normal faults and tilted blocks. Stability is another advantage of this method in which it is fast and robust to provide travel times [18]. Furthermore, by comparing the travel times, it is obvious that the anisotropic wavefronts laterally move faster than the isotropic ones; however, the isotropic and anisotropic wavefronts propagate in a same speed vertically. This difference is due to the parameter η that affects the wave propagation mainly laterally. Waheed et al. [6] demonstrated that the maximum influence of η on the propagation of the wave is in the direction orthogonal to the symmetry axis. The errors introduced to the anisotropy parameter affected the travel times which below 30 percent is not too much but above 30 percent causes a noticeable difference. The gap between isotropic and VTI warfronts increases while the wavefronts propagate away from the source. To know how these variations influence imaging, the Marmousi synthetic data was migrated by using these travel times.

Figure 5 illustrates the images of VTI and isotropic PSDM for Marmousi model as well as the overlay of the images with the reflectivity model which was derived from the velocity model and its calculated acoustic impedance. Comparison of Figures 5(a) and 5(b) demonstrates several differences.

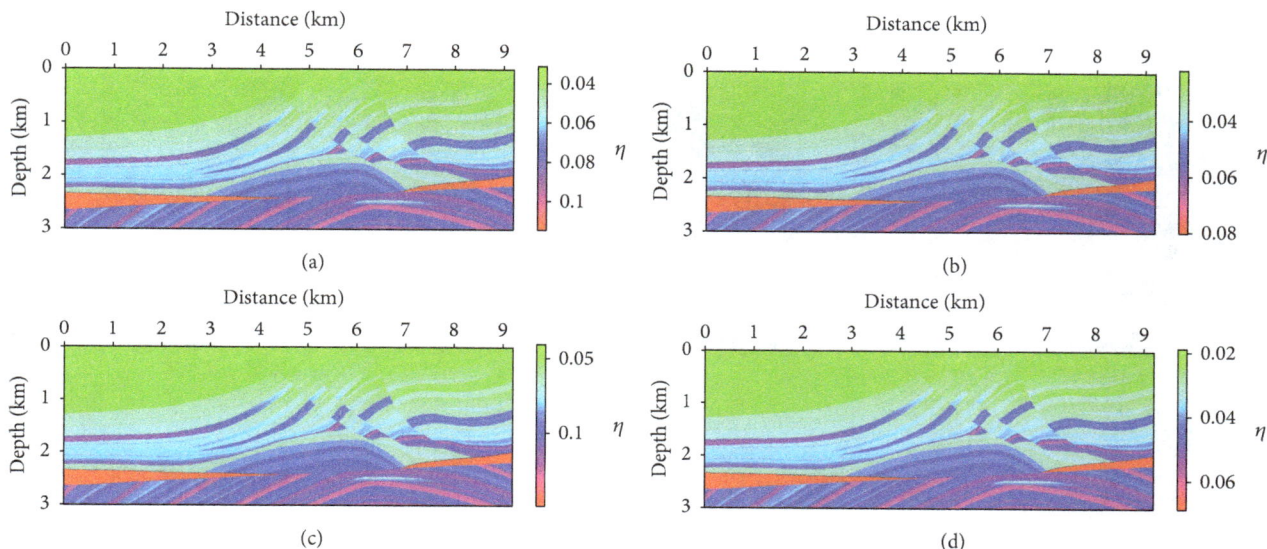

FIGURE 3: Marmousi anisotropy model: (a) true η model, (b) model with 30% reduced η value, (c) model with 30% added η value, and (d) model with 40% reduced η value.

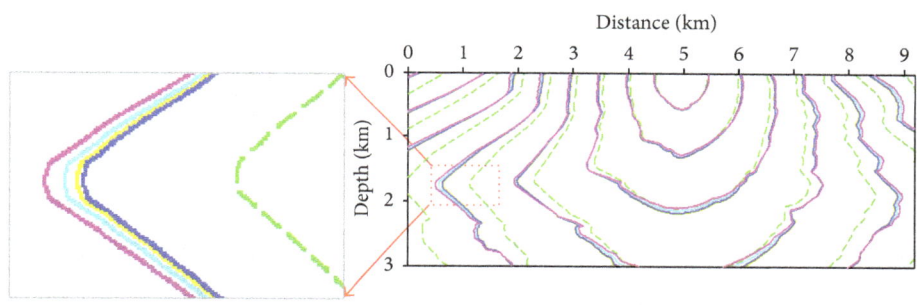

FIGURE 4: Fast marching travel time results of Marmousi model for isotropy (dashed green line), anisotropy with true η value (light blue line), anisotropy with 30% error less than true η value (yellow line), anisotropy with 30% error more than true η value (magenta line), and anisotropy with 40% error less than true η value (dark blue line) for a source at $x = 5$ km.

Circles, number 1, show misshaping of pinchout and a wrong dip for the reflector in the isotropic image. Also, the spacing between the isotropic interfaces is more than the anisotropic ones.

The resolution is another effect where, in the anisotropic image, layers appear to be more clear and sharp, but, in the isotropic image, they are poorly imaged.

In rectangles, number 2, more fractures are detected by VTI imaging, and different positioning is obvious. Number 3 again indicates changing in the reflector's dip where the isotropic layer has steep tilt compared to anisotropic layer. Comparing the images with the model confirms that the VTI result is much more accurate, and it is better matched.

Considering Figure 5(d) as a reference, which is imaged by true η model and by comparing it with the images from the modified η models, it can be concluded that the images difference for those with below 30 percent errors is not remarkable. In contrast, for those above 30 percent errors, whether higher or lower than true η value, there is more

mispositioning in the deeper sections. Hence, if the estimated anisotropy parameter has error lower than 30 percent, the result of our VTI algorithm is reliable for interpretation and other applications. However, by increasing error in η model, mispositioning becomes more apparent that geophysicists need to consider it.

4. Conclusions

A prestack VTI depth imaging algorithm was designed and applied for the isotropic and anisotropic media. A VTI fast marching eikonal solver based on anelliptic was utilized to compute the seismic travel times. The Marmousi velocity model and its seismic dataset are used for imaging. After Kirchhoff imaging, results showed that the VTI imaging algorithm generates images with higher resolution than the isotropic ones specifically in deeper parts, and the subsurface features are imaged correctly regarding their pattern and location. Moreover, study on error in anisotropy estimation

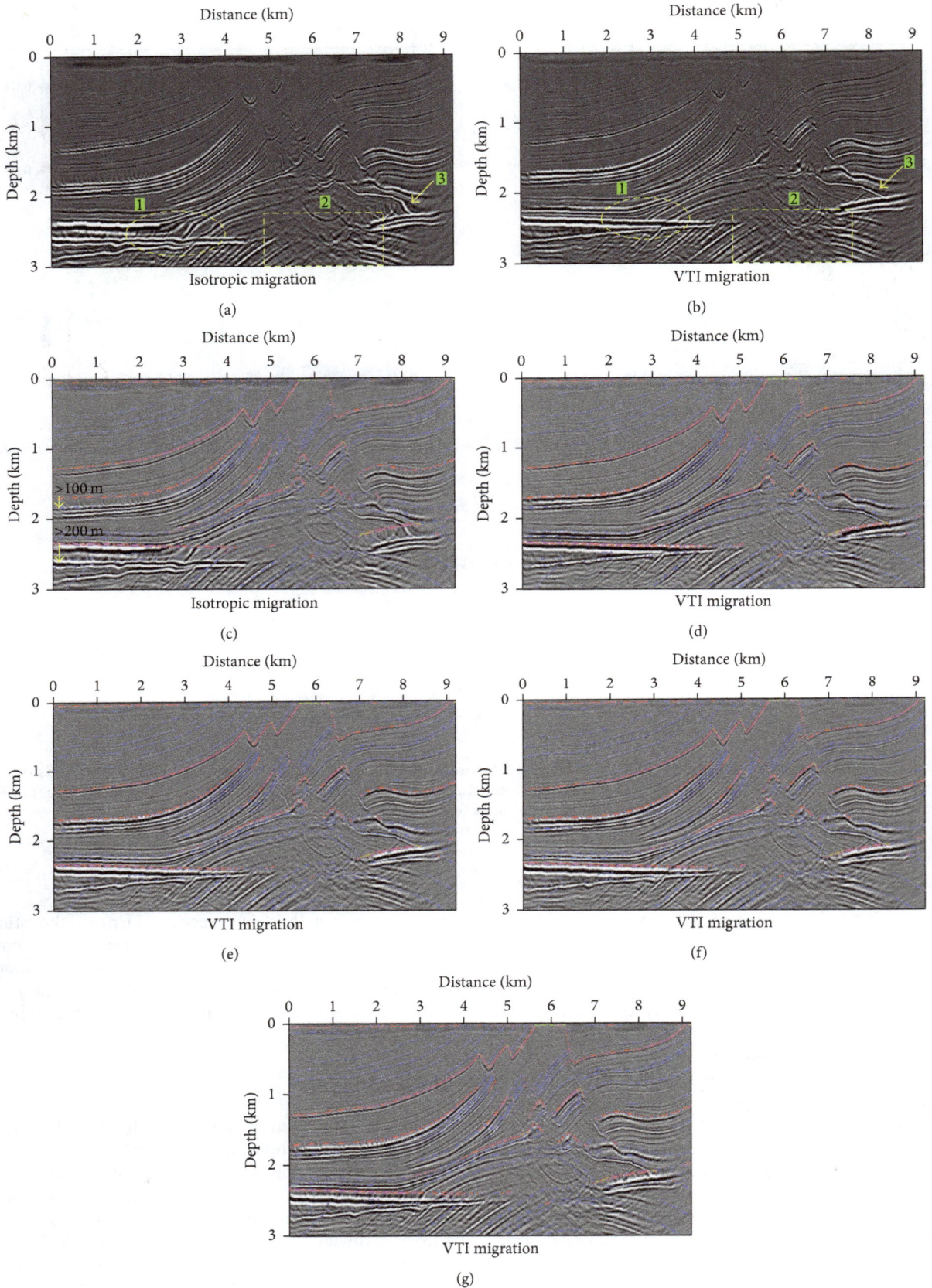

FIGURE 5: Comparison between (a) isotropic and (b) VTI Marmousi images. Superposition of reflectivity on images generated by (c) isotropic model, (d) true eta model, (e) eta model with 30% decreased value, (f) eta model with 30% increased value, and (g) eta model with 40% decreased value.

demonstrated that up to 30 percent error can be tolerated in imaging but beyond that leads to considerable mispositioning which may affect the interpretation of the subsurface.

Competing Interests

The authors declare that they have no competing interests.

Acknowledgments

The authors gratefully acknowledge members of Centre of Seismic Imaging (CSI) at UTP for their helpful discussions. This work is funded by PETRONAS. Seismic package of Madagascar was used in developing the algorithm.

References

[1] G. E. Backus, "Long-wave elastic anisotropy produced by horizontal layering," *Journal of Geophysical Research*, vol. 67, no. 11, pp. 4427–4440, 1962.

[2] L. Thomsen, "Weak elastic anisotropy," *Geophysics*, vol. 51, no. 10, pp. 1954–1966, 1986.

[3] R. M. Pereira, J. C. R. Cruz, and J. D. S. Protázio, "Anelliptic rational approximations of traveltime P-wave reflections in VTI media," in *Proceedings of the 14th International Congress of the Brazilian Geophysical Society & EXPOGEF*, pp. 945–949, Rio de Janeiro, Brazil, August 2015.

[4] A. Stovas and S. Fomel, "Generalized nonelliptic moveout approximation in t-p domain," *Geophysics*, vol. 77, no. 2, pp. U23–U30, 2012.

[5] J. A. Sethian and A. M. Popovici, "3-D traveltime computation using the fast marching method," *Geophysics*, vol. 64, no. 2, pp. 516–523, 1999.

[6] U. Waheed, T. Alkhalifah, and H. Wang, "Efficient traveltime solutions of the acoustic TI eikonal equation," *Journal of Computational Physics*, vol. 282, pp. 62–76, 2015.

[7] U. Waheed, C. E. Yarman, and G. Flagg, "An iterative, fast-sweeping-based eikonal solver for 3D tilted anisotropic media," *Geophysics*, vol. 80, no. 3, pp. C49–C58, 2014.

[8] P. Sava and S. Fomel, "Huygens wavefront tracing: a robust alternative to ray tracing," in *SEG Technical Program Expanded Abstracts*, pp. 1961–1964, Society of Exploration Geophysicists, 1998.

[9] J. Dellinger, "Anisotropic finite–difference traveltimes," in *SEG Technical Program Expanded Abstracts*, pp. 1530–1533, Society of Exploration Geophysicists, Tulsa, Okla, USA, 1991.

[10] H. Zhao, *A Fast Sweeping Method for Eikonal Equation Mathematics of Computation*, American Mathematics of Society, Providence, RI, USA, 2004.

[11] Y.-T. Zhang, H.-K. Zhao, and J. Qian, "High order fast sweeping methods for static Hamilton-Jacobi equations," *Journal of Scientific Computing*, vol. 29, no. 1, pp. 25–56, 2006.

[12] J. A. Sethian and A. Vladimirsky, "Ordered upwind methods for static Hamilton-Jacobi equations," *Proceedings of the National Academy of Sciences of the United States of America*, vol. 98, no. 20, pp. 11069–11074, 2001.

[13] E. Cristiani, "A fast marching method for Hamilton-Jacobi equations modeling monotone front propagations," *Journal of Scientific Computing*, vol. 39, no. 2, pp. 189–205, 2009.

[14] S. Fomel, "On anelliptic approximations for qP velocities in VTI media," *Geophysical Prospecting*, vol. 52, no. 3, pp. 247–259, 2004.

[15] F. Muir and J. Dellinger, "A practical anisotropic system," Stanford Exploration Project SEP-44, 1985, vol. 55, p. 58.

[16] S. Fomel, "A variational formulation of the fast marching eikonal solver," in *SEP-95: Stanford Exploration Project*, pp. 127–147, 1997.

[17] B. Biondi, *3D Seismic Imaging*, Society of Exploration Geophysicists, Tulsa, Okla, USA, 2006.

[18] S. Y. Moussavi Alashloo, D. Ghosh, Y. Bashir, and W. I. Wan Yusoff, "A comparison on initial-value ray tracing and fast marching eikonal solver for VTI traveltime computing," in *Proceedings of the IOP Conference Series: Earth and Environmental Science (AeroEarth '15)*, vol. 30, no. 1, Jakarta, Indonesia, 2015.

11

Localized Increment and Decrement in the Total Electron Content of the Ionosphere as a Response to the April 20, 2018, Geomagnetic Storm

Carlos Sotomayor-Beltran (iD)

Image Processing Research Laboratory (INTI-Lab), Universidad de Ciencias y Humanidades, Lima 39, Peru

Correspondence should be addressed to Carlos Sotomayor-Beltran; csotomayor@uch.edu.pe

Academic Editor: Angelo De Santis

A moderate geomagnetic storm occurred on April 20, 2018. Using vertical total electron content (VTEC) maps provided by the Center for Orbit Determination in Europe, ionospheric responses to the geomagnetic storm could be identified in generated two-dimensional differential VTEC maps. During the day of the storm the enhancement of the equatorial ionization anomaly (EIA), product of the super-fountain effect was identified. A localized TEC enhancement (LTE) was also observed to the south of the EIA on April 20, 2018. It was also possible to visualize this LTE in a longitudinal section of the EIA as a third crest. The maximum increment of VTEC for the LTE was 204%. This LTE is quite unique because it happened during the expected solar cycle 24 and 25 minimum, and according to a previous study no LTE observation could be done for the last solar two-cycle minimum. The origin of the observed LTE is suggested to be partly product of the super-fountain effect. Finally, a localized TEC decrement (LTD) was observed towards the end of the day, April 20, 2018. Because this LTD consisted in the disappearance of the northern and southern crests of the EIA and this occurred during the recovery phase of the geomagnetic storm, it can be suggested that the LTD origin is due to the westward disturbance electric field. This mechanism was put forward by a past study that also analyzed the responses to a geomagnetic storm (the 2015 St. Patrick's day storm), being one of the responses the inhibition of both crests of the EIA.

1. Introduction

Geomagnetic storms can be very disruptive. For instance, they can induce currents in main electricity grids, can adversely affect global positioning system by considerably reducing its accuracy and also can cause the loss of high-frequency radio communications (e.g., [1, 2]). The ionosphere, which is a layer of the atmosphere located between ~60 and 1000 km in altitude, responds to geomagnetic storms in the form of ionospheric storms. Ionospheric storms can be classified as positive or negative (e.g., [3–5]), i.e., when there is an increment or decrement in the total electron density (TEC), respectively.

At low latitudes, a noticeable feature in the F layer of the ionosphere is the equatorial ionization anomaly (EIA; [6]). This anomaly consists of two regions of enhanced plasma located at $\pm15°$ of the magnetic equator. This phenomenon is mainly due to eastward zonal electric field which in combination with the Earth's magnetic field at equatorial latitudes produces an upward $\mathbf{E}$ x $\mathbf{B}$ drift of the plasma. Once high altitudes are reached, the plasma diffuses along the geomagnetic field lines due to pressure gradient forces in conjunction with the action of gravity [7]. This whole process is also known as the equatorial fountain effect. On the other hand when a geomagnetic storm sets in, it has been previously observed in several studies (e.g., [8–12]) that the EIA is also considerably enhanced, which is due to the penetration of the interplanetary electric field from high to low latitudes and which lasts for several hours [13]. As a consequence of the electric field penetration, there is an upward displacement of plasma to higher altitudes and also an expansion of the EIA in the direction of the poles (e.g., [12]). In this case, this phenomenon is known as the super-fountain effect.

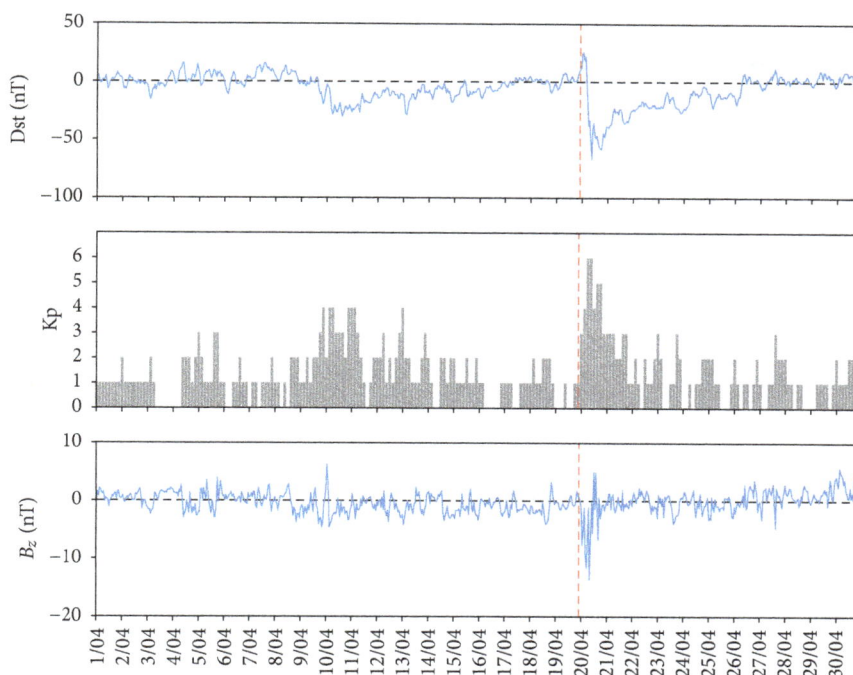

FIGURE 1: Dst and Kp geomagnetic indices during April 2018. Additionally, the vertical component of the interplanetary magnetic field (B_z) is shown. The vertical red dashed line indicates the storm sudden commencement (SSC).

Besides the expected diurnal response of the EIA to geomagnetic storms in the low and midlatitude ionosphere, there are still not well-defined physical mechanisms that can account for the anomalous increase or decrease of TEC outside the extent of the EIA. Very recently, Edemskiy et al. [14] have observed that the ionospheric response to the August 15, 2015, geomagnetic storm presented not only the regular increment of TEC in the EIA, but also two regions of localized TEC enhancements (LTEs or localized positive ionospheric storms) at central-middle (~35°S) and lower-high latitudes (~62°S) were identified. The aforementioned study suggested that the excursion of the vertical interplanetary magnetic field (B_z) was necessary but not a definite requirement for the appearance of the observed LTEs.

In this paper, publicly available vertical TEC (VTEC) data from the Center for Orbit Determination in Europe (CODE) were used to investigate the ionospheric responses to the April 20, 2018, geomagnetic storm that happened during the expected time of the solar cycle 24 and 25 minimum [15].

2. Data and Method

2.1. Ionospheric Data. In order to detect ionospheric anomalies due to the geomagnetic storm of April 20, 2018, global maps of VTEC provided by CODE were used. As indicated by previous studies [16, 17], global ionospheric maps (GIMs) are good indicators of space weather events.

GIMs are routinely produced every hour and they come in daily IONosphere Map EXchange files (IONEX; [18]). CODE makes IONEX files publicly available via their ftp

site (ftp://ftp.aiub.unibe.ch/CODE/). GIMs provide VTEC in a global geographical grid with a resolution of 5° x 2.5° (longitude and latitude, respectively). The total number of cells in latitude is 71 and in longitude is 73. The VTEC values are given in TEC Units (TECU), where 1 TECU = 10^{16} electrons/m^2. The accuracy of the GIMs is between 2 to 9 TECU. Due to the bit of structural complexity of the IONEX files, a program using the library NumPy from Python was written. NumPy has the ability to work efficiently with N-dimensional arrays [19]. The result after this preprocessing stage was to obtain 3D cubes of VTEC per each IONEX file.

Ionospheric disturbances can be properly identified from GIMs assuming that the VTEC within a certain range of days follows a Gaussian distribution (e.g., [20–23]). Under this consideration, the upper bound (UB) and lower bounds (LB) can be defined as $\mu + 2\sigma$ and $\mu - 2\sigma$, respectively (where μ is the mean and σ the standard deviation). In this way, a 3D cube of differential VTEC (ΔVTEC) at a confidence level of 95% was constructed based on the 3D VTEC cubes. Positive and negative ionospheric anomalies are identified in the 3D ΔVTEC cube when

$$\Delta\text{VTEC} = \begin{cases} +, & \text{if VTEC} > UB \\ 0, & \text{if } UB \geq \text{VTEC} \geq LB \\ -, & \text{if VTEC} < LB \end{cases} \quad (1)$$

2.2. Geomagnetic Conditions. In Figure 1, the variability of two important geomagnetic indices, Dst and Kp, during the

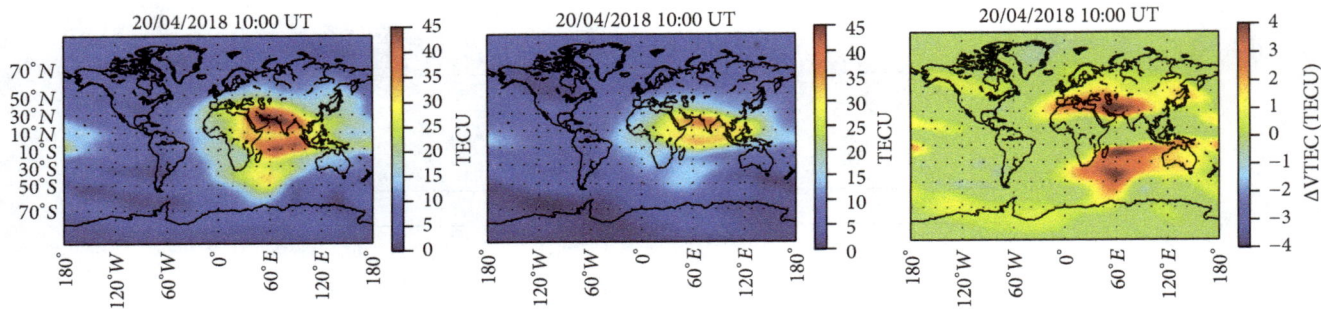

FIGURE 2: Left: GIM for April 20, 2018 at 10:00 UT. Center: map of the mean (μ) VTEC for April 20, 2018, at 10:00 UT. Right: differential VTEC map for April 20, 2018, at 10:00 UT.

month of April 2018 can be observed. While the Kp index indicates the severity of a storm, the Dst index is a proxy for the strength of westward terrestrial ring current [24]. As seen in Figure 1, the Kp index for April 20, 2018, between 06:00 and 12:00 has a value of Kp = 6, which according to the National Oceanic and Atmospheric Administration (NOAA) space weather scales points to a moderate G2 storm. On April 19, 2018, at 22:00 UT the Dst index rises to a value of 11 nT, which marks the storm sudden commencement (SSC), and it also indicates the start of the initial phase of the storm [25, 26]. On April 20, 2018, at 05:00 UT the Dst decreased drastically to a negative value of -19 nT. The Dst index reached its minimum of -66 nT at 09:00 UT, time in which the main phase finished. Kp and Dst data were downloaded from the German Research Center for Geosciences (https://www.gfz-potsdam.de/en/kp-index/) and the World Data Center for Geomagnetism in Kyoto (http://wdc.kugi.kyoto-u.ac.jp/wdc/Sec3.html), respectively.

In addition to the geomagnetic indices, a very crucial parameter that drives the development of a geomagnetic storm is the vertical interplanetary magnetic field (B_z). Gonzalez and Tsurutani [27] have indicated that, for strong storms, the observed southward B_z is lower than -10 nT and it remains below this threshold for at least three hours. On the other hand, Gonzalez et al. [25] pointed that, for moderate storms (-100 nT < Dst $\leq$ -50 nT), $B_z \leq$ -5 nT for at least 2 hours. It is the latter what is observed during the geomagnetic storm of April 20, 2018. In the bottom plot of Figure 1, it can be observed that $B_z \leq$ -5 nT, most of the time ranging within April 20, 2018, at 01:00 until 16:00 UT. Data for B_z in geomagnetic solar magnetospheric coordinate system was retrieved as hourly averages from the OMNI database (https://omniweb.gsfc.nasa.gov/form/dx1.html).

3. Results and Discussion

In order to better appreciate how maps of ΔVTEC improve the detection of ionospheric anomalies due to geomagnetic storms, Figure 2 shows a GIM (left), a map containing the mean VTEC (center), and the ΔVTEC map (right) for April 20, 2018, at 10:00 UT. It can be observed in the ΔVTEC map that the two enhanced plasma regions that form the EIA are

displaced towards the poles (~30°N and ~25°S), confirming what it was noticed for other geomagnetic storms in past studies (e.g., [8–12, 28, 29]). Furthermore in the ΔVTEC map of Figure 2, a well-defined LTE in the southern hemisphere at higher-middle latitudes (~44°S) can be clearly identified. The GIM is also showing this LTE; however, there is no doubt when looking at the ΔVTEC map that the LTE is properly displayed showing clearly its complete spatial distribution.

3.1. Localized TEC Enhancement. After carefully inspecting the generated ΔVTEC cube, as previously seen in Figure 2, a LTE was observed to appear on April 20, 2018, at ~06:00 UT (Figure 3). The LTE appeared about 8 hours after the SSC to the south of the EIA (~44°S) right over the Indian Ocean. At later times the LTE started to drift westward (same as the direction of the Sun). Edemskiy et al. [14] have also observed such displacement for the two LTEs they detected during the August 15, 2015, geomagnetic storm. At ~10:00 UT the LTE reached its maximum intensity with a ΔVTEC of 4.07 TECU. At this hour the EIA reached also its peak ΔVTEC intensity of 4.33 TECU in the northern plasma region. As seen in Figure 3, the enhancement of the EIA lasted only for about 6 hours (08:00 until 14:00 UT), whereas the LTE is almost visible until April 21, 2018, at 00:00 UT. At this time the LTE is located over the south Pacific to finally fade away. It is worth mentioning that approximately two hours after (16:00 UT) the enhanced EIA disappears, B_z returns to its regular variability. A day before (April 19, 2018) and after (April 21, 2018) the occurrence of the geomagnetic storm, the ΔVTEC maps for these days revealed that there was no EIA enhancement or appearance of LTEs.

In order to appreciate the shape of the EIA and the LTE, a meridional profile along the 58.5°E at 10:00 UT (14:00 local time for this longitudinal coordinate) is shown in Figure 4. At this time and longitudinal coordinate, which is pretty much the same for the capital city of Oman, Muscat, the LTE reached, as previously mentioned, its highest intensity. The EIA shape is also shown at the same UT for 2 days before and after the geomagnetic storm. From Figure 4 it can be noticed that, for April 18, 19, 21, and 22, 2018, at 10:00 UT, the EIA shape has a well-defined double-crest with a trough form [6], which is expected during daytime

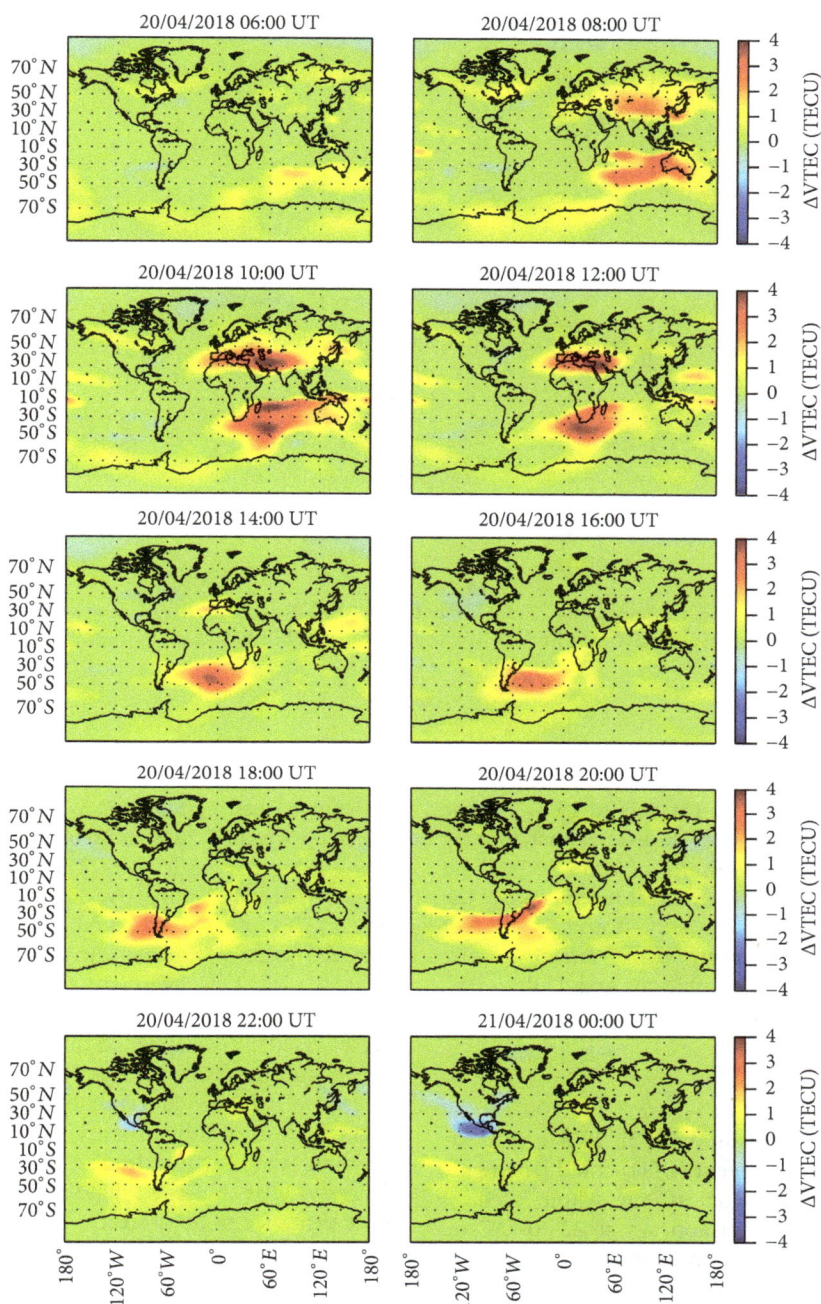

FIGURE 3: Differential VTEC maps between April 20, 2018 (06:00 UT), and April 21, 2018 (00:00 UT).

because of the equatorial fountain effect. During the day of the storm at 10:00 UT, Figure 4 also shows that the EIA is enhanced by showing a considerably increase of the VTEC in the northern hemisphere of ~133% and at the same time a poleward displacement of the crests, this being product of the super-fountain effect (e.g., [8–12, 29]). On the other hand, an unexpected response to the storm in the EIA shape is the additional presence of a third crest (~44°S), which corresponds to the LTE observed in the southern hemisphere in Figure 3. However, it can be seen that, during the days

before the storm (April 18 and 19, 2018) and after (April 21 and 22, 2018), the meridional profiles show as well a slight increment of their TEC in the same position of the observed third crest, for the LTE the peak increment in VTEC is of ~204%. Here it could be suggested that the super-fountain effect, which is caused due to southern flip of B_z, with some other physical mechanism may be playing an important role in the generation of this LTE.

Edemskiy et al. [14] have not only detected the two LTEs observed during the geomagnetic storm of August

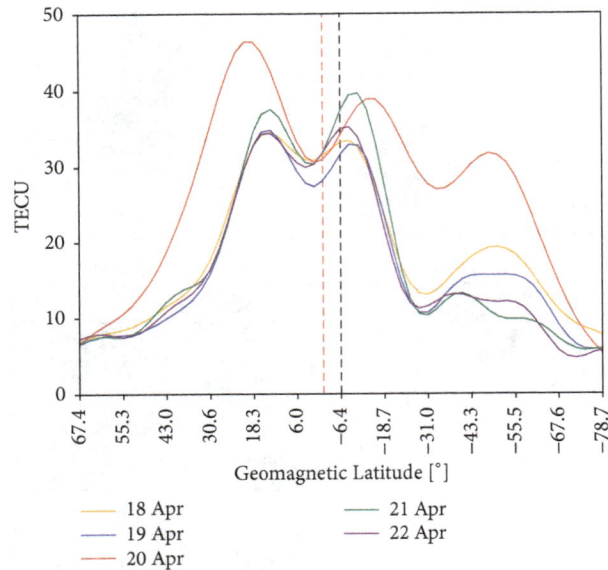

FIGURE 4: Shape of the VTEC for the 58.5°E meridian at 10:00 UT between April 18 and 22, 2018. A relevant range of geomagnetic latitudes is shown, 67.4°N - 78.7°S (75°N - 75°S in geographic latitudes). The vertical red and black dashed lines indicate the magnetic equator and the equator (geographic latitude = 0°), respectively.

15, 2015, but they also discovered further 26 LTEs for the period of 2002-2015. However, they indicated that, during the time of the solar activity minimum of 2006-2009, there was no detection of LTEs. Thus in the present study, the detected LTE during the April 20, 2018, storms is the first identification of such type of event (LTE) during a solar cycle minimum. According to Hathaway and Upton [15], between 2016 and 2019 the solar cycle 24 and 25 minimum is expected.

3.2. Localized TEC Decrement. In Figure 3 at 22:00 UT on April 20, 2018, and at 00:00 UT on April 21, 2018, a negative storm or a localized TEC decrement (LTD) could be already observed. However, for completeness, the whole event is presented in Figure 5. This LTD appears at 22:00 UT on April 20, 2018, right over the southern half of Mexico. At 00:00 UT on April 21, 2018, the LTD moved to the west and reached its minimum peak intensity with a ΔVTEC of -3.21 TECU. This negative storm was quite transitory; by 04:00 UT it completely disappeared. On the other hand the GIMs for this period of hours, also presented in Figure 5, show apparently that the northern plasma region of the EIA disappears, although from the GIMs alone this is not completely clear.

In order to have a better view of what is occurring with the EIA, meridional profiles for 100°W at 00:00 UT between April 19 and 23, 2018, are presented in Figure 6. The 100°W longitudinal coordinate is about the same as the one for Mexico City (capital of Mexico), where the local time is 19:00 hours. On April 19, 20, 22, and 13, 2018, at 00:00 UT (19:00 local time) the EIA shape shows a double-crest with a trough, which is still expected after sunset. However, what was unforeseen, as

to what is seen in Figure 5, was the disappearance of both crests on April 21, 2018, at 00:00 UT, time when the recovery phase is already under way. The disappearance of the crests corresponds to the negative ionospheric storm observed in the ΔVTEC maps of Figure 5. In Figure 6 it can be seen that the peak of the meridional profile for April 21, 2018, at 00:00 UT falls nearly in the location of the magnetic equator. A similar situation, where the northern and southern crests faded away, happened during the 2015 St. Patrick's Day storm on March 18, 2015, at the beginning of the recovery phase [30]. In the aforementioned study it was indicated that the westward disturbance electric field is the most likely mechanism that could inhibit the formation of both crests. Thus it can be suggested that the same process is causing the disappearance of both crests observed for the April 20, 2018 storm during the recovery phase.

4. Conclusions

Analysis of differential VTEC content maps during the G2 (moderate) geomagnetic storm that occurred on April 20, 2018, has revealed different ionospheric responses. The expected response to the super-fountain effect was observed for a period of 6 hours with a maximum increment of VTEC of 133% in the northern crest of the equatorial ionization anomaly.

A localized TEC enhancement has also been observed during the storm, located to the south of the EIA. The maximum increment of VTEC for the LTE is of 204%. This LTE appeared two hours before the EIA was enhanced and also lasted until nearly the end of the day. The LTE shows itself as a third crest in the longitudinal profile of the EIA.

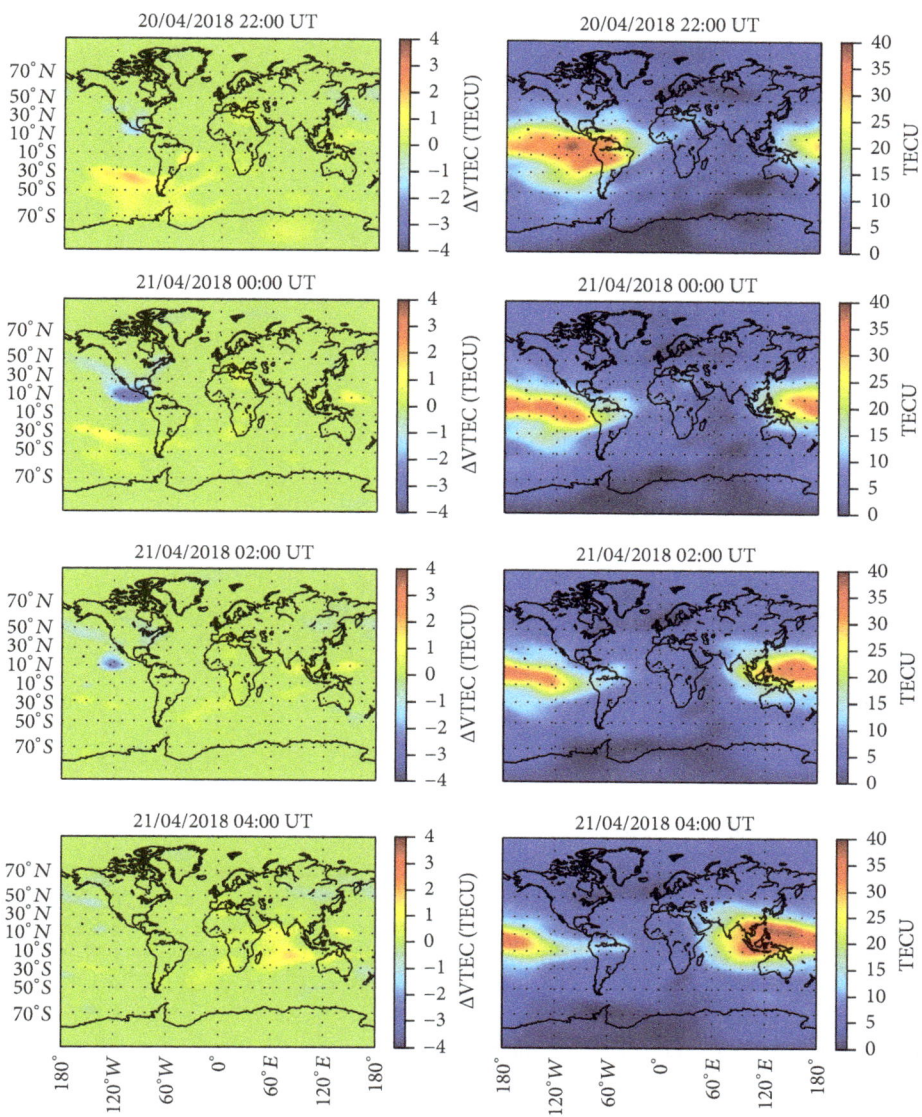

FIGURE 5: Left column: differential VTEC maps between April 20, 2018 (22:00 UT) and April 21, 2018 (04:00 UT). Right column: GIMs between April 20, 2018 (22:00 UT), and April 21, 2018 (04:00 UT).

It is suggested that the super-fountain effect may be playing an important role in the origin of the LTE. Moreover, this LTE is quite unique because, according to a previous study which detected 26 LTEs during 2002-2015, no LTEs have been observed during the solar two-cycle minimum of 2006-2009, and the April 20, 2018, storm occurred during the solar cycle 24 and 25 expected minimum.

Finally, a localized TEC decrement (or negative ionospheric storm) was also identified. This LTD revealed itself in the longitudinal profile of the EIA as the depletion of both its northern and southern crest. The LTD was observed to occur during the recovery phase of the storm. This similar behavior was observed for the 2015 St. Patrick's Day geomagnetic storm. The authors of that study attributed the inhibition of both crests to the westward disturbance dynamo electric field. Hence, the origin of the LTD during the April 20, 2018, storm is considered to be due to the same mechanism. However, more observations of such types of LTDs are needed in order to understand better their origin.

Acknowledgments

The author would like to acknowledge the Center for Orbit Determination in Europe for making IONEX files available to the public.

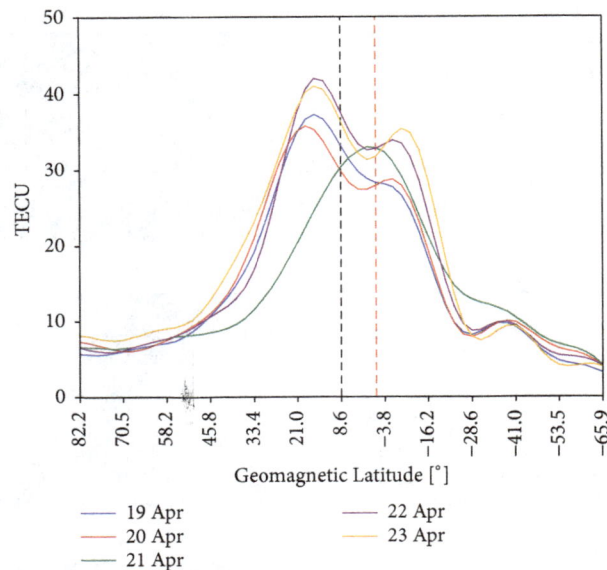

FIGURE 6: Shape of the VTEC for the 100°W meridian at 00:00 UT between April 19 and 23, 2018. A relevant range of geomagnetic latitudes is shown, 82.2°N - 65.9°S (75°N - 75°S in geographic latitudes). The vertical red and black dashed lines indicate the magnetic equator and the equator (geographic latitude = 0°), respectively.

References

[1] J. J. Love and C. A. Finn, "The USGS geomagnetism program and its role in space weather monitoring," *Space Weather Journal*, vol. 9, no. 7, 2011.

[2] M. H. MacAlester and W. Murtagh, "Extreme space weather impact: An emergency management perspective," *Space Weather Journal*, vol. 12, no. 8, pp. 530–537, 2014.

[3] I. Tsagouri, A. Belehaki, G. Moraitis, and H. Mavromichalaki, "Positive and negative ionospheric disturbances at middle latitudes during geomagnetic storms," *Geophysical Research Letters*, vol. 27, no. 21, pp. 3579–3582, 2000.

[4] I. Horvath and B. C. Lovell, "Positive and negative ionospheric storms occurring during the 15 May 2005 geomagnetic superstorm," *Journal of Geophysical Research: Space Physics*, vol. 120, no. 9, pp. 7822–7837, 2015.

[5] P. R. Fagundes, F. A. Cardoso, B. G. Fejer, K. Venkatesh, B. A. G. Ribeiro, and V. G. Pillat, "Positive and negative GPS-TEC ionospheric storm effects during the extreme space weather event of March 2015 over the Brazilian sector," *Journal of Geophysical Research: Space Physics*, vol. 121, no. 6, pp. 5613–5625, 2016.

[6] E. V. Appleton, "Two anomalies in the ionosphere [1]," *Nature*, vol. 157, article 691, 1946.

[7] C. Stolle, C. Manoj, H. Lühr, S. Maus, and P. Alken, "Estimating the daytime Equatorial Ionization Anomaly strength from electric field proxies," *Journal of Geophysical Research: Space Physics*, vol. 113, no. A9, 2008.

[8] T. Tanaka and K. Ohtaka, "Ionospheric disturbances during low-latitude auroral events and their association with magnetospheric processes," *Journal of Geophysical Research: Space Physics*, vol. 101, no. A8, pp. 17151–17159, 1996.

[9] B. Tsurutani, A. Mannucci, B. Iijima et al., "Global dayside ionospheric uplift and enhancement associated with interplanetary electric fields," *Journal of Geophysical Research: Space Physics*, vol. 109, no. A8, 2004.

[10] C. H. Lin, A. D. Richmond, R. A. Heelis et al., "Theoretical study of the low- and midlatitude ionospheric electron density enhancement during the October 2003 superstorm: Relative importance of the neutral wind and the electric field," *Journal of Geophysical Research: Atmospheres*, vol. 110, no. A12, 2005.

[11] A. J. Mannucci, B. T. Tsurutani, B. A. Iijima et al., "Dayside global ionospheric response to the major interplanetary events of October 29-30, 2003 'Halloween Storms'," *Geophysical Research Letters*, vol. 32, no. 12, 2005.

[12] E. Astafyeva, "Effects of strong IMF Bz southward events on the equatorial and mid-latitude ionosphere," *Annales Geophysicae*, vol. 27, no. 3, pp. 1175–1187, 2009.

[13] C.-S. Huang, J. C. Foster, and M. C. Kelley, "Long-duration penetration of the interplanetary electric field to the low-latitude ionosphere during the main phase of magnetic storms," *Journal of Geophysical Research: Space Physics*, vol. 110, no. A11, 2005.

[14] I. Edemskiy, J. Lastovicka, D. Buresova, J. Bosco Habarulema, and I. Nepomnyashchikh, "Unexpected Southern Hemisphere ionospheric response to geomagnetic storm of 15 August 2015," *Annales Geophysicae*, vol. 36, no. 1, pp. 71–79, 2018.

[15] D. H. Hathaway and L. A. Upton, "Predicting the amplitude and hemispheric asymmetry of solar cycle 25 with surface flux transport," *Journal of Geophysical Research: Space Physics*, vol. 121, no. 11, pp. 744–753, 2016.

[16] A. Coster and A. Komjathy, "Space weather and the global positioning system," *Space Weather Journal*, vol. 6, no. 6, 2008.

[17] M. Hernández-Pajares, J. M. Juan, J. Sanz et al., "The IGS VTEC maps: a reliable source of ionospheric information since 1998," *Journal of Geodesy*, vol. 83, no. 3-4, pp. 263–275, 2009.

[18] S. Schaer, W. Gurtner, and J. Feltens, "Ionex: The ionosphere map exchange format version 1," 233–247.

[19] T. E. Oliphant, "Python for scientific computing," *Computing in Science & Engineering*, vol. 9, no. 3, Article ID 4160250, pp. 10–20, 2007.

[20] J. Y. Liu, Y. J. Chuo, S. J. Shan et al., "Pre-earthquake ionospheric anomalies registered by continuous GPS TEC measurements," *Annales Geophysicae*, vol. 22, no. 5, pp. 1585–1593, 2004.

[21] F. Zhu, Y. Wu, J. Lin, and Y. Zhou, "Temporal and spatial characteristics of VTEC anomalies before Wenchuan Ms8.0 earthquake," *Geodesy and Geodynamics*, vol. 1, no. 1, pp. 23–28, 2010.

[22] Y. B. Yao, P. Chen, S. Zhang, J. J. Chen, F. Yan, and W. F. Peng, "Analysis of pre-earthquake ionospheric anomalies before the global M = 7.0+ earthquakes in 2010," *Natural Hazards and Earth System Sciences*, vol. 12, no. 3, pp. 575–585, 2012.

[23] J. Li, G. Meng, X. You, R. Zhang, H. Shi, and Y. Han, "Ionospheric total electron content disturbance associated with May 12, 2008, Wenchuan earthquake," *Geodesy and Geodynamics*, vol. 6, no. 2, pp. 126–134, 2015.

[24] M. Sugiura, "Hourly value of equatorial Dst for the IGY," *Annals of the International Geophysical Year*, vol. 35, 1964.

[25] W. D. Gonzalez, J. A. Joselyn, Y. Kamide et al., "What is a geomagnetic storm?" *Journal of Geophysical Research: Atmospheres*, vol. 99, no. A4, pp. 5771–5792, 1994.

[26] W. Wang, J. Lei, A. G. Burns et al., "Ionospheric response to the initial phase of geomagnetic storms: Common features," *Journal of Geophysical Research: Space Physics*, vol. 115, no. A7, 2010.

[27] W. D. Gonzalez and B. T. Tsurutani, "Criteria of interplanetary parameters causing intense magnetic storms (Dst < -100 nT)," *Planetary and Space Science*, vol. 35, no. 9, pp. 1101–1109, 1987.

[28] B. Zhao, W. Wan, and L. Liu, "Responses of equatorial anomaly to the October-November 2003 superstorms," *Annales Geophysicae*, vol. 23, no. 3, pp. 693–706, 2005.

[29] E. Astafyeva, I. Zakharenkova, and M. Förster, "Ionospheric response to the 2015 St. Patrick's Day storm: A global multi-instrumental overview," *Journal of Geophysical Research: Space Physics*, vol. 120, no. 10, pp. 9023–9037, 2015.

[30] L. Spogli, C. Cesaroni, D. Di Mauro et al., "Formation of ionospheric irregularities over Southeast Asia during the 2015 St. Patrick's Day storm," *Journal of Geophysical Research: Space Physics*, vol. 121, no. 12, pp. 211–233, 2016.

12

TEM Response of a Large Loop Source over the Multilayer Earth Models

A. K. Tiwari, S. P. Maurya, and N. P. Singh (iD)

Department of Geophysics, Faculty of Science, Banaras Hindu University, Varanasi 221005, India

Correspondence should be addressed to N. P. Singh; singhnpbhu@yahoo.co.in

Academic Editor: Rudolf A. Treumann

The general expression of TEM response of large loop source over the layered earth models is not available in the literature for arbitrary source-receiver positions, except for the case of central loop and coincident loop configurations over the homogeneous earth model. In the present study, an attempt is made to present the TEM response of a large loop source over the layered earth model for arbitrary receiver positions. The frequency domain responses of large loop source over the layer earth model for arbitrary receiver positions are converted into the impulse (time derivative of magnetic field) TEM response using Fourier cosine or sine transform. These impulse TEM responses in turn are converted into voltage responses for arbitrary receiver positions, namely, central loop, arbitrary in-loop, and offset-loop TEM responses over the layered earth models. For checking the accuracy of the method, results are compared with the results obtained using analytical expression over a homogeneous earth model. The complete matching of both of the results suggests that the present computational technique is capable of computing TEM response of large loop source over the homogeneous earth model with high accuracy. Thereafter, the technique is applied for computation of TEM response of a large loop source over the layered earth (2-layer, 3-layer, and 4-layer) models for the central loop, in-loop, and offset-loop configurations and the results are presented in voltage decay form. The results depict their characteristic variations. These results would be useful for modeling and inversion of large loop TEM data over the layer earth models for all the possible configurations resulting from a large loop source.

1. Introduction

A large horizontal loop on or above the earth is one of the most widely used sources in TEM and airborne TEM (ATEM) methods. Using a large loop source, one can measure at the center of the loop, at any arbitrary point inside the loop, coincident loop, and at arbitrary offset-loop points. A number of studies on methods of computation of TEM response and modeling can be found in the literature [1–13]. All these studies are based on considering transmitter as a large circular loop and receiver either at the center of the loop and/or coincident with the loop, over the simple layer/homogenous earth models, and majority of them are based on the computation of TEM responses from their frequency responses only for the central and/or coincident loop configurations, because of the computational intricacies associated with the frequency domain response computation of large loop TEM source for arbitrary receiver positions

(except for central and coincident loop configurations) over the layer earth models which involve product of two Basel functions that make the integral unstable and nonconvergent. Singh and Mogi [14, 15] for the first time presented a new computational method for computation of EM responses of large loop source over the layer earth models for arbitrary receiver positions. Thereafter, a number of researches appeared in the literature [16–24]. Thus, there was scarcity of EM modeling due to large loop sources over the layer earth models even in frequency domain, which resulted in a lack of studies in TEM response of large loop sources over the layer earth models for arbitrary receiver positions, except for the case of central loop and coincident loop configurations that were too over the homogeneous half-space because of nonavailability of analytical expression for arbitrary receiver positions, and sophisticated computational methods for computing the frequency domain responses from which TEM responses are usually derived. Moreover, during the process of converting the frequency domain EM response into time domain, even a small error of less than 1% in frequency response can produce a considerable difference in time domain EM response of large loop source over the layered earth models for all the configurations. Therefore, with the view of overcoming this drawback and filling the gap in the existing literature for TEM response of large loop sources, in this study, an attempt is made to present a reliable method for computation of TEM response of a large loop source over the layer earth models for any arbitrary receiver positions using the frequency response computation method described in [14, 15]. The computed TEM voltage response is compared with the available TEM responses over the homogeneous earth model for checking the accuracy and reliability of the method.

2. Theoretical Background

The plan view of large loop TEM method with central loop, in-loop, and offset-loop configurations over a layer earth is shown in Figure 1. The large loop presents a source loop and the small loop at the center of the source loop (P_1) represents receiver position corresponding to the central loop configuration. Similarly, the large loop presents a source loop and small loop at arbitrary position inside the source loop (P_2) represents receiver position corresponding to the in-loop configuration and small loop at arbitrary point outside the source loop (P_3) represents receiver position corresponding to arbitrary offset-loop configuration.

The frequency domain expressions of EM field components at a point on or above the surface of an n-layered earth due to a finite horizontal circular loop of radius a, carrying a current $Ie^{i\omega t}$ and placed at the height $z = -h$ above the surface of layered earth model, can be found in [9]. The expression of H_z field component at a measurement point on the surface of n-layered earth (i.e., at $z = 0$) can be written as

$$H_z(\omega, \rho, h)$$
$$= \frac{Ia}{2} \int_0^\infty \frac{\left[e^{-u_0 h}(1 + r_{TE})\right] \lambda^2}{u_0} J_1(\lambda a) J_0(\lambda r) d\lambda, \quad (1)$$

where $r_{TE} = (Y_0 - \hat{Y}_1)/(Y_0 + \hat{Y}_1)$ with $Y_0 = u_0/i\omega\mu_0$ (intrinsic admittance of free space) and $\hat{Y}_1 = H_y^{TE}/E_x^{TE} = -H_x^{TE}/E_y^{TE}$ (surface admittance at $z = 0$).

For an n-layer case, the surface admittances are given by the recurrence relation

$$\hat{Y}_1 = Y_1 \frac{\hat{Y}_2 + Y_1 \tanh(u_1 h_1)}{Y_1 + \hat{Y}_2 \tanh(u_1 h_1)},$$
$$\hat{Y}_n = Y_n \frac{\hat{Y}_{n+1} + Y_n \tanh(u_n h_n)}{Y_n + \hat{Y}_{n+1} \tanh(u_n h_n)}, \quad (2)$$

and $\hat{Y}_n = Y_n$ with $Y_n = u_n/i\omega\mu_n$, $u_n = (k_x^2 + k_y^2 - k_n^2)^{1/2} = (\lambda^2 - k_n^2)^{1/2}$, and $k_n^2 = \omega^2 \mu_n \varepsilon_n - i\omega\mu_n\sigma_n$.

Central loop configuration $OP_1 = 0.0$
In-loop configuration $OP_2 = a/2$
Offset loop configuration $OP_3 = 2a$
Loop radius $= a$

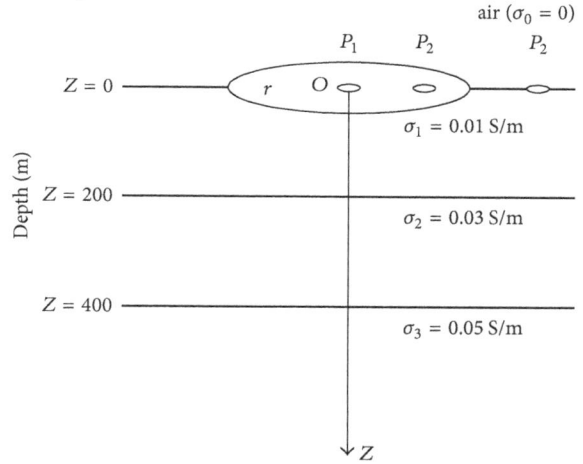

FIGURE 1: Schematic diagram showing all the possible configurations of large loop TEM system over a 3-layer earth model.

Here, r is source-receiver offset (measured from the center of the loop). For calculation purposes, $\tanh(u_n h_n)$ is used in its exponential form for stability reasons [5].

Thereafter, the time derivative of vertical magnetic field ($\partial h_z/\partial t$) is obtained by transforming the frequency domain solution of vertical magnetic field computed using [14] into the time domain solution using the Fourier cosine and/or sine transform as given in [25].

$$\frac{\partial h_z(t)}{\partial t} = \frac{-2}{\pi} \int_0^\infty \text{Re}\left[H_z(\omega, \rho, h)\right] \cos(\omega t) d\omega,$$
$$\frac{\partial h_z(t)}{\partial t} = \frac{2}{\pi} \int_0^\infty \text{Im}\left[H_z(\omega, \rho, h)\right] \sin(\omega t) d\omega, \quad (3)$$

where $\text{Re}[H_z(\omega, \rho, h)]$ and $\text{Im}[H_z(\omega, \rho, h)]$ are the real and imaginary parts of the vertical magnetic field over a layered earth model in frequency domain. The components of vectors ρ and h are the resistivities and thickness of different layers of the layered earth model, and ω is the angular frequency.

In general, the data collected from a large loop TEM system consist of vertical voltage measurements made at various time intervals after the current in the transmitter is turned off. These voltage measurements are related to the time derivatives of vertical magnetic field ($\partial h_z/\partial t$) in accordance with the following relation:

$$V(t) = -\mu_0 \left(\frac{\partial h_z}{\partial t}\right) M, \quad (4)$$

where M is the area-turns product of the receiver coil. These voltage data can be further transformed into the apparent resistivity because sometimes it is preferable to use apparent resistivity transformation to have a direct relation with the geoelectrical section, as well as an initial estimate of the layer resistivities, which are often required in nonlinear inversion for interpretation of TEM data.

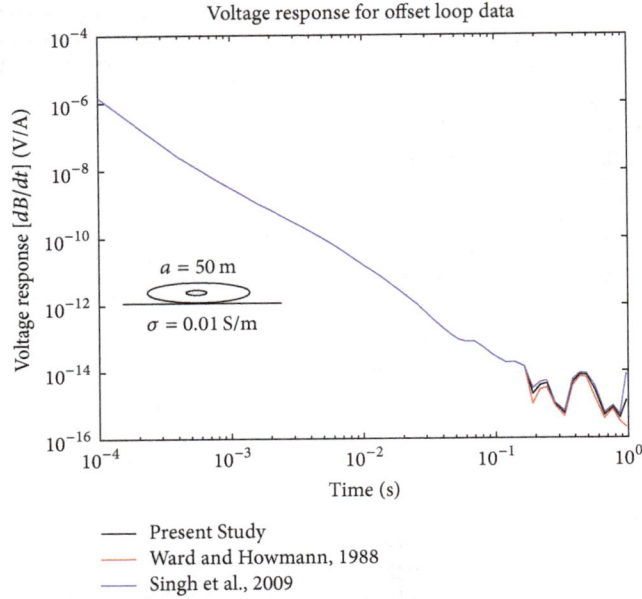

FIGURE 2: Comparison of results computed using the present method for voltage response of vertical magnetic field at the center of a large circular loop source of radius, $a = 50$ m, placed over the surface of a homogeneous earth model of conductivity ($\sigma = 0.01$ S/m) with the results generated using the analytical expression given by ward and Hohmann (1988) and Singh et al. (2009).

Therefore, starting with the computation of $H_z(\omega, \rho, h)$ field (see (1)), using the method and algorithm described in [14, 15], we have computed the time derivative of vertical magnetic field using the Fourier cosine and sine transforms (see (3)) [8, 25–27]. Thereafter, the transient voltage response is computed using (4). Computations are performed for the homogeneous, 2-layer, 3-layer, and 4-layer resistive and conductive earth models and the results are presented in the following section.

3. Results and Discussions

3.1. Check for Validity of the Method. For checking the validity and accuracy of the method, we applied it for the computation of TEM (impulse) response of a large circular loop source over the surface of a homogeneous earth model for central loop configurations, and the result is compared with the published results for TEM response of a large loop source generated using the central loop analytical expression for impulse response [8, 9]. The analytic expression for the impulse response of the h_z field at the center of a large loop source of radius (a) over the homogeneous earth model of conductivity σ can be written as [8, 9]

$$h_{z|\text{Impulse}} = \frac{\partial h_{z|\text{Step}}}{\partial t}$$

$$= \frac{I}{\mu_0 \sigma a^3} \left[3 \, \text{erf}(\theta a) - \frac{2}{\pi^{1/2}} \theta a \left(3 + 2\theta^2 a^2 \right) e^{-\theta^2 a^2} \right], \quad (5)$$

where $\theta = (\mu_0 \sigma/4t)^{1/2}$, erf indicates the error function, and t means the delay time after the current is turned off.

3.2. Illustration of TEM Response over Layer Earth Models. For illustrating the accuracy, applicability, and efficiency of the program for generating voltage response due to large loop TEM methods over the layer earth models for arbitrary receiver positions, we applied it for computing the TEM response over the homogeneous earth model of conductivity 0.01 S/m and compared the results with the published results [8, 9] (Figure 2). From Figure 2, it is evident that the computed results match very well the published results computed using the analytical expressions [9], thereby indicating the accuracy and validity of the method for computation of TEM response of large loop source over the homogeneous earth models.

Further, the method is applied for computation of TEM responses of a large loop source over the 2-layer, 3-layer, and 4-layer earth models for arbitrary receiver positions, that is, central loop, in-loop, and offset-loop configurations, for different loop sizes, and the results are presented in Figures 3–5.

Figure 3 presents TEM responses of a large loop source over a two-layer earth model (as shown in the inset of the figure) for central loop, in-loop, and offset-loop configurations, respectively, for source loop radii 100 m and 200 m. The computed results depict characteristic features of TEM response over the two-layer earth models.

Figure 4 presents TEM response of a large loop source over the three-layer earth model (as shown in the inset of the figure) for loop sizes of radii 100 m and 200 m, for central loop, in-loop, and offset-loop configurations. The computed results depict characteristic response over the three-layer earth models.

Figure 5 depicts TEM response of a large loop source over the four-layer earth model (as shown in the inset

(a)

(b)

(c)

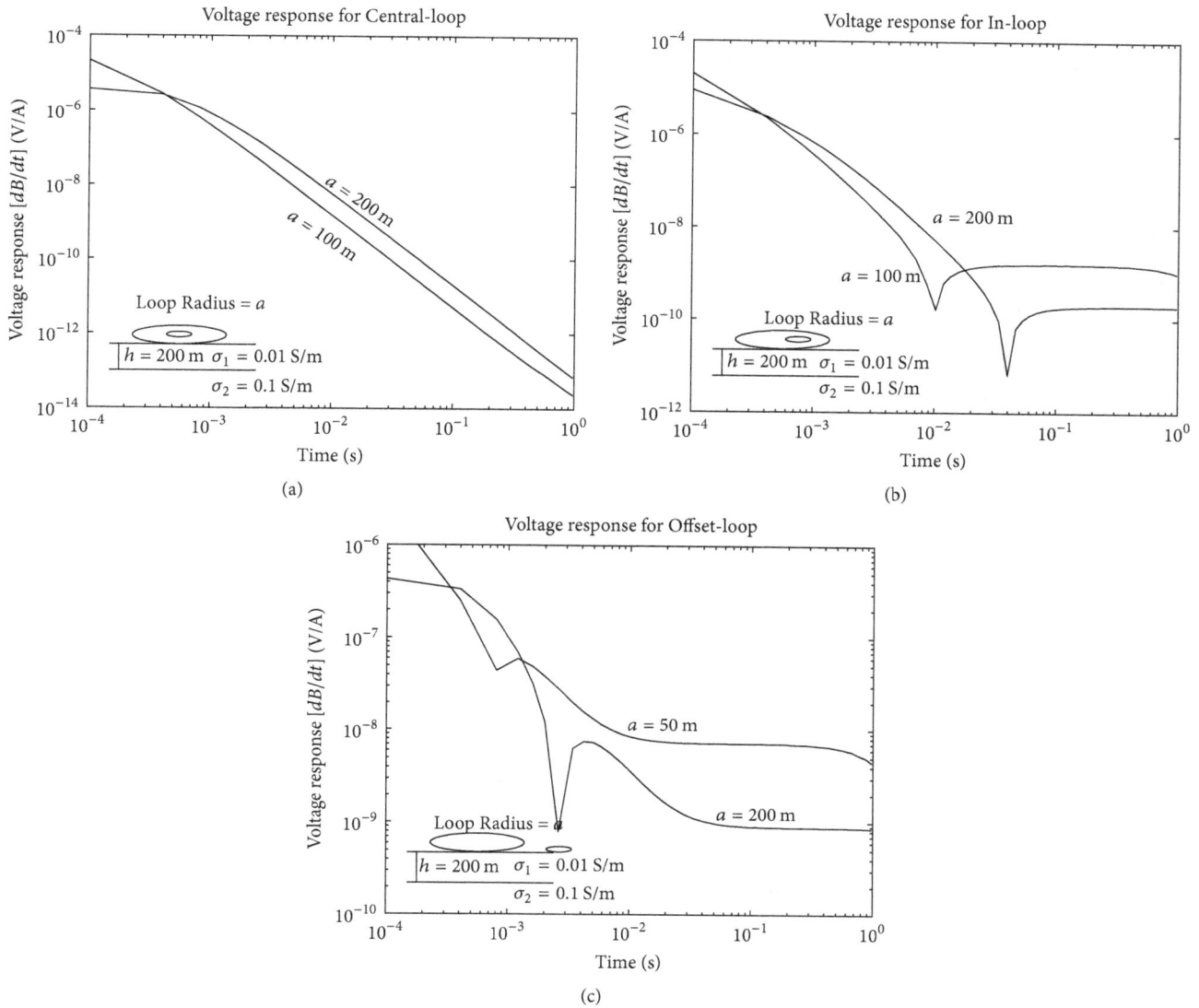

FIGURE 3: Transient voltage response due to large circular loop source of radii 100 m and 200 m over the 2-layer earth model (as shown in the inset of the figure) for (a) central loop configuration, $r = 0$, (b) arbitrary in-loop configuration, $r = a/2$, and (c) arbitrary offset-loop configuration, $r = 2a$.

of the figure) for source loop sizes of radii 100 m and 200 m, for central loop, in-loop, and arbitrary offset-loop configurations, respectively. The results depict characteristic response of large loop sources over the four-layer earth models.

From these figures (Figures 3–5), it is clear that, with the increase in loop size, the TEM response shows smooth and well-defined characteristics. Moreover, it is also noticed that the central loop TEM responses are more regular as compared to the offset-loop and arbitrary in-loop responses.

4. Conclusion

The present article describes a simple and sophisticated method for computation of TEM response of a large loop source over layered earth model for arbitrary receiver

positions, that is, at the center of the loop, at arbitrary in-loop, and at arbitrary offset-loop points. The method is based on conversion of frequency domain results computed using EMLCLTR program of Singh and Mogi [15] into the time derivative of the vertical magnetic field, that is, impulsive TEM responses. The method is simple and suitable for computation of TEM response of a large loop source at any arbitrary receiver locations, that is, central loop, in-loop, and offset-loop configurations. During its frequency domain computation, it has the option of computing frequency domain responses with or without the inclusion of displacement current factor and hence is a more reliable and accurate one.

For illustrating the nature and characteristics of large loop TEM responses over the layer earth models at arbitrary receiver positions, results are presented for the TEM response

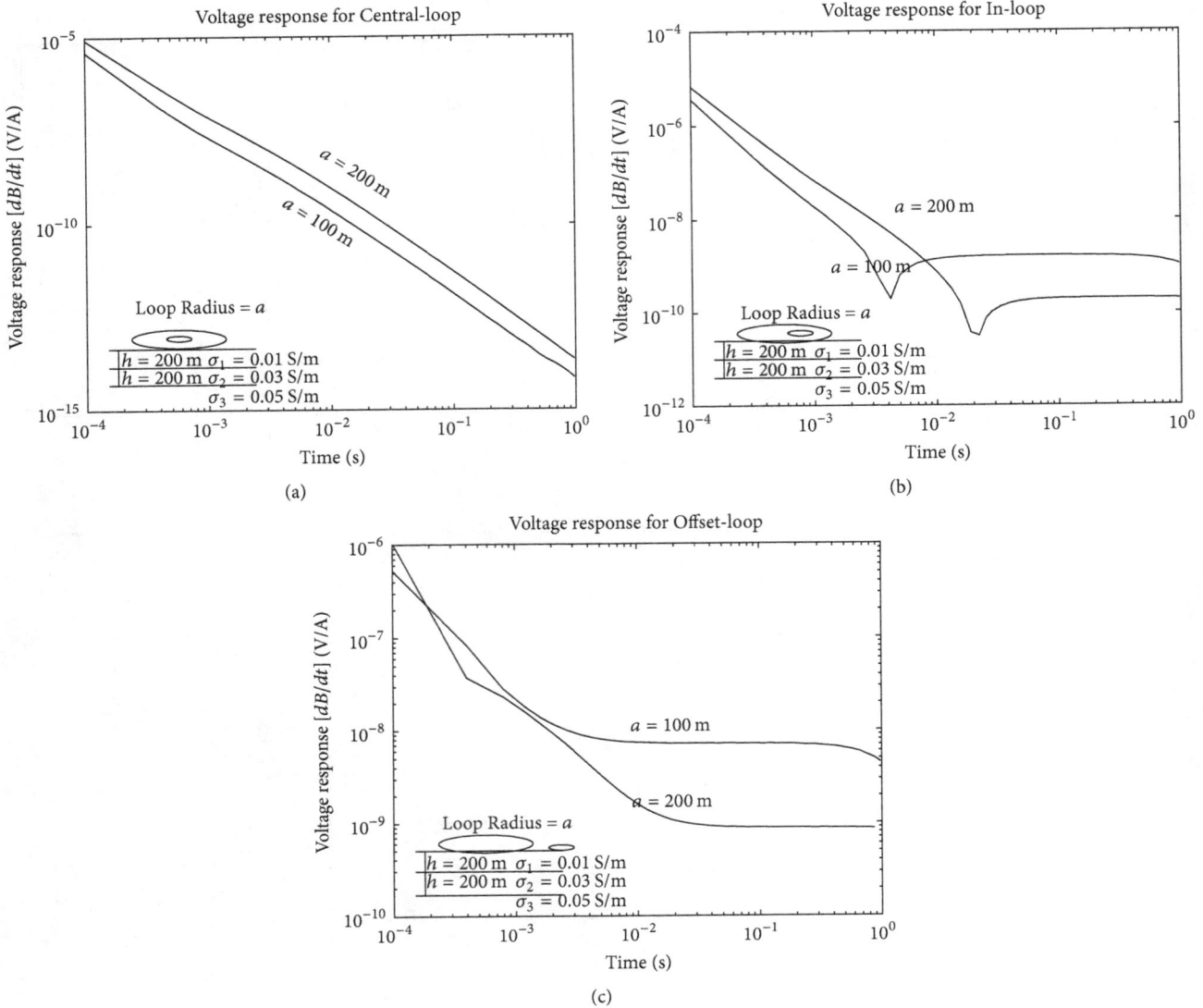

Voltage response for Central-loop

Voltage response for In-loop

(a)

(b)

Voltage response for Offset-loop

(c)

FIGURE 4: Transient voltage response of large circular loop source of radii 100 m and 200 m over the 3-layer earth model (as shown in the inset of the figure) for (a) central loop configuration, $r = 0$, (b) arbitrary in-loop configuration, $r = a/2$, and (c) arbitrary offset-loop configuration, $r = 2a$.

at source-receiver offsets $r = 0$, $r = a/2$, and $r = 2a$ pertaining to the central loop, in-loop, and offset-loop configurations over 2-layer, 3-layer, and 4-layer earth models. The results depict their characteristic variations. From the results, it is noticed that the voltage response curves are more regular for central loop, followed by offset loop, and the least for the in-loop configurations. Further, it is also observed that, with the increase in source loop size, the voltage responses become smoother.

This study would enable the prospect, development, and use of loop-in-loop method (in-loop method) and loop-offset-loop method (offset-loop method) along with the well-developed central loop and coincident loop methods using large loop sources, and thus it would enhance the applicability and cost-effectiveness of the large loop source transient electromagnetic method for exploration and applied geophysics applications.

Acknowledgments

One of the authors (A. K. Tiwari) is thankful to the Department of Geophysics, Banaras Hindu University, Varanasi, for providing the opportunity and facility for completion of this work.

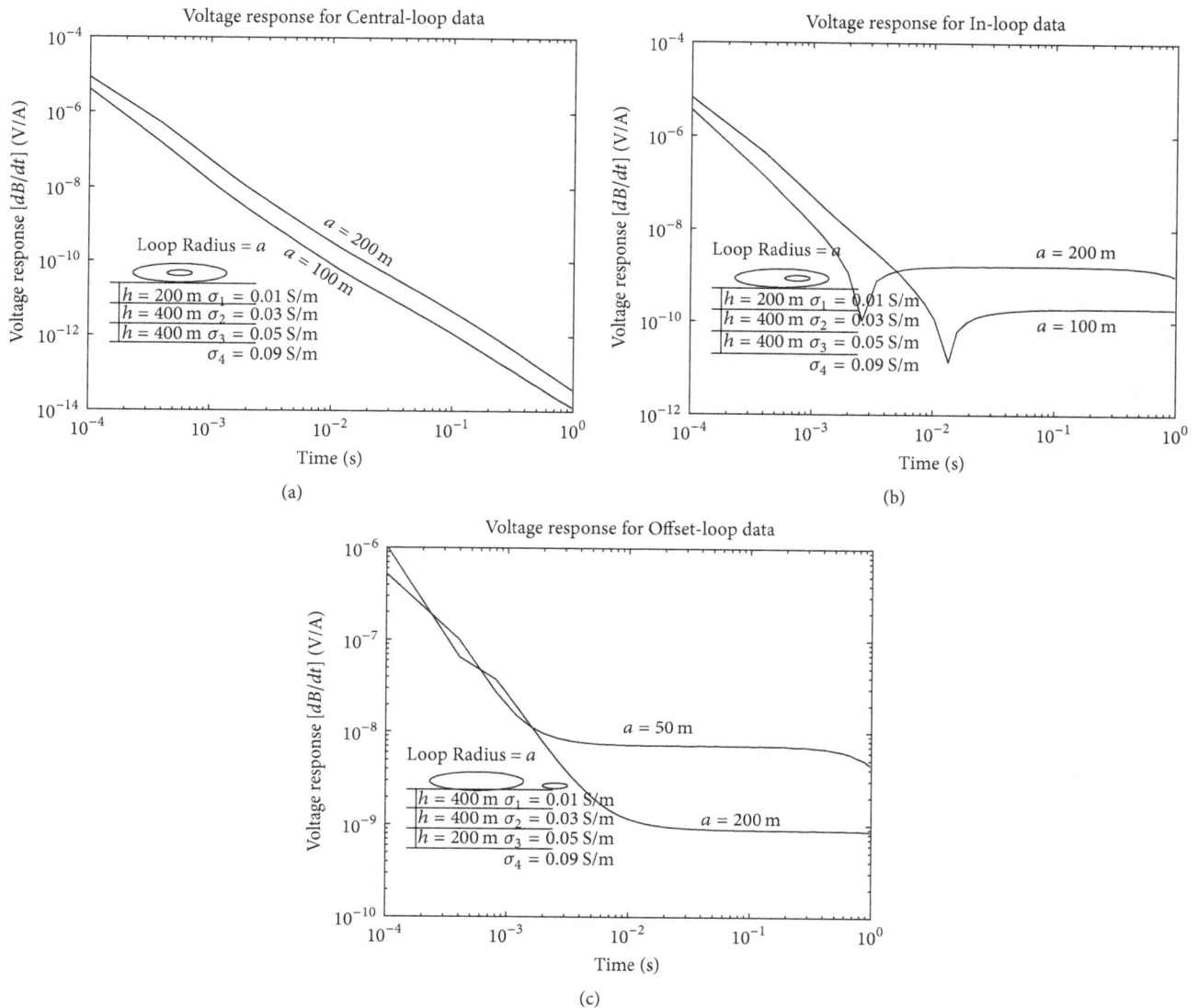

FIGURE 5: Transient voltage response of large circular loop source of radii 100 m and 200 m over the 4-layer earth model (as shown in the inset of the figure) for (a) central loop configuration, $r = 0$, (b) arbitrary in-loop configuration, $r = a/2$, and (c) arbitrary offset-loop configuration, $r = 2a$.

References

[1] A. V. Christiansen and N. B. Christensen, "A quantitative appraisal of airborne and ground-based transient electromagnetic (TEM) measurements in Denmark," *Geophysics*, vol. 68, no. 2, pp. 523–534, 2003.

[2] H. Huang and G. J. Palacky, "Damped least-squares inversion of time-domain airborne EM data based on singular value decomposition," *Geophysical Prospecting*, vol. 39, no. 6, pp. 827–844, 1991.

[3] T. Ingeman-Nielsen and F. Baumgartner, "CR1Dmod: A Matlab program to model 1D complex resistivity effects in electrical and electromagnetic surveys," *Computers & Geosciences*, vol. 32, no. 9, pp. 1411–1419, 2006.

[4] H. Jang, H. Jang, K. H. Lee, and H. J. Kim, "Step-off, vertical electromagnetic responses of a deep resistivity layer buried in marine sediments," *Journal of Geophysics and Engineering*, vol. 10, no. 2, Article ID 025011, 2013.

[5] J. H. Knight and A. P. Raiche, "Transient electromagnetic calculations using the Gaver- Stehfest inverse Laplace transform method.," *Geophysics*, vol. 47, no. 1, pp. 47–50, 1982.

[6] H. F. Morrison, R. J. Phillips, and D. P. Obrien, "Quantitative interpretation of transient electromagnetic fields over a layered half space," *Geophysical Prospecting*, vol. 17, no. 1, pp. 82–101, 1969.

[7] A. P. Raiche, "Transient electromagnetic field computations for polygonal loops on layered earths," *Geophysics*, vol. 52, no. 6, pp. 785–793, 1987.

[8] N. P. Singh, M. Utsugi, and T. Kagiyama, "TEM response of a large loop source over a homogeneous earth model: A generalized expression for arbitrary source-receiver offsets," *Pure and Applied Geophysics*, vol. 166, no. 12, pp. 2037–2058, 2009.

[9] S. H. Ward and G. W. Hohmann, "Electromagnetic theory for geophysical applications," in *Electromagnetic Methods in Applied Geophysics*, M. N. Nabighian, Ed., vol. 1, pp. 131–311,

Society of Exploration Geophysicists, Tulsa, Oklahoma, 1988.

[10] G.-Q. Xue, C.-Y. Bai, and X. Li, "Extracting the virtual reflected wavelet from TEM data based on regularizing method," *Pure and Applied Geophysics*, vol. 169, no. 7, pp. 1269–1282, 2012.

[11] G. Q. Xue, S. Yan, and N. N. Zhou, "Theoretical study on the errors caused by dipole hypothesis of large-loop TEM response," *Diqiu Wuli Xuebao*, vol. 54, no. 9, pp. 2389–2396, 2011.

[12] M. S. Zhdanov, "Electromagnetic geophysics: Notes from the past and the road ahead," *Geophysics*, vol. 75, no. 5, pp. 75A49–75A66, 2010.

[13] M. S. Zhdanov, D. A. Pavlov, and R. G. Ellis, "Localized S-inversion of time-domain electromagnetic data," *Geophysics*, vol. 67, no. 4, pp. 1115–1125, 2002.

[14] N. P. Singh and T. Mogi, "EMLCLLER-a program for computing the EM response of a large loop source over a layered earth model," *Computers & Geosciences*, vol. 29, no. 10, pp. 1301–1307, 2003.

[15] N. P. Singh and T. Mogi, "Electromagnetic response of a large circular loop source on a layered Earth: a new computation method," *Pure and Applied Geophysics*, vol. 162, no. 1, pp. 181–200, 2005.

[16] S. Cristina and M. Parise, "Fast calculation of theoretical response curves for induction depth sounding," in *Proceedings of the European Microwave Week 2009, EuMW 2009: Science, Progress and Quality at Radiofrequencies - 39th European Microwave Conference, (EuMC '09)*, pp. 1567–1570, October 2009.

[17] M. Jamie, "Note on: 'EMLCLLER—A program for computing the EM response of a large loop source over a layered earth model' by N.P. Singh and T. Mogi, Computers & Geosciences 29 (2003) 1301–1307," *Computers & Geosciences*, vol. 96, pp. 236–246, 2016.

[18] M. Parise, "Fast computation of the forward solution in controlled-source electromagnetic sounding problems," *Progress in Electromagnetics Research*, vol. 111, pp. 119–139, 2011.

[19] M. Parise, "Full-wave analytical explicit expressions for the surface fields of an electrically large horizontal circular loop antenna placed on a layered ground," *IET Microwaves, Antennas & Propagation*, vol. 11, no. 6, pp. 929–934, 2017.

[20] Y. Qi, L. Huang, X. Wu, G. Fang, and G. Yu, "Effect of loop geometry on TEM response over layered earth," *Pure and Applied Geophysics*, vol. 171, no. 9, pp. 2407–2415, 2014.

[21] J. T. Ratnanather, J. H. Kim, S. Zhang, A. M. Davis, and S. K. Lucas, "Algorithm 935: IIPBF, a MATLAB toolbox for infinite integral of products of two Bessel functions," *ACM Transactions on Mathematical Software (TOMS)*, vol. 40, no. 2, p. 14, 2014.

[22] N. P. Singh and T. Mogi, "EMDPLER: A F77 program for modeling the EM response of dipolar sources over the non-magnetic layer earth models," *Computers & Geosciences*, vol. 36, no. 4, pp. 430–440, 2010.

[23] J. Deun and R. Cools, "Note on electromagnetic response of a large circular loop source on a layered earth: A new computation method," in *Pure and Applied Geophysics*, N. P. Singh and T. Mogi, Eds., vol. 164, pp. 1107–1111, 2007.

[24] G.-Q. Xue, X. Li, L. J. Gelius, Z.-P. Qi, N.-N. Zhou, and W.-Y. Chen, "A new apparent resistivity formula for in-loop fast sounding TEM theory and application," *Journal of Environmental & Engineering Geophysics*, vol. 20, no. 2, pp. 107–118, 2015.

[25] G. A. Newman, G. W. Hohmann, and W. L. Anderson, "Transient electromagnetic response of a three-dimensional body in a layered earth," *Geophysics*, vol. 51, no. 8, pp. 1608–1627, 1986.

[26] W. L. Anderson, "Numerical integration of related Hankel transforms of orders 0 and 1 by adaptive digital filtering," *Geophysics*, vol. 44, no. 7, pp. 1287–1305, 1979.

[27] W. L. Anderson, "Nonlinear least-squares inversion of transient soundings for a central induction loop system (Program NLSTCI)," U.S. Geological Survey 82, 1982.

13

Planetary Sciences, Geodynamics, Impacts, Mass Extinctions, and Evolution: Developments and Interconnections

Jaime Urrutia-Fucugauchi and Ligia Pérez-Cruz

Programa Universitario de Perforaciones en Océanos y Continentes, Departamento de Geomagnetismo y Exploración Geofísica, Instituto de Geofísica, Universidad Nacional Autónoma de México, Delegación Coyoacán, 04510 Mexico City, DF, Mexico

Correspondence should be addressed to Jaime Urrutia-Fucugauchi; juf@geofisica.unam.mx

Academic Editor: Robert Tenzer

Research frontiers in geophysics are being expanded, with development of new fields resulting from technological advances such as the Earth observation satellite network, global positioning system, high pressure-temperature physics, tomographic methods, and big data computing. Planetary missions and enhanced exoplanets detection capabilities, with discovery of a wide range of exoplanets and multiple systems, have renewed attention to models of planetary system formation and planet's characteristics, Earth's interior, and geodynamics, highlighting the need to better understand the Earth system, processes, and spatio-temporal scales. Here we review the emerging interconnections resulting from advances in planetary sciences, geodynamics, high pressure-temperature physics, meteorite impacts, and mass extinctions.

1. Introduction

In the 16th and 17th centuries, physics encompassed a wide field of inquiry with significant advances coming from many widely separated endeavors in what are now astronomy, optics, mechanics, gas chemistry, thermodynamics, and so forth. They included development of the heliocentric model for the solar system, formulation of the laws of planetary motion, experimental and mathematical descriptions of pendular and parabolic motion, the law of universal gravitation, inertial reference frame and laws of motion, the pressure-volume Boyle law, and the ideal gas law, among many other discoveries. Modern research in physics continues to encompass a wide field of inquiry, which is reflected into the different disciplines and emerging frontiers.

Increased awareness on the role of interactions among the Earth's components of the atmosphere, hydrosphere, lithosphere, ionosphere, and biosphere (Figure 1) leads to integrative approaches in Earth system science. This has led to better understanding of interactions, component flow, and feedback mechanisms acting with distinct spatial-temporal scales and manifested in the geochemical cycles, surface processes, and Earth's climate. Recent advances in the study of the Earth's interior and external processes in the near Earth environment and the solar system are resulting in broad integrative approaches.

Recent and long standing questions are being investigated using technological and theoretical developments, which include high performance computing, big data analysis, satellite observation system, instrumental networks, and planetary missions. Planetary missions to the solar system and the discovery of exoplanets and multiple systems provide a broad context for Earth's studies, integrating studies and challenging models and theories. Here, we review developments in geodynamics, high pressure mineral physics, meteorite impacts, mass extinctions, and planetary sciences and in the emerging interconnections as fields develop.

2. Geodynamics and Earth's Deep Interior

In the 1960s and early 1970s, development of plate tectonics provided a new paradigm for the Earth sciences, with Earth's upper layer divided into several plates undergoing large-scale plate motions [1]. Plate tectonics integrated surface

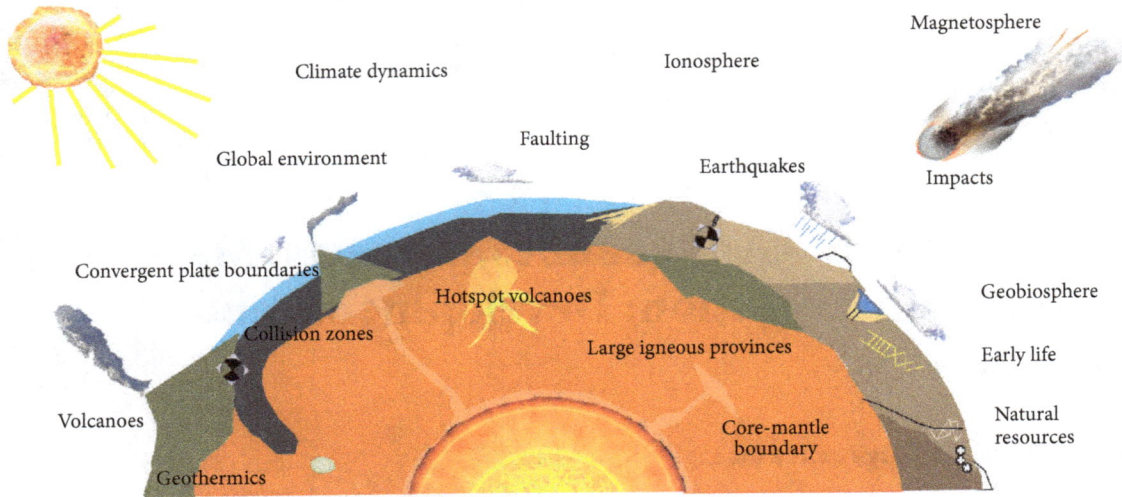

FIGURE 1: Earth system components connecting the Earth's deep interior, core, mantle, lithosphere, atmosphere, oceans, magnetosphere, and ionosphere with external processes like asteroid impacts, cosmic radiation, and solar winds (credits: International Scientific Continental Drilling Program (ICDP) website: http://www.icdp-online.org).

tectonic processes with the Earth's interior and deep energy sources, linking magmatism, seismicity, mountain building, and metallogeny in a unified way. The theory integrated the rich long-held archive of geological and geophysical data on the continents with the more recently acquired information on the oceans, particularly on the mid-ocean ridges, fracture zones, and trenches [2–4]. Plate tectonics provides a kinematic framework building on a global synthesis of geologic mapping, structural geology, stratigraphy, paleontology, petrology, geochemistry, seismology, paleomagnetism, geodesy and marine geology, and geophysics.

In the past three decades, plate tectonics has proved highly successful, prompting multi- and interdisciplinary studies. Understanding how the planet works has however remained a challenge, with the dynamics of deep and surface processes, mechanisms, and energy sources only partly investigated. Key aspects of plate dynamics, mantle convection, hotspot magmatism, mantle layered structure, convection, core-mantle, intraplate deformation, vertical motions, polar wandering, and plate driving forces still remain only partly understood.

Plate tectonics provide a global model for the lithosphere, which is broken into several plates undergoing relative motion at plate boundaries (Figure 2(a)). Oceanic lithosphere is created at ridges and recycled back into the mantle at subduction zones. The advent of international and regional broadband seismological and GPS networks has provided instantaneous plate motion data. This has opened new ways to study plate kinematics, with improved spatial and temporal resolution. Plate models integrating geological and geodetic data are being constructed, which permit analyzing plate reorganizations for the past few million years and evaluating plate deformation and diffuse plate boundaries. The recent synthesis by DeMets et al. [5] incorporates 27 plates, including six small plates not directly linked to the ridge system, and gives a high resolution plate kinematic model

(Figure 2(b)). Their results confirm the rigid plate assumption and provide constraints on plate deformation resulting from thermal contraction and wide plate boundaries.

Over long time scales, plate motions have undergone major changes and plate reorganizations with ocean basins closing and opening, which relate to deep processes and mantle convection [6]. Geological estimates of plate motion using the marine magnetic anomalies, fracture zones, hotspot tracks, and paleomagnetic directions have been used to reconstruct plate kinematics for the past 200 Ma. Studies of oceanic plateaus, igneous provinces, orogenic belts, and volcanic arcs provide tight constraints on plate motions and mantle convection for the Phanerozoic and Precambrian, with formation of supercontinent assemblies and continental breakup [7]. The role of deep mantle structures in plate tectonics can be observed on the residual geoid long wavelength characteristics and shear wave velocity zones in the deep mantle and core-mantle boundary.

The challenge is how to use the improved resolution on plate kinematics for modeling plate dynamics [8]. The forces that control plate motion and the relation to deep processes in the mantle, the nature of hotspots, fate of subducted lithosphere, and processes at the core-mantle D'' zone are in general poorly constrained. Earth's deep structure, mineral composition, convection, high pressure/temperature physics, and energy sources remain as a major frontier [9, 10].

Advances are being made from seismological analyses imaging velocity anomalies, wave polarization, and seismic anisotropy features in the mantle and core [11]. Seismic wave attenuation anomalies have been documented with depth, which correlate with estimates from mantle viscosity from geodynamic modeling. Measurements of attenuation and other anelastic properties have been linked to rheological properties, which are investigated in theoretical models and laboratory experiments. The layered structure of Earth's interior is characterized by increase of pressure and temperature,

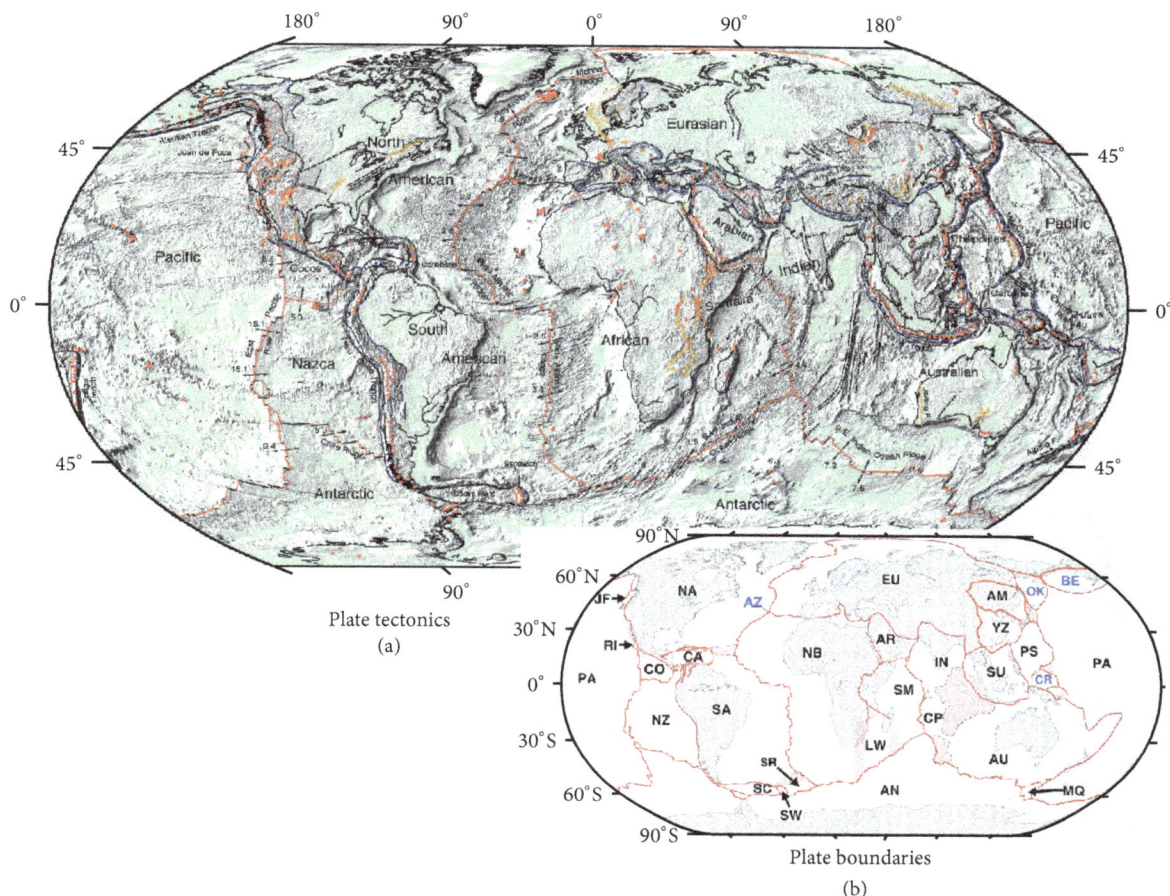

FIGURE 2: The Earth's lithosphere is divided into several tectonic plates that undergo relative motion. Plate boundaries are divergent boundaries (seafloor spreading ridges), convergent boundaries (subduction zones), and transform boundaries (transform faults). (a) Plate tectonic boundaries. Structural information on normal and reverse faults and volcanic centers is added (credits: NASA Earth Observatory and Goddard Space Flight Center website: http://earthobservatory.nasa.gov). (b) Global plate model incorporating 27 plates in a high resolution plate kinematic model (adapted from DeMets et al. [5]).

with variation of physical properties and mineralogy and phase changes. Pressure increases from about 24 GPa in the crust to 364 GPa in the inner core (Figure 3). Recently, physical and compositional structural mineral properties are being determined at increasing pressure and temperature using diamond-anvil cells, laser beams, noble gas graphite furnaces, and synchrotron sources. MgSiO-rich perovskite is the main constituent of the lower mantle down to 2900 km. This mantle mineral undergoes a phase transformation to denser postperovskite at core-mantle conditions, characterizing the physical properties at the D'' layer. Iron and iron-silica alloys are investigated at simulated outer and inner core conditions, with pressures and temperatures up to 257 GPa and 2400 K [12] and 364 GPa and 5500 K [13]. Experiments on high pressure mineral physics are providing novel data on the mineralogy and physical properties like anelasticity and plasticity, which are coupled from first principles calculations in constraining phase transformations and depth variations [14–16].

Computer modeling of convection permits testing different boundary conditions, property contrasts, and geometries, including those long explored of whole-mantle and double-layer convection. Dynamo modeling for geomagnetic field generation simulates short- and long-term variations observed at the surface in secular variation and regional anomalies, including polarity reversals. Thermal boundary conditions play major roles in dynamo behavior. Increasing computational power permits simulating fine mesh geometries with higher resolution. The field of geodynamic modeling, coupled to deep interior models for layered convection, mantle viscosity, and physical property contrast regional anomalies, has greatly expended in recent years showing large potential for further developments [17].

Plate boundaries are locations for active exchange interactions from the deep mantle to the surface, which manifest in seismicity, heat flow, and magmatic activity (Figure 4). Regional instrumental networks, geophysical surveys, and modeling on zones like the San Andreas transform fault in western United States, the Dead Sea and Anatolian faults in the Middle East, or the Honshu subduction zone in Japan are providing fresh high resolution data. Studies also address the economic implications, where significant mineral and energy resources concentrate at plate boundaries and

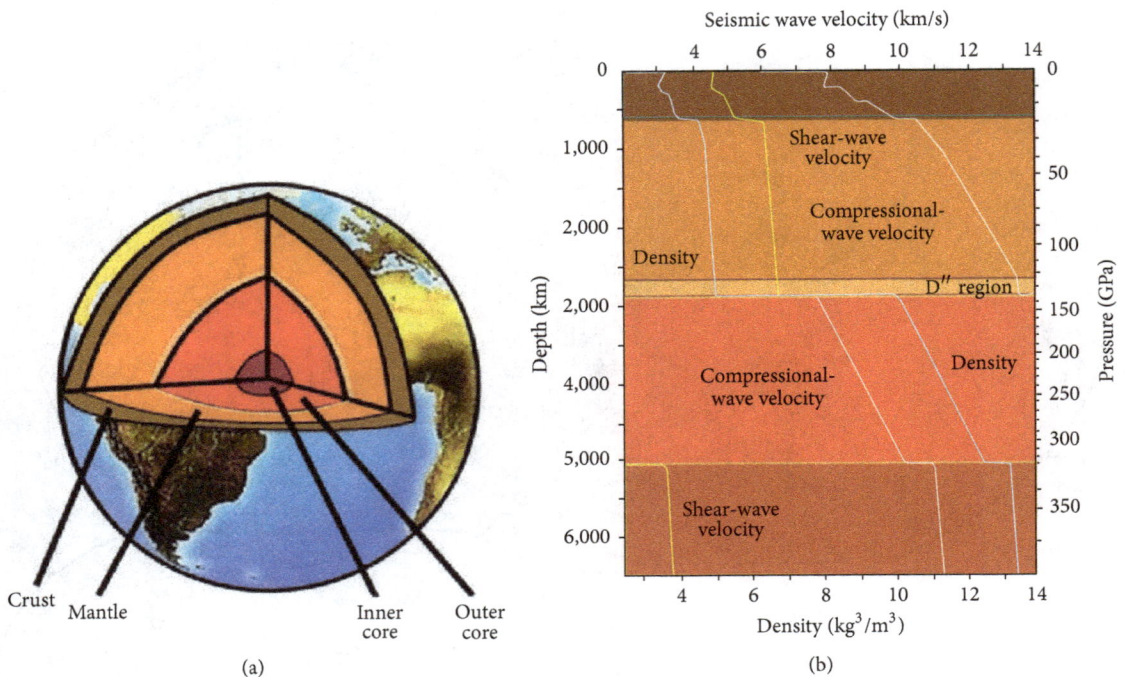

FIGURE 3: Density and seismic velocity variation with depth in Earth's interior (adapted from Romanowicz [9]).

related hazards associated with earthquakes and volcanic eruptions [18, 19]. Research on earthquakes, slow slip events, and volcanic eruptions provides enhanced understanding of mechanisms and developing new monitoring tools. Studies are addressing megathrust earthquakes like the Tokai-Oki magnitude 9.0 earthquake and the plate subduction process [20]. Active volcanoes present special challenges, particularly to model magma inside the conduits and deep connections in the mantle, which has prompted development of a range of methods of remote sensing, GPS, tiltmeters, broadband seismic networks, and integrated potential field and electromagnetic surveys. New tools being added include muons tomography exploiting secondary cosmic rays produced in the upper atmosphere with enhanced capabilities for imaging deep volcano structures [21, 22].

3. Impacts, Mass Extinctions, and Evolution

The evolution of life had been mostly studied from the fossil record, which provides evidence on past living organisms preserved along Earth's history. Paleontological studies have built a broad picture of life evolution from the single-celled organisms in the Precambrian to the multicellular organism in the Phanerozoic, providing a spatial-temporal reference system incorporated into the geological time scale. The field moved from stratigraphic, fossilization, and taxonomic based studies to exploring the ecosystems, physiology, reproductive traits, organism diseases, climate and environmental interactions, and feedbacks. With the introduction of isotope geochemistry and molecular studies the paleobiology field is being expanded, becoming increasingly multi- and interdisciplinary.

The extinction rates have been climbing as a result of the effects of climate and environmental changes and anthropogenic activity. The global warming, ocean acidification, deforestation, and pollution are affecting the ecosystems, with the extinction of species in the land and marine realms. Over a longer time span from the last deglaciation at the Late Pleistocene and Holocene transition, a large number of species including many land and marine vertebrates has disappeared. The extinction rates and magnitude has increased interest in studying past extinction events, particularly those associated with the five mass extinctions in the Phanerozoic (Figure 5). Mass extinctions are characterized by being above the rates of background extinction levels, occurring over a relatively short time [23, 24]. Barnosky et al. [25] have analyzed the recent extinctions in a geological context and compared them with the past five events. Most of the species that have ever developed are extinct, so studies of extinction rates and mechanisms are critical for understanding the evolution processes.

The end-Cretaceous mass extinction, the second in severity in the Phanerozoic and most recent one, is being intensely studied. It affected significant numbers of species and genera, with extinction of the dinosaurs, pterosaurs, ammonites and numerous marine microorganisms, causing the disappearance of about 75% of the species. The mass extinction marks the end of the Mesozoic Era. The Cretaceous/Paleogene (K/Pg) boundary is recognized by a globally distributed thin clay layer (Figure 6), which represents the fine-grain-sized fraction of the ejecta from the Chicxulub impact [26–28]. The K/Pg boundary layer is a global stratigraphic marker, which permits unprecedented temporal resolution and lateral correlation of events.

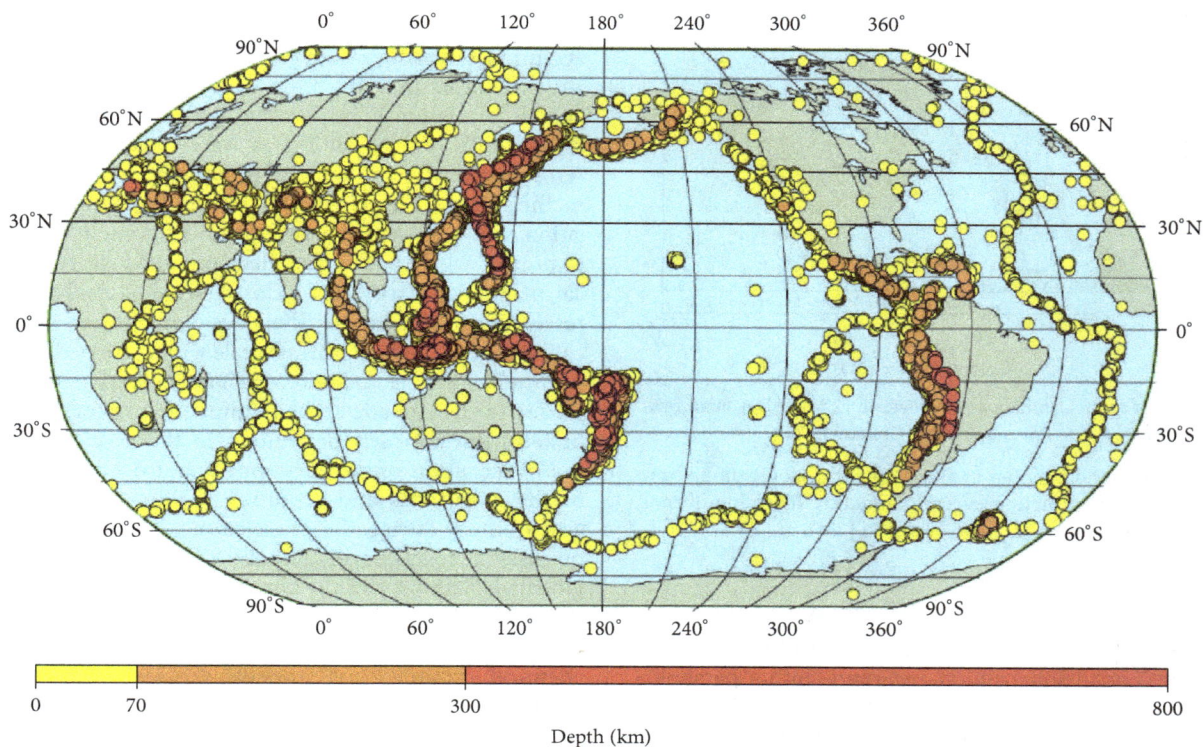

(a)

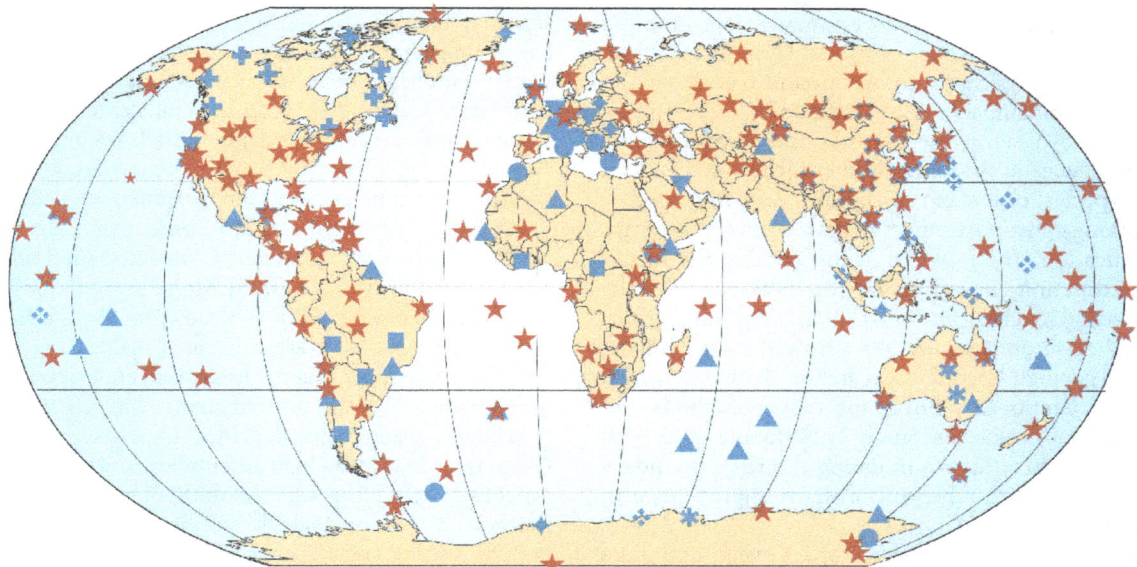

★ GSN   ● Italy
✳ Australia   ⬩ Japan
✚ Canada   ◼ USA
▲ France   ▼ Other
◆ Germany

(b)

FIGURE 4: (a) Seismicity and plate boundaries, with focal depth distribution. (b) Global seismic networks (adapted from Romanowicz [9]).

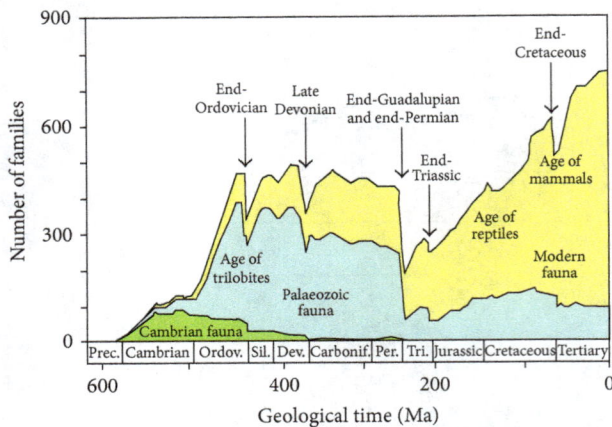

Percentage of extinct families: late Ordovician 12%, late Devonian 14%, late Permian 52%, late Triassic 12%, late Cretaceous 11%

FIGURE 5: Number of families as a function of geologic time, showing the five major extinction events marked by sharp biodiversity decrease (adapted from Raup and Sepkoski [23]).

The K/Pg layer is a few millimeter-to-centimeters thick, formed by a basal spherulitic layer, representing parabolic-emplaced melted droplets or condensates from a high temperature ejecta cloud, and the clay representing the fine-grained ejecta emplaced in the upper stratosphere (Figure 6). In the Gulf of Mexico-Caribbean Sea region, it has a more complex structure with high-energy tsunami deposit and a high temperature layer. Analyses of the layer distribution, composition, and physical properties permit reconstructing the dynamics of the impact event. Studies of K/Pg boundary sections provide data on the climatic and environmental changes and effects on the biota. Studies include analyses on the extinct species, ecosystem disruption, surviving species, short- and long-term postimpact effects, recovery patterns, and diversification. The problem in interpreting the mechanisms of extinction and effects on the biota has been the precision needed in dating and correlation. Separating events on the scale of seconds to months involved in the impact event in the geologic record are a major challenge, which has sparked attempts in refining the dating methods and stratigraphy. The most recent analysis by Renne et al. [29] has reduced the uncertainties in dating the K/Pg boundary to within ~30 ka, which represents a sharp improvement in dating capabilities.

Studies on the K/Pg boundary, impact event, and mass extinction are expanding, addressing life evolution at short and long time scales. One of the processes investigated addresses the evolution on maximum body size of terrestrial mammals, which coexisted with dinosaurs during most of the Mesozoic. For about 140 Ma mammals coexisted with the dinosaurs, restricted to small body sizes and ecosystems. Following the extinction of dinosaurs, first the birds increased their size, including some large predators. Later, mammals started to diversify and increase their maximum body size during the Paleocene and early Eocene. Smith et al. [30] have analyzed the evolution of maximum body size for terrestrial

mammals showing that the groups increase their body mass by the late Eocene, irrespective of the landmass.

The fossil record provides a punctuated view of life evolution, biased to certain geological settings, environments, and life forms that are more easily preserved. Dating and lateral correlation of rock strata present a further complication, with less resolution as we go back in time. High resolution stratigraphic methods, making use of multiproxy methods integrating statistical, spectral, and numerical simulation analyses, are being developed. Radiometric dating has improved, which is being applied combined with astronomical, magnetic polarity, and cyclostratigraphy, resulting in high resolution chronologies. The developments are applied to calibrating the geological time scale with increased precision.

Studies of the fossil record and evolution are closely related to the climatic and environmental factors, which are linked from the early beginnings in the Precambrian with the oxygenation of the atmosphere and oceans, the advent of the eukaryotes and evolution of life, and climate and environment during the Phanerozoic. Studies are focusing on early life forms, formation of the iron banded formations, global glaciations, and the construction of the life tree. New tools for climate reconstruction with increased high resolution are being developed using a wide range of biological, chemical, isotopic, and physical proxies. In Mexico and North and Central America, studies assess the effects, mechanisms, and interconnections of the Inter-tropical Convergence Zone latitudinal migration, North American monsoon, El Niño-Southern Oscillation, Pacific Decadal Oscillation, solar irradiance, and teleconnections [31, 32]. The studies are addressing climate evolution at different spatial and temporal scales, which are coupled with computational simulations and theoretical models for millennial, centennial to decadal resolution. Recent studies explore the links and influence of climatic and environmental factors on evolutionary patterns and the interconnections of the biosphere with climate [33].

A major development has come from the molecular clocks, which have significantly impacted methods to calibrate evolutionary time [34, 35]. Modeling tools for molecular tree analysis have rapidly evolved, providing estimates for branching events that are calibrated against the minimum ages from the fossil record. Improved understanding of the different genomes and rates of change has remained a major challenge in using molecular clocks to provide absolute dates. Given the advances in instrumentation and methods that are capable of providing vast amounts of data and processing power, the molecular clock will provide higher resolution in investigating evolutionary time. Multigene clocks applied to multitaxa are already giving unprecedented details in branching points, integrating phylogenetic reconstructions, the fossil record, and constraints on genome evolutionary rates [36].

Molecular analysis is well suited for studying macroevolutionary evolution, for instance, the appearance of eukaryotes, which in the fossil record appear at about 800 Ma when global changes in the oceans and climate were occurring. The molecular estimates for the early eukaryotic diversification are younger at around 1866 to 1679 Ma [36]. This older date is consistent with reports on eukaryotic microfossils,

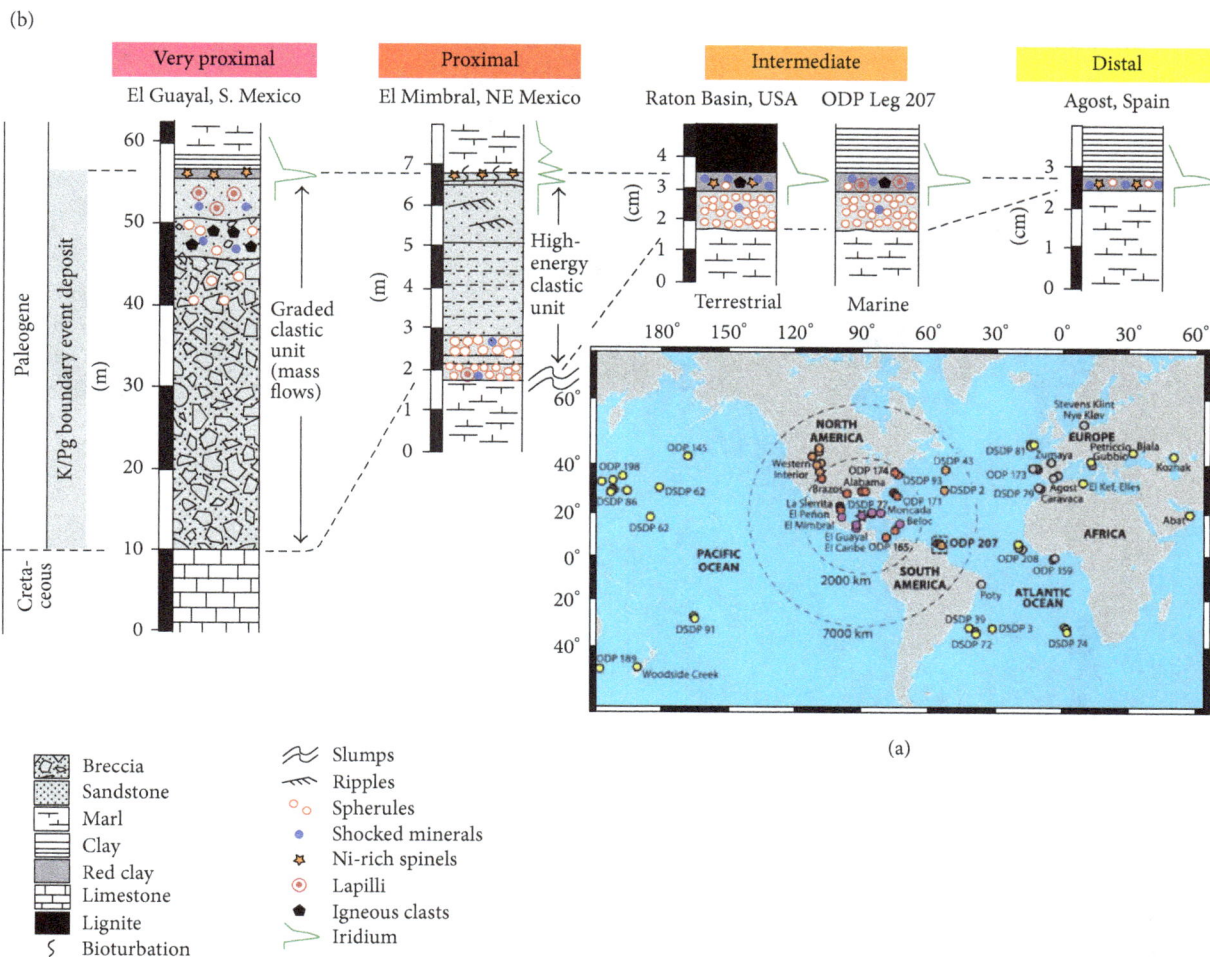

FIGURE 6: Cretaceous/Paleogene (K/Pg) boundary sections for distal, intermediate, proximal, and very proximal sites. Schematic K/Pg boundary sections (b). (a) Distribution of K/Pg boundary sites (Schulte et al. [26]).

indicating a long time span in the diversification of the major eukaryotic lineages [33, 36]. Studies are addressing evolutionary traits at genomic level, investigating eukaryotic evolution over million-year periods across species. Organism complexity is related to genomic features such as cell type number, gene contents, protein length, proteome disorder, and protein interactivity, which are being quantified [37, 38]. In the 1.4 Ga evolution of eukaryotes, alternative splicing has steadily increased with organism complexity [38].

4. Planetary Sciences

Exploration of the solar system using Earth based multispectral remote sensing and space probes has opened new research frontiers. Planetary missions to the terrestrial planets and moons of the gas giant planets have provided data on the structure, surface morphology, magmatic activity, tectonic styles, and deep interiors.

Observations of the surfaces of the inner planets and moons show that they are characterized by craters of different sizes and morphologies. They have been formed by collision of asteroid and cometary fragments over time, from small sized impacts to the large peak ring and multiring basin impacts. Large impacts produce deep transient excavation cavities in the curst, fragmenting and removing large volumes of rock and redistributing crustal material. On Earth, the active tectonic environment and erosion have effectively erased the record of impacts, with a relatively small number of craters documented and only three large multiring basins [39]. The Chicxulub crater, with a ~200 km rim diameter formed at the K/Pg boundary, is the youngest of the multiring basins and the only one with the ejecta preserved [26, 27, 29]. The other two structures formed in Precambrian times: Sudbury at about 1.8 and Vredefort at about 2 Ga ago. Chicxulub crater is located in the Yucatan platform in the southern Gulf of Mexico. The structure is covered by carbonate sediments and is being investigated by geophysical methods and deep drilling (Figure 7) [40, 41].

Impacts produce deformation at various depths, generating thermal anomalies and forming long-lived hydrothermal systems. The craters showing hydrothermal alteration are being investigated for manifestations of life forms, forming part of the exobiology programs. Studies of impact craters in the terrestrial record and elsewhere are enhancing

(a)   (b)

(c)   (d)

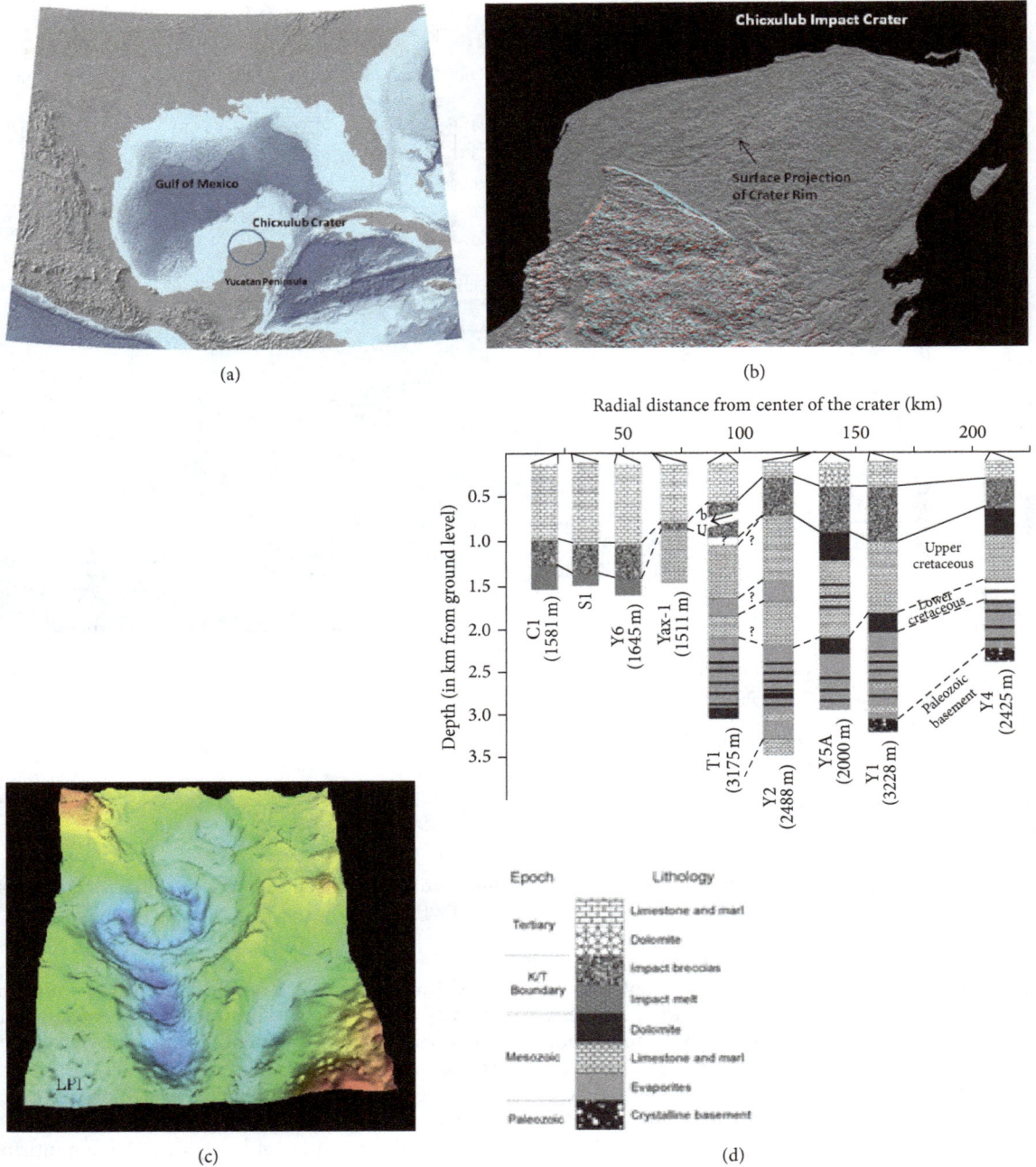

FIGURE 7: Chicxulub impact crater. (a) Gulf of Mexico and location of Chicxulub crater in the Yucatan platform. (b) Satellite interferometric radar image of Yucatan peninsula (credits: JPL-Caltech NASA), showing surface features associated with the buried crater structure. (c) Bouguer gravity anomaly of the Chicxulub crater (Sharpton et al. [40]). (d) Schematic lithological columns and lateral correlation for deep boreholes in the Chicxulub crater area, plotted as a function of relative distance to crater center (Urrutia-Fucugauchi et al. [27, 41]).

understanding of these highly energetic phenomena in shaping planetary surfaces, including those in the asteroid belt.

Analysis of frequency, density, and size distribution of craters permits estimating the age of the planetary surfaces, with ancient surfaces marked by high density of craters, often including the large multiring basins [39]. The size-frequency crater relationships are also related to the geodynamics and deep structure. Plate tectonics appears restricted to Earth

[10, 42]. Magmatic activity is observed in other bodies, including Mars, Venus, and Io. Mars lithosphere appears not being fragmented and under relative motion. Venus shows intense deformation and experienced a catastrophic resurfacing event about 500 Ma ago.

Evidence on the deep structure, thermal state, and convection comes from studies of meteorites, magnetic fields, and core dynamos. Meteorites have long been used for

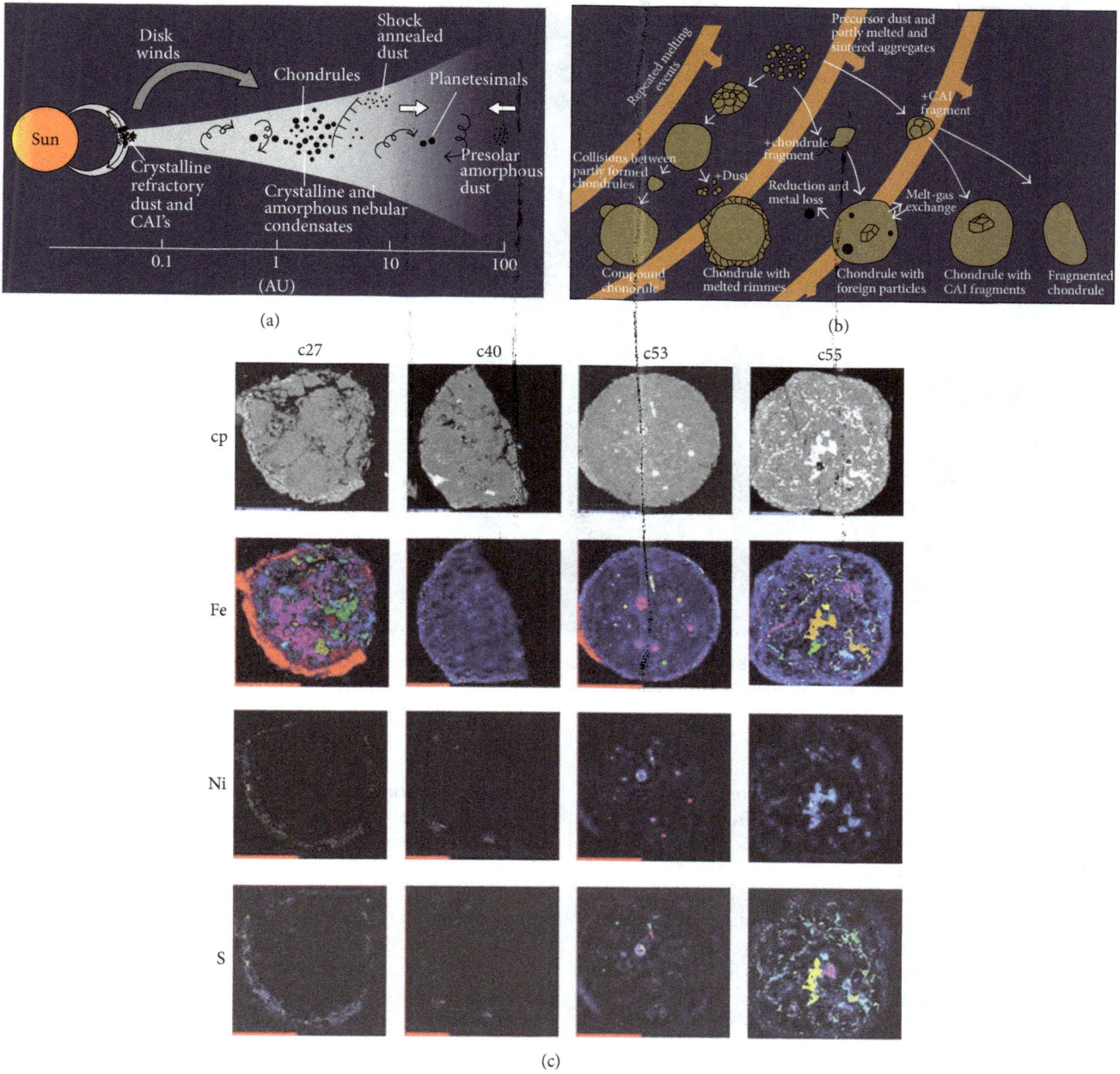

FIGURE 8: Schematic model of formation of chondrules and calcium-aluminium inclusions CAIs. (a) Protoplanetary disk. (b) Chondrule types with different morphologies and internal structures (adapted from Scott [43]). (c) Scanning electron microscopy images of individual chondrules from the Allende meteorite, showing the different morphologies, internal structures, and Fe, Ni, and S compositions. Numbers refer to laboratory sample identifications (Urrutia-Fucugauchi et al. [45]).

studying the origin and early stages of evolution of the planetary system (Figure 8). Analyses of chondrites and other primitive meteorites have documented the age of the first solids represented by refractory inclusions and chondrules, chemical composition of the solar nebula, and formation of planetesimals [43]. Studies are providing increasing resolution on the evolutionary stages (e.g., [43, 44]). Studies on chondrites and iron and stony-iron meteorites support that

their planetesimals had differentiated iron cores capable of sustaining dynamo action for ~10 Ma periods [45–49]. The paleomagnetic record of main group pallasites supports the fact that they come from near the core-mantle boundary of differentiated planetesimals that sustained internal magnetic fields [47]. Partly differentiated planetesimals might have been relatively abundant in the early stages of the solar system [48]. Many were destroyed by energetic collisions, and

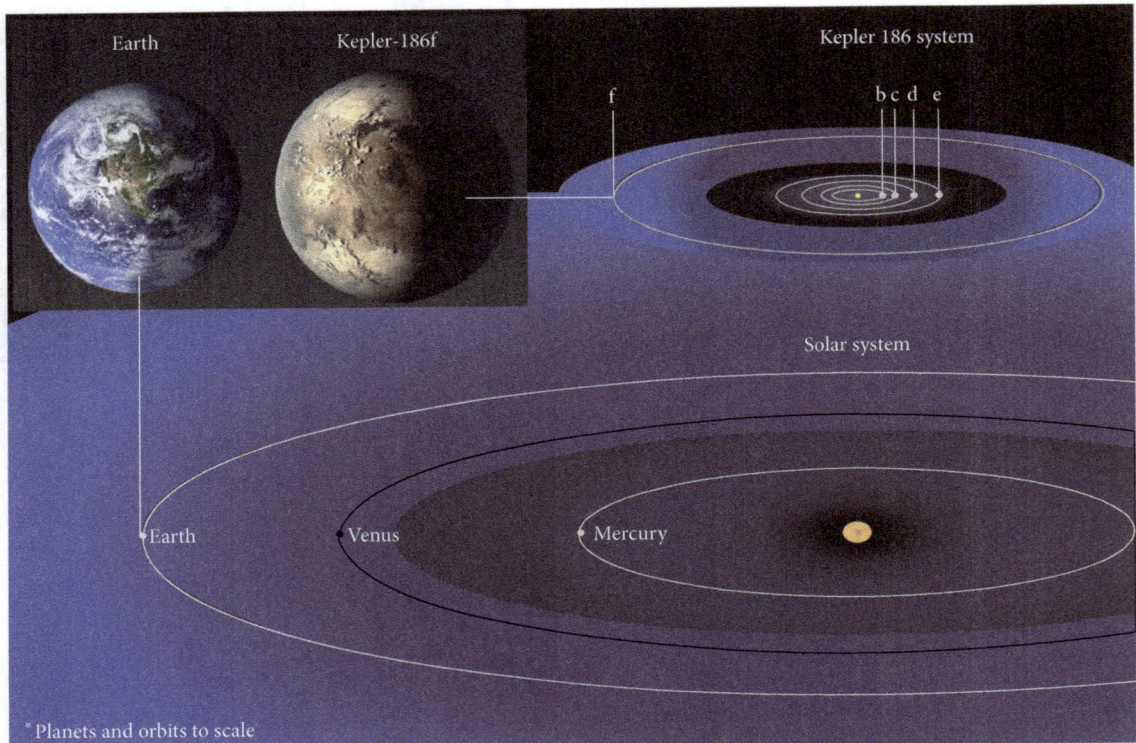

FIGURE 9: Schematic artistic representation of Kepler-186 multiple system compared with the inner solar system. Kepler-186 is a five-planet system located ~500 light-years away orbiting an M star half the Sun mass (Quintana et al. [51]) (credits: NASA Ames/SETI Institute/JPL-Caltech).

a fraction of them are preserved in the asteroid belt. Recent analyses show that asteroid Vesta had a convecting iron core in the early stages [49].

Planetary exploration is one of the most rapidly expanding frontiers in geophysics, with new data coming from the solar system missions and new exciting findings of exoplanets and planetary systems. The recent discoveries of exoplanets and multiple systems challenge the models for formation and early evolution of planetary systems based on observations of our solar system [50]. The large number of exoplanets discovered revives interest in planetary models with distinct formation zones for gas-icy giants and rocky planets within given regions of the accretion disk and models involving large-scale planet migration.

With increasing resolution and detection capacity, smaller Earth-sized planets are being detected. The Kepler space-based telescope mission is currently analyzing thousands of candidates, including several small mass planets. Recently, Quintana et al. [51] reported the finding of Kepler-186f, a 1.11 Earth-radius exoplanet in an orbit within the habitable zone around a M1-type dwarf star of the main sequence (Figure 9). Kepler-186f is the outermost planet of a five-planet system characterized by coplanar orbits. The multiplanet system is compatible with formation in a protoplanetary disk, with planets formed from accretion of local material and/or collisional growth of planetesimals. Numerical simulations conducted by Quintana and coauthors [51] for the Kepler-186 system show that too

steep density configurations, with dense accretion disk close to the star, are required. These results suggest that planets underwent inward migration while forming or a late stage perturbation.

Detection methods focus mainly on large planets close to the star, so most discoveries are large gas planets in orbits close to their stars. Detecting small Earth-like planets remains a challenge. Robertson et al. [52] analyzed the system around the M dwarf Gliese 581 star, showing that stellar activity might cause interference resulting in false exoplanet detection. Their results show that the signal for GJ 581 g, one of the four exoplanets in the system, depends on the eccentricity assumed for the companion GJ 581 d.

A major challenge in studying exoplanets lies in constraining the mass, density, composition, and orbital parameters. Recent developments start to provide new tools and data. Rocky planets are expected to have smaller sizes than gas and icy planets, but additional observations are required, which can be explored from the star metallicity. Buchhave et al. [53] analyzed the abundance of elements heavier than hydrogen and helium for 405 exoplanet host stars, finding that the exoplanet sizes separate into three metallicity regions. The three populations are interpreted in terms of rocky, gas dwarf, and gas-icy giant exoplanets. Another field of intense scrutiny is the detection of atmospheres for the super-Earths, gas dwarfs, and icy-gas giants [54]. Recent studies using transmission spectroscopy data report absorption features giving details on the atmosphere properties, confirming clouds in a super-Earth [55]. Considering that a significant fraction of exoplanets so far detected range in size between Earth and Neptune, the new studies open an interesting research field.

Determining the orbital parameters and spin provides important constraints on the planet ambient characteristics. Many exoplanets detected show orbits close to the stars, which are easier to detect with current methods. Spectroscopy observations can provide data on the spin velocity, which has been recently reported for gas giant planet β Pictoris b [56]. The exoplanet is located far from the star, about twice the distance of Jupiter in our system, and is quite bright. The spin determination comes from (blue) shifted carbon monoxide spectral signals from the planet, which gives an estimate of 25 km/s. In the solar system, spin correlates with the mass, showing a broad trend with the exception of Mercury and Venus. The fast rotation velocity, about 2 and 50 times greater than Jupiter's and Earth's, fits well with the planet mass. The study adds an interesting tool for characterizing multiplanet systems, which can provide constraints for models of planetary formation.

Interest in extraterrestrial life, which for a long time remained limited to theoretical analyses, has led to studies of organisms in extreme environments. Studies of extremophile communities from the deep crust, ocean thermal vents, hyperarid deserts, or polar caps have expanded understanding on food webs, energy sources, reproductive strategies, and metabolic states. Planetary missions are being directed to extraterrestrial life searches. Several missions have been directed to Mars, since the Viking missions experiments have tested the properties of the soils and atmosphere looking for evidence on liquid water and organic compounds. Recent missions are expanding the characterization of surface liquid water, hydrothermal activity, organic compounds, and fossil clues. New missions and spectroscopy observations use remote sensing clues of life activity in the planetary atmospheres.

Until the mid-1990s, the only planetary system known was our own. Models for evolution of planetary nebula predicted the formation of planets from planetary disks, but no observational evidence was available. The recent reports of hundreds of exoplanets and multiple planet systems and the observations on their sizes, orbits, and star characteristics are drastically changing and expanding theories and models for formation of planets and planetary systems [57–60].

5. Conclusions

New tools like the Earth observation satellite network, the global positioning system, planetary missions, high pressure/temperature experiments, high resolution tomography, and high performance computing play a major role in expanding research frontiers in geophysics. Increased interest in understanding Earth processes and new developments in instrumentation, modeling, and observation capabilities also comes from population growth and demographic changes, which increase global demand for minerals, water, and energy resources, resulting in pollution, land use changes, deforestation, environmental degradation, organism extinction,

changes in atmospheric gas composition, and global warming. In this context, understanding Earth's subsystems of the atmosphere, oceans, continents, ionosphere, magnetosphere, biosphere, and deep interior, their interconnections, cycles, spatio-temporal scales, and feedback mechanisms has become a major priority. The anthropogenic induced changes are comparable to those caused by geologic forces on the planet, highlighting the importance of integrated research. This has prompted global approaches in Earth system science and development of research fields, many of them at cross-disciplinary borders like biogeosciences, environmental geophysics, exobiology, and planetary sciences.

In a broad general context, the developments in high performance computing power, personal computers, telecommunications, electronics, and advent of the internet are profoundly changing the scientific research enterprise. The developments touch practically every area related to research with electronic databases, publications, electronic archives, search engines, software, and personal and group interactions. The capacity for analyzing massive data sets using supercomputers and computer networks facilitates using numerical methods and complex simulations. High performance computing allows modeling of the complex climate system, core and mantle tomography, Earth observational satellite multispectral data, or exoplanet detection systems with the massive data sets from the space-telescope Kepler and other search missions.

Studies in widely different fields are interconnected with the recent developments, opening bridges across previously separated endeavors. Studies on the origin and evolution of the solar system are linked to the new areas of planetary sciences, which challenge current models opening new questions. Most of the exoplanets discovered are in the size range between Earth and Neptune, for which there are no analogs in the solar system. Studies of the structure and properties, orbital characteristics, and formation mechanisms for the super-Earths and gas giants are giving fresh insights on planetary evolution [58]. Studies are addressing finer details in the characteristics of exoplanets in addition to size, orbit, and mass, such as the spin, surface temperature, and presence and composition of atmospheres and clouds [59, 60]. The mass-spin relation in the solar system is related to the breakup velocity and impacts added angular momentum. The estimation of the fast spin for β Pictoris b, which fits with the trend for fast spin and large mass, opens the link of impacts in the formation of planets [57, 59]. β Pictoris b is a young planet still contracting and cooling, towards a size comparable to Jupiter. Determination of the spin characteristics for a larger group of exoplanets will allow investigating how planets form and evolve in different protoplanetary disks environments.

Exoplanet research and planetary missions connect with investigation of the cratering record on Earth and in other bodies of the solar system, including the large impacts during the early stages of planet formation. Satellites in the solar system show different characteristics of the rocky and gasicy planets, with small satellites in large planets and larger satellites in small planets. Studies on the tectonics and deep structure on Earth are now related to planetary research on the planet interiors, planet formation models, and thermal states [42]. Results from high pressure and temperature mineral physics [11–15] relate and constrain models of formation of super-Earth and giant icy-gas exoplanets [51–60], as well as the planets in the solar system [50]. We have similar links between studies of life on extreme terrestrial environments, origin, and evolution of life in the young Earth and studies of exobiology [61]. Studies are uncovering relationships and exploring new questions and interconnections.

Competing Interests

The authors declare that there is no conflict of interests regarding the publication of this paper.

Acknowledgments

The authors thank Ana Escalante and Miguel Angel Diaz for assistance with the figures. This study forms part of National University of Mexico Programs on the Chicxulub Impact, the Cretaceous/Paleogene Boundary, and MeteorPlan. Partial support comes from Papiit IG-101115 and Conacyt grants.

References

[1] X. LePichon, J. Francheatau, and J. Bonin, *Plate Tectonics*, Elsevier, Amsterdam, The Netherlands, 1973.

[2] J. T. Wilson, "A new class of faults and their bearing on continental drift," *Nature*, vol. 207, no. 4995, pp. 343–347, 1965.

[3] W. J. Morgan, "Rises, trenches, great faults, and crustal blocks," *Journal of Geophysical Research*, vol. 73, no. 6, pp. 1959–1982, 1968.

[4] D. P. McKenzie and R. L. Parker, "The North Pacific: an example of tectonics on a sphere," *Nature*, vol. 216, no. 5122, pp. 1276–1280, 1967.

[5] C. DeMets, R. G. Gordon, and D. F. Argus, "Geologically current plate motions," *Geophysical Journal International*, vol. 181, no. 1, pp. 1–80, 2010.

[6] K. Burke, "Plate tectonics, the wilson cycle, and mantle plumes: geodynamics from the top," *Annual Review of Earth and Planetary Sciences*, vol. 39, pp. 1–29, 2011.

[7] R. N. Mitchell, T. M. Kilian, and D. A. D. Evans, "Supercontinent cycles and the calculation of absolute palaeolongitude in deep time," *Nature*, vol. 482, no. 7384, pp. 208–211, 2012.

[8] D. L. Turcotte and G. Schubert, *Geodynamics: Applications of Continuum Physics to Geological Problems*, John Wiley & Sons, New York, NY, USA, 1982.

[9] B. Romanowicz, "Using seismic waves to image Earth's internal structure," *Nature*, vol. 451, no. 7176, pp. 266–268, 2008.

[10] G. Schubert, D. Turcotte, and P. Olson, *Mantle Convection in the Earth and Planets*, Cambridge University Press, Cambridge, UK, 2001.

[11] S. A. Karato, A. M. Forte, R. C. Liebermann, G. Masters, and L. Stixrude, Eds., *Earth's Deep Interior: Mineral Physics and Tomography from the Atomic to the Global Scale*, vol. 117 of *AGU Geophysical Monograph*, American Geophysical Union, 2000.

[12] H. Asanuma, E. Ohtani, T. Sakai et al., "Phase relations of Fe-Si alloy up to core conditions: implications for the Earth inner core," *Geophysical Research Letters*, vol. 35, no. 12, Article ID L12307, 2008.

[13] S. Tateno, K. Hirose, Y. Ohishi, and Y. Tatsumi, "The structure of iron in Earth's inner core," *Science*, vol. 330, no. 6002, pp. 359–361, 2010.

[14] M. Murakami, K. Hirose, K. Kawamura, N. Sata, and Y. Ohishi, "Post-perovskite phase transition in $MgSiO_3$," *Science*, vol. 304, no. 5672, pp. 855–858, 2004.

[15] D. C. Rubie, T. Duffy, and E. Ohtani, "New developments in high pressure mineral physics and applications to the Earth's interior," *Physics of the Earth and Planetary Interiors*, vol. 143-144, pp. 1–3, 2004.

[16] J.-F. Lin, W. Sturhahn, J. Zhao, G. Shen, H.-K. Mao, and R. J. Hemley, "Sound velocities of hot dense iron: Birch's Law revisited," *Science*, vol. 308, no. 5730, pp. 1892–1894, 2005.

[17] L. Hwang, T. Jordan, L. Kellog, J. Tromp, and R. Wiellemann, *Advancing Solid Earth System Science Through High-Performance Computing*, Computational Infrastructure for Geodynamics, University of California, Davis, Calif, USA, 2014.

[18] ICSU, *Earth System Science for Global Sustainability: The Grand Challenges*, International Council for Science, Paris, France, 2010.

[19] A. Ismail-Zadeh, J. Urrutia-Fucugauchi, A. Kijko, K. Takeuchi, and I. Zialapin, Eds., *Extreme Natural Hazards, Disaster Risks and Societal Implications*, Cambridge University Press, Cambridge, UK, 2014.

[20] M. Simons, S. E. Minson, A. Sladen et al., "The 2011 magnitude 9.0 Tohoku-Oki earthquake: mosaicking the megathrust from seconds to centuries," *Science*, vol. 332, no. 6036, pp. 1421–1425, 2011.

[21] H. K. M. Tanaka, T. Uchida, M. Tanaka, H. Shinohara, and H. Taira, "Cosmic-ray muon imaging of magma in a conduit: degassing process of Satsuma-Iwojima Volcano, Japan," *Geophysical Research Letters*, vol. 36, no. 1, Article ID L01304, 2009.

[22] V. Grabski, R. Nuñez, S. Aguilar et al., "Use of horizontal cosmic muons to study density distribution variations in the Popocatepetl volcano," in *Proceedings of the 33rd International Cosmic Ray Conference (ICRC '13)*, vol. 33, pp. 1–4, Rio de Janeiro, Brazil, July 2013.

[23] D. M. Raup and J. J. Sepkoski Jr., "Mass extinctions in the marine fossil record," *Science*, vol. 215, no. 4539, pp. 1501–1503, 1982.

[24] J. J. Sepkoski Jr., "Patterns of phanerozoic extinction: a perspective from global data bases," in *Global Events and Event Stratigraphy in the Phanerozoic*, O. H. Walliser, Ed., pp. 35–51, Springer, New York, NY, USA, 1996.

[25] A. D. Barnosky, N. Matzke, S. Tomiya et al., "Has the Earth's sixth mass extinction already arrived?" *Nature*, vol. 471, no. 7336, pp. 51–57, 2011.

[26] P. Schulte, L. Alegret, I. Arenillas et al., "The Chicxulub asteroid impact and mass extinction at the Cretaceous-paleogene boundary," *Science*, vol. 327, no. 5970, pp. 1214–1218, 2010.

[27] J. Urrutia-Fucugauchi, A. Camargo-Zanoguera, and L. Pérez-Cruz, "Discovery and focused study of the Chicxulub impact crater," *Eos*, vol. 92, no. 25, pp. 209–210, 2011.

[28] L. W. Alvarez, W. Alvarez, F. Asaro, and H. V. Michel, "Extraterrestrial cause for the Cretaceous-Tertiary extinction," *Science*, vol. 208, no. 4448, pp. 1095–1108, 1980.

[29] P. R. Renne, A. L. Deino, F. J. Hilgen et al., "Time scales of critical events around the cretaceous-paleogene boundary," *Science*, vol. 339, no. 6120, pp. 684–687, 2013.

[30] F. A. Smith, A. G. Boyer, J. H. Brown et al., "The evolution of maximum body size of terrestrial mammals," *Science*, vol. 330, no. 6008, pp. 1216–1219, 2010.

[31] G. H. Haug, K. A. Hughen, D. M. Sigman, L. C. Peterson, and U. Röhl, "Southward migration of the intertropical convergence zone through the holocene," *Science*, vol. 293, no. 5533, pp. 1304–1308, 2001.

[32] L. Pérez-Cruz, "Hydrological changes and paleoproductivity in the Gulf of California during middle and late Holocene and their relationship with ITCZ and North American Monsoon variability," *Quaternary Research*, vol. 79, no. 2, pp. 138–151, 2013.

[33] J. L. Blois and E. A. Hadly, "Mammalian response to cenozoic climatic change," *Annual Review of Earth and Planetary Sciences*, vol. 37, pp. 181–208, 2009.

[34] S. Kumar, "Molecular clocks: four decades of evolution," *Nature Reviews Genetics*, vol. 6, no. 8, pp. 654–662, 2005.

[35] S. Kumar and S. B. Hedges, "A molecular timescale for vertebrate evolution," *Nature*, vol. 392, no. 6679, pp. 917–920, 1998.

[36] L. W. Parfrey, D. J. G. Lahr, A. H. Knoll, and L. A. Katz, "Estimating the timing of early eukaryotic diversification with multigene molecular clocks," *Proceedings of the National Academy of Sciences of the United States of America*, vol. 108, no. 33, pp. 13624–13629, 2011.

[37] E. Schad, P. Tompa, and H. Hegyi, "The relationship between proteome size, structural disorder and organism complexity," *Genome Biology*, vol. 12, article R120, 2011.

[38] L. Chen, S. J. Bush, J. M. Tovar-Corona, A. Castillo-Morales, and A. O. Urrutia, "Correcting for differential transcript coverage reveals a strong relationship between alternative splicing and organism complexity," *Molecular Biology and Evolution*, vol. 31, no. 6, pp. 1402–1413, 2014.

[39] J. Urrutia-Fucugauchi and L. Perez-Cruz, "Multiring-forming large bolide impacts and evolution of planetary surfaces," *International Geology Review*, vol. 51, no. 12, pp. 1079–1102, 2009.

[40] V. L. Sharpton, K. Burke, A. Camargo-Zanoguera et al., "Chicxulub multiring impact basin: size and other characteristics derived from gravity analysis," *Science*, vol. 261, no. 5128, pp. 1564–1567, 1993.

[41] J. Urrutia-Fucugauchi, A. Camargo-Zanoguera, L. Pérez-Cruz, and G. Pérez-Cruz, "The Chicxulub multi-ring impact crater, yucatan carbonate platform, Gulf of Mexico," *Geofisica Internacional*, vol. 50, no. 1, pp. 99–127, 2011.

[42] C. O'Neill, A. M. Jellinek, and A. Lenardic, "Conditions for the onset of plate tectonics on terrestrial planets and moons," *Earth and Planetary Science Letters*, vol. 261, no. 1-2, pp. 20–32, 2007.

[43] E. R. D. Scott, "Chondrites and the protoplanetary disk," *Annual Review of Earth and Planetary Sciences*, vol. 35, pp. 577–620, 2007.

[44] J. N. Connelly, M. Bizzarro, A. N. Krot, Å. Nordlund, D. Wielandt, and M. A. Ivanova, "The absolute chronology and thermal processing of solids in the solar protoplanetary disk," *Science*, vol. 338, no. 6107, pp. 651–655, 2012.

[45] J. Urrutia-Fucugauchi, L. Pérez-Cruz, and D. Flores-Gutiérrez, "Meteorite paleomagnetism—from magnetic domains to planetary fields and core dynamos," *Geofisica Internacional*, vol. 53, no. 3, pp. 343–363, 2014.

[46] L. T. Elkins-Tanton, B. P. Weiss, and M. T. Zuber, "Chondrites as samples of differentiated planetesimals," *Earth and Planetary Science Letters*, vol. 305, no. 1-2, pp. 1–10, 2011.

[47] J. A. Tarduno, R. D. Cottrell, F. Nimmo et al., "Evidence for a dynamo in the main group pallasite parent body," *Science*, vol. 338, no. 6109, pp. 939–942, 2012.

[48] B. P. Weiss and L. T. Elkins-Tanton, "Differentiated planetesimals and the parent bodies of chondrites," *Annual Review of Earth and Planetary Sciences*, vol. 41, pp. 529–560, 2013.

[49] R. R. Fu, B. P. Weiss, D. L. Shuster et al., "An ancient core dynamo in asteroid Vesta," *Science*, vol. 338, no. 6104, pp. 238–241, 2012.

[50] A. Morbidelli, J. I. Lunine, D. P. O'Brien, S. N. Raymond, and K. J. Walsh, "Building terrestrial planets," *Annual Review of Earth and Planetary Sciences*, vol. 40, pp. 251–275, 2012.

[51] E. V. Quintana, T. Barclay, S. N. Raymond et al., "An Earth-sized planet in the habitable zone of a cool star," *Science*, vol. 344, no. 6181, pp. 277–280, 2014.

[52] P. Robertson, S. Mahadevan, M. Endl, and A. Roy, "Stellar activity masquerading as planets in the habitable zone of the M dwarf Gliese 581," *Science*, vol. 345, no. 6195, pp. 440–444, 2014.

[53] L. A. Buchhave, M. Bizzarro, D. W. Latham et al., "Three regimes of extrasolar planet radius inferred from host star metallicities," *Nature*, vol. 509, no. 7502, pp. 593–595, 2014.

[54] H. A. Knutson, B. Benneke, D. Deming, and D. Homeier, "A featureless transmission spectrum for the Neptune-mass exoplanet GJ436b," *Nature*, vol. 505, no. 7481, pp. 66–68, 2014.

[55] L. Kreidberg, J. L. Bean, J.-M. Désert et al., "Clouds in the atmosphere of the super-Earth exoplanet GJ 1214b," *Nature*, vol. 505, no. 7481, pp. 69–72, 2014.

[56] I. A. G. Snellen, B. R. Brandl, R. J. De Kok, M. Brogi, J. Birkby, and H. Schwarz, "Fast spin of the young extrasolar planet β Pictoris b," *Nature*, vol. 508, no. 7498, pp. 63–65, 2014.

[57] A. W. Howard, "Observed properties of extrasolar planets," *Science*, vol. 340, no. 6132, pp. 572–576, 2013.

[58] T. Barman, "Astronomy: a new spin on exoplanets," *Nature*, vol. 508, no. 7498, pp. 41–42, 2014.

[59] X. Dumusque, F. Pepe, C. Lovis et al., "An Earth-mass planet orbiting α Centauri B," *Nature*, vol. 491, no. 7423, pp. 207–211, 2012.

[60] R. M. Canup and W. R. Ward, "A common mass scaling for satellite systems of gaseous planets," *Nature*, vol. 441, no. 7095, pp. 834–839, 2006.

[61] C. S. Cockell, *Astrobiology: Understanding Life in the Universe*, Wiley-Blackwell, 2015.

14

Assessment of Surface Runoff for Tank Watershed in Tamil Nadu using Hydrologic Modeling

Marykutty Abraham ⓘ[1] **and Riya Ann Mathew** ⓘ[2]

[1]*Sathyabama University, Chennai, India*
[2]*Anna University, Chennai, India*

Correspondence should be addressed to Marykutty Abraham; marykutty_ab@yahoo.co.in

Academic Editor: Yun-tai Chen

Providing safe and wholesome water in sufficient quantity on a sustainable basis remains elusive for large population especially in semiarid regions and hence water balance estimation is vital to assess water availability in a watershed. The water balance study is formulated to assess the runoff that can be harvested for effective utilization. The study area is Urapakkam watershed with a chain of 3 tanks having an aerial extent of $4.576 \, km^2$ with hard rock formation underneath and thus has limited scope for groundwater recharge. Hence surface water is the main water source in this area. Runoff computed for the watershed using USDA-NRCS model varied from 94.95 mm to 2324.34 mm and the corresponding rainfall varied from 575.7 mm to 3608.0 mm, respectively. A simple regression model was developed for the watershed to compute runoff from annual rainfall. Average annual runoff estimated for the watershed was around 37% of the rainfall for the study period from 2000-01 to 2013-14. Statistical analysis and test of significance for runoff obtained by NRCS model and regression model did not show any significant difference thus proving that regression model is efficient in runoff computation for ungauged basins. The volume of water accessible for fifty percent dependable flow year is obtained as 2.46 MCM and even if 50% of it can be effectively harnessed the water available in the watershed is 1.23 MCM. The water demand of the area is estimated as 0.148 MCM for domestic purpose and 0.171 MCM for irrigation purpose, which is much lower than the available runoff that can be harnessed from the watershed. Thus there is scope to harvest 1.23 MCM of water which is more than the demand of the watershed. The study reveals that it is feasible to harvest and manage water effectively if its availability and demand are computed accurately.

1. Introduction

The global annual per capita water supply is around $7000 \, m^3$ [1], which shows that the earth has sufficient freshwater for the needs of the population. According to Food and Agriculture Organisation [2] average per capita water availability ranges from near zero in Kuwait to around $10,000 \, m^3$ in Tajikistan. In addition to the spatial variability, there is seasonal variability in these areas, which results in extreme events. Cosgrove and Rijsberman [3] predicted that by the year 2025 about 60% of the world's population will face acute water shortage. Groundwater table is highly influenced by urbanization of an area.

Water demand is increasing in most of the cities as per capita water use of an urban citizen is almost twice than that of a rural citizen and India is rapidly urbanizing. As the Indian cities are facing water problems, groundwater exploitation has become inevitable which leads to the decline of groundwater table. To sustain urban cities there is an urgent need to augment groundwater resources. The per capita water availability of India is $1545 \, m^3$ per annum [4].

Monsoonal climate prevalent in India causes half of the precipitation to occur within a period of 15 days and more than 90% of the annual runoff during four monsoon months. The large regional and seasonal variation in rainfall and runoff in India have prompted human interferences in hydrological cycle. Options that combine augmentation of surface water and groundwater resources can ensure sufficient water availability in a sustainable manner. As the study area is on the threshold for rapid development and urbanization there will be an impetus for drastic increase in water demand. The villages in the study area are already experiencing rapid urbanization.

Water balance estimates are essential to strengthen the decision-making pertaining to water management. Rainwater harvesting and water conservation measures deserve the highest priority in water management compared to other measures because of the huge investment required and longer time period involved [5]. So estimation of water potential is very important for the effective utilization of the available water resources in the context of fast urbanization.

The widely accepted empirical method to estimate runoff for an ungauged watershed is Natural Resource Conservation Service (NRCS) model, formerly known as Soil Conservation Service (SCS) model, which is based on water balance equation developed by United States Department of Agriculture [6]. Rainfall-runoff relation was first proposed by Sherman [7]. Surface runoff based on land use, soil, rainfall, duration of storm, and annual temperature for watersheds was developed by Mockus [8]. Mishra and Singh [9] conducted extensive research on Soil Conservation Service-curve number (SCS-CN) method. Mishra et al. [10] investigated SCS model for US fields with size ranging from 0.3 to 30352 hectare using rainfall-runoff events and found that the model is appropriate for high rainfall (>50.8 mm) areas. SCS method was successfully applied for a forest watershed having permeable soils [11]. Rainfall-runoff relationships using various empirical and numerical methods have been developed by various researchers [12–16]. Applicability of NRCS model was successfully demonstrated by many researchers worldwide for computing runoff [11, 17–19]. NRCS method is widely accepted and is proved useful for surface runoff estimation by several authors for various watersheds of Tamil Nadu [20, 21]. Modifications were made for SCS model for steep slopes by Gupta et al. [22]. Water balance computation of Noyil Basin in Tamil Nadu was presented by Chatterjee et al. [23]. Central Ground Water Board [24] conducted water balance studies for upper Yamuna basin and the recharge estimated was compared with water table fluctuation method.

Thakuriah and Saikia [25] successfully demonstrated methodology for runoff estimation by integrating remote sensing and geographic information system (GIS) in Buriganga watershed of Assam. Remote sensing data along with ground truth verification is widely utilized for mapping various resources in an economic and efficient way including soil mapping, land use mapping, and fertility mapping [26, 27]. The applicability of SCS model was improved by accounting for spatial and temporal variability in soil and land use with remote sensing and geographic information system [28–34].

The literature study reveals that NRCS model is used for hydrologic simulation of small watersheds worldwide for a long period and can also give realistic results even for larger watersheds. Hence NRCS model is used in the present study for runoff computation owing to its efficiency, precision, and simplicity for ungauged watersheds.

2. The Study Area and Data Collection

The study area is Urapakkam watershed which consists of three tanks, namely, Urapakkam tank, Karanaipuducheri tank, and Periyar Nagar tank. Urapakkam watershed with an area of 4.576 km^2 is part of Guduvanchery watershed in the Adyar basin, in Kanchipuram district of Tamil Nadu, India. The area falls under geographical coordinates 12° 52′ N to 12° 47′ N and 80° 42′ E to 80° 52′ E. Elevation of the district varies from 0.5 m to 230 m above mean sea level. The temperature ranges between 37.1°C and 20.5°C [35]. Index map of study area is presented in Figure 1. Rainfall occurs mainly during October to December through Northeast monsoon. The area receives an average rainfall of around 1150 mm per annum. Rainfall data was collected from India Meteorological Department (IMD) Chennai for fourteen years from 2000 to 2014. Agriculture is the main livelihood of the watershed community. Paddy is the major crop and groundnut, sugarcane, cereals, millets, pulses, etc. are also cultivated. Water from Palar river, tanks, and wells is used for irrigation in the watershed [36].

3. Methodology

3.1. Runoff Computation. Watershed boundary for Urapakkam watershed was delineated using District Soil & Watershed Atlas [37] and verified with Survey of India toposheet and satellite imagery. Land use map was prepared for the tank watershed in GIS platform using ArcGIS-9.8 software and the area under each land use is calculated.

Surface runoff was assessed by applying NRCS-Hydrologic model using daily time step [38]. The runoff equation is

$$Q = \frac{(P - I_a)^2}{(P - I_a + S)} \tag{1}$$

$$S = \frac{25400}{CN} - 254, \tag{2}$$

where

P is Rainfall, in mm

CN is curve number which ranges between 0 and 100

S is storage index, depending on the curve number

I_a is initial abstraction, a fraction of S

Q is runoff, in mm

Using $I_a = 0.2S$, (2) takes the form

$$Q = \frac{(P - 0.2S)^2}{(P - 0.8S)} \tag{3}$$

Daily rainfall data, land use map, hydrological soil group, and infiltration study were used to arrive at runoff curve numbers for NRCS model. The watershed curve number is closely related to soil storage index. According to infiltration characteristics, soil is grouped under four hydrologic soil groups, namely, A, B, C, and D, where A has the highest infiltration rate and D has the least infiltration rate [39].

Surface conditions of a watershed are taken care of by curve numbers. The curve numbers corresponding to various land uses and hydrologic soil groups can be obtained from

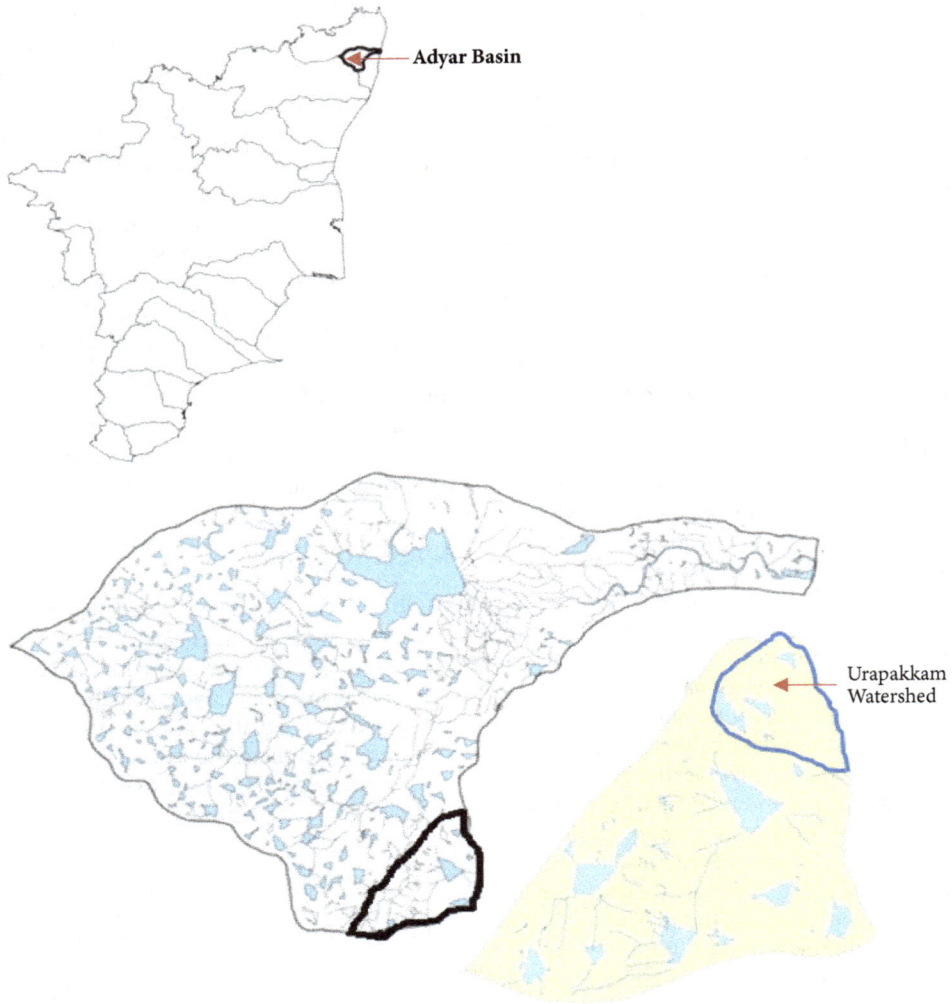

FIGURE 1: Index map of study area.

Table 1 [40]. From this, the weighted curve number of the watershed can be determined which corresponds to the average Antecedent Moisture Condition (AMC II). AMC for dry and wet situations is determined using five-day antecedent rainfall which corresponds to less than 13 mm and more than 28 mm rainfall, respectively, and gives rise to AMC I and AMC III. The curve number determined for AMC II can be converted for other conditions, namely, AMC I and AMC III using the following relations [38]:

$$CN_I = \frac{4.2CN_{II}}{10 - 0.0584CN_{II}} \qquad (4)$$

$$CN_{III} = \frac{23CN_{II}}{10 + 0.13CN_{II}}, \qquad (5)$$

where

CN$_I$ is curve number for AMC I

CN$_{II}$ is curve number for AMC II

CN$_{III}$ is curve number for AMC III

A simple regression model was developed to compute annual runoff from rainfall which was then calibrated and validated with NRCS model. Such simpler equations can be used for computation of runoff for ungauged watersheds instead of arduous and data intensive computation methods. Performance of developed regression model was evaluated using statistical methods such as coefficient of determination, standard error of estimate, and mean absolute error. The statistical test of significance used was t-test.

Coefficient of Determination (R^2) shows the degree of linear relation between the two variables. R^2 value close to unity indicates a high degree of association between the two variables [41].

Standard Error of Estimate (SEE) is an estimate of the average deviation of the regression value from the observed data.

$$SEE = \sqrt{\frac{\sum (x_i - x)^2}{n - 2}} \qquad (6)$$

TABLE 1: Runoff curve numbers.

Description of Land use	Hydrologic Soil Group			
	A	B	C	D
Paved parking lots, roofs, driveways	98	98	98	98
Streets and Roads:				
Paved with curbs and storm sewers	98	98	98	98
Gravel	76	85	89	91
Dirt	72	82	87	89
Cultivated (Agricultural Crop) Land*:				
Without conservation treatment (no terraces)	72	81	88	91
With conservation treatment (terraces, contours)	62	71	78	81
Pasture or Range Land:				
Poor (<50% ground cover or heavily grazed)	68	79	86	89
Good (50–75% ground cover; not heavily grazed)	39	61	74	80
Meadow (grass, no grazing, mowed for hay)	30	58	71	78
Brush (good, >75% ground cover)	30	48	65	73
Woods and Forests:				
Poor (small trees/brush destroyed by over-grazing or burning)	45	66	77	83
Fair (grazing but not burned; some brush)	36	60	73	79
Good (no grazing; brush covers ground)	30	55	70	77
Open Spaces (lawns, parks, golf courses, cemeteries, etc.):				
Fair (grass covers 50–75% of area)	49	69	79	84
Good (grass covers >75% of area)	39	61	74	80
Commercial and Business Districts (85% impervious)	89	92	94	95
Industrial Districts (72% impervious)	81	88	91	93
Residential Areas:				
1/8 Acre lots, about 65% impervious	77	85	90	92
1/4 Acre lots, about 38% impervious	61	75	83	87
1/2 Acre lots, about 25% impervious	54	70	80	85
1 Acre lots, about 20% impervious	51	68	79	84

Source: United States Department of Agriculture, Soil Conservation Service, 1986 [40]. Technical Release 55.

Mean Absolute Error (MAE) is the average of all absolute errors which can be expressed as

$$MAE = \frac{1}{n}\sum_{i=1}^{n}|x_i - x|, \qquad (7)$$

where

x is runoff by NRCS model

x_i is runoff by regression model

n is number of data sets

Test for significance: t-test was used to test whether there is significant difference between runoff computed by NRCS model and those estimated by regression model.

The assumptions used in this test were samples which are independent and random in nature and samples which are normally distributed.

Two-independent sample t-test statistics is defined as

$$t = \frac{\left(\sum |x_i - x|\right)/n}{\sqrt{\left(\left(\sum |x_i - x|\right)^2 - \left(\left(\sum |x_i - x|\right)^2 /n\right)\right)/(n-1)(n)}} \qquad (8)$$

3.2. Water Balance Study. Water availability and demand for domestic use and irrigation in the watershed were computed separately. Domestic water demand was obtained as the product of per capita water requirement and the population. The difference between water availability and water demand will give a comprehensive understanding on the total utilizable water.

Estimation of Irrigation Demand. Irrigation water demand was calculated from the method given by Brouwer and Heibloem, 1986 [42]. Potential evapotranspiration was computed from Blaney–Criddle equation [43] and is given by

$$ET_o = p\left(0.46 T_{mean} + 8\right), \qquad (9)$$

where

T_{mean} is mean daily temperature (°C)

p is mean daily percentage of annual daytime hours.

Crop coefficient (K_c) for paddy for various growth stages was estimated and crop water requirement was determined from

$$ET_{crop} = ET_o \times K_c \qquad (10)$$

TABLE 2: Land use details and the corresponding curve numbers.

Land use category	Area km^2	Curve Number
Forest /Forest Plantations	1.089	79
Crop land	0.193	81
Water bodies	0.600	89
Barren land	0.167	80
Settlement/Residential area	2.527	87

Other water needs such as water to saturate the soil (SAT), percolation and seepage losses (PSL), and water needed for standing water layer (WL) were used to compute the irrigation need.

Effective rainfall (P_e) was calculated from the formulae

$$P_e = 0.8P - 25 \quad \text{if } P > 75 \, \text{mm/month}$$
$$P_e = 0.6P - 10 \quad \text{if } P < 75 \, \text{mm/month}, \tag{11}$$

where

P is rainfall in mm/month

Irrigation Need (IN) is calculated as

$$\text{IN} = \text{ET}_{\text{crop}} + \text{SAT} + \text{PSL} + \text{WL} - P_e \tag{12}$$

4. Result

The runoff for Urapakkam watershed in Adyar basin was computed using USDA-NRCS model. A simple regression model was also developed and validated for its suitability for runoff computation of the watershed. By knowing the utilizable water available in a watershed, water supply for irrigation and other purposes can be planned and alternative cropping patterns can be adopted to avoid losses.

4.1. Runoff Computation

4.1.1. NRCS Model. Watershed boundary for the chain of tanks was taken from District Soil & Watershed Atlas and verified with Survey of India toposheet and satellite imagery. Land use map for Urapakkam watershed was prepared from satellite imagery with field verification in GIS platform using ArcGIS and the area under various land uses was determined (Figure 2).

The watershed is in the hydrologic soil group D as per the soil details. Hydrological modeling considers water year from July to June of ensuing year and uses daily time step. Urapakkam tank has both free catchment and intercepted catchment. The hydrological model estimated the runoff for fourteen water years from 2000-01 to 2013-14 with daily time step. Land use details and corresponding curve numbers are given in Table 2.

Weighted curve number for the watershed with a drainage area of 4.576 km^2 is 84.85. The curve numbers for Antecedent Moisture Conditions I, II, and III are 70.17, 84.85, and 92.80, respectively.

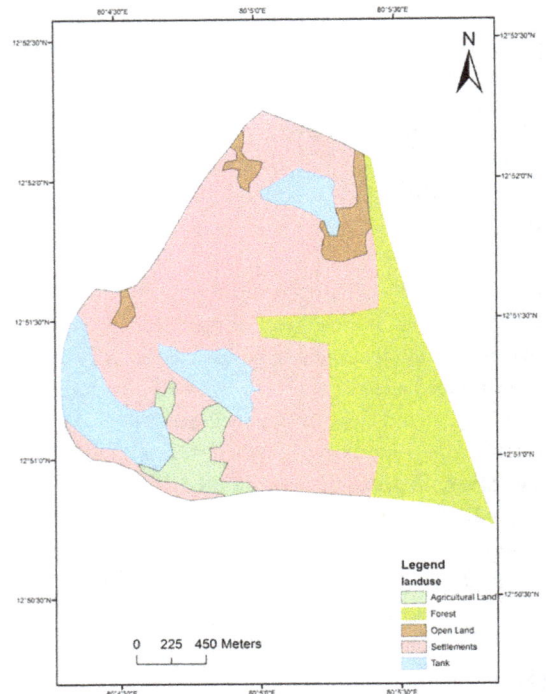

FIGURE 2: Land use map of Urapakkam watershed.

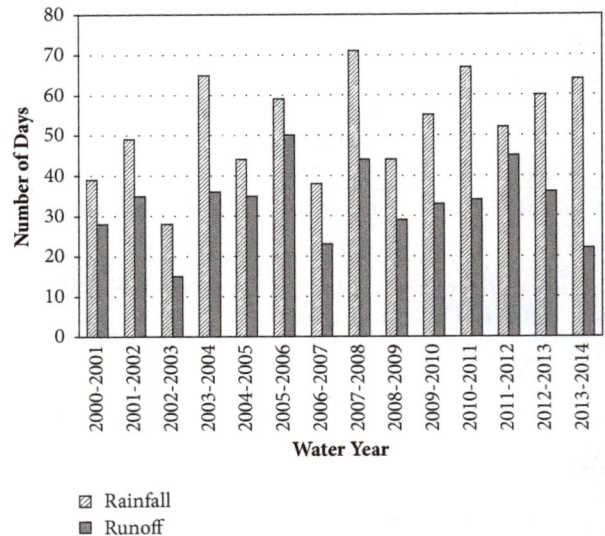

Rainfall
Runoff

FIGURE 3: Number of days with rainfall and runoff by NRCS model.

Days with rainfall and runoff for the study period are given in Figure 3. Monthly rainfall and the corresponding runoff as estimated from NRCS model are given in Figure 4. Table 3 gives monthly rainfall and runoff for the water year 2008–2009, when fifty percent dependable flow was observed.

Annual rainfall for the watershed varied from 575.7 mm (2002–03) to 3608.0 mm (2005–06). The runoff as estimated from NRCS model varied from 94.95 mm to 2324.34 mm and the corresponding percentage varied from 16.49 to 64.42 with an average value of 37.09 percent for the fourteen water years:

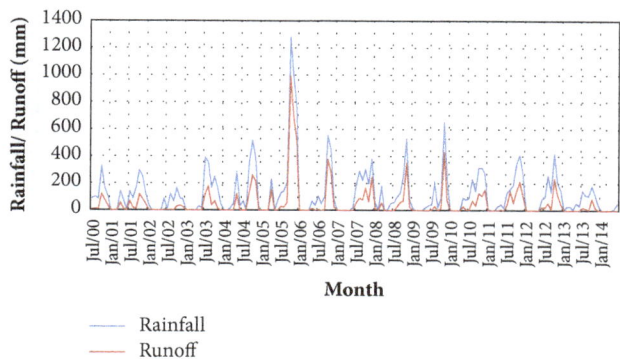

FIGURE 4: Monthly rainfall and runoff by NRCS model for Urapakkam tank watershed.

TABLE 3: Monthly rainfall and runoff for Urapakkam watershed using NRCS model for the water year 2008-2009.

Month	Rainfall mm	Runoff mm
Jul-08	61.00	0.05
Aug-08	101.00	27.49
Sep-08	129.00	63.11
Oct-08	254.00	69.09
Nov-08	534.00	352.89
Dec-08	84.00	26.33
Jan-09	15.00	0.00
Feb-09	0.00	0.00
Mar-09	0.00	0.00
Apr-09	0.00	0.00
May-09	24.00	0.00
Jun-09	36.80	0.00

2000-01 to 2013-14. Bar chart showing annual rainfall and runoff is given in Figure 5. Volume of runoff for the watershed varied between 0.434 MCM and 10.637 MCM during the fourteen years of study. Annual runoff for a representative water year (2008-09) is obtained as 2.46 MCM (Table 4). When the rainfall exceeds the threshold limit, the proportion of rainfall contributing to runoff was found to be higher. Wide variations were observed in the quantum of runoff from rainfall of same magnitude due to varying antecedent moisture conditions in the soil. Figure 6 shows the average annual rainfall and runoff estimated for Urapakkam watershed.

4.1.2. Regression Model. A Linear regression model was developed for the watershed to compute annual runoff from the annual rainfall. Ten-year data from NRCS model was used for model calibration and four-year data was used for validation. The regression relationship obtained between annual rainfall (mm) and annual runoff (mm) is given as

$$Runoff = 0.736 \times rainfall - 467.8 \qquad (13)$$

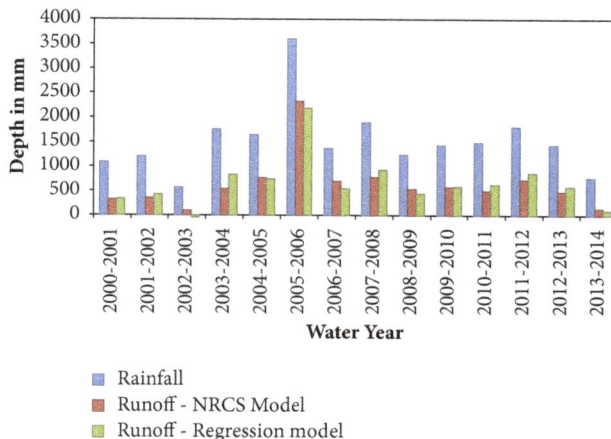

FIGURE 5: Annual rainfall and runoff for Urapakkam watershed.

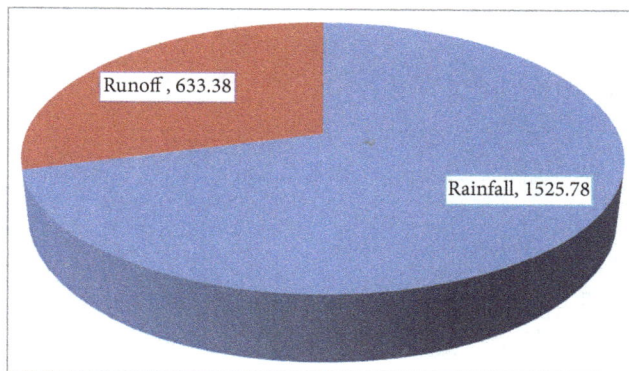

FIGURE 6: Average annual rainfall and runoff estimated for Urapakkam watershed.

R^2 value was obtained as 0.944 during calibration and 0.988 during validation period. The plots of rainfall versus runoff as obtained from NRCS model during calibration and validation period are shown in Figure 7.

4.1.3. Validation of Regression Model. Figure 8 shows the comparison between the annual runoff by NRCS model and the developed regression model which illustrates that the regression model could be used to effectively predict runoff for the watershed even during the validation phase. The coefficient of determination (R^2) for runoff by NRCS model versus runoff by regression model was 0.939 at 95% confidence level (Figure 9). The statistical tests such as t-test, MAE, and SEE obtained for the two runoff estimates (NRCS model and regression model) were 0.55, 10.60, and 13.90, respectively. The average runoff for the watershed by linear regression model was obtained as 37.17 percent of the rainfall which ranges from 12.79% to 60.63%. It shows no considerable deviation from runoff computed by NRCS model. Thus the statistical analysis proves that tedious and data intensive computation methods can be eliminated and the volume of runoff from a watershed can be efficiently and easily computed using simple regression model.

TABLE 4: Annual rainfall and runoff estimated for Urapakkam watershed using NRCS model for the period 2000-01 to 2013–14.

Water Year	Rainfall		Runoff			
	No. of days	Depth mm	No. of days	Depth mm	Percentage of Rainfall	Volume MCM
2000-2001	39	1092.30	28	323.96	29.66	1.483
2001-2002	49	1211.10	35	355.39	29.34	1.626
2002-2003	28	575.70	15	94.95	16.49	0.434
2003-2004	65	1765.80	36	537.50	30.44	2.460
2004-2005	44	1648.00	35	768.59	46.64	3.517
2005-2006	59	3608.00	50	2324.34	64.42	10.637
2006-2007	38	1379.00	23	701.20	50.85	3.209
2007-2008	71	1899.50	44	780.43	41.09	3.571
2008-2009	44	1238.80	29	538.97	43.51	2.466
2009-2010	55	1439.00	33	582.24	40.46	2.665
2010-2011	67	1489.00	34	499.09	33.52	2.284
2011-2012	52	1810.00	45	734.11	40.56	3.359
2012-2013	60	1435.40	36	484.12	33.73	2.215
2013-2014	64	769.30	22	142.48	18.52	0.652
Minimum	**28**	**575.7**	**15**	**94.95**	**16.49**	**0.434**
Maximum	**71**	**3608.0**	**44**	**2324.34**	**64.42**	**10.637**

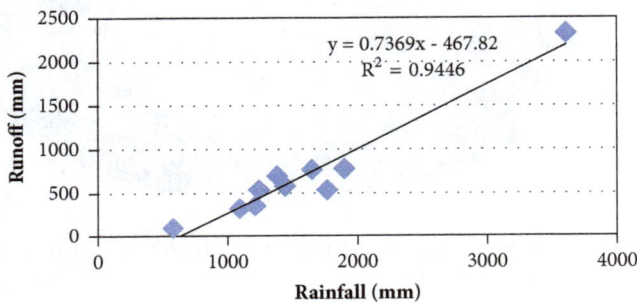

(a) Calibration

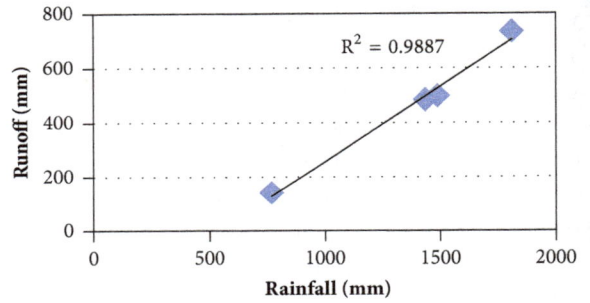

(b) Validation

FIGURE 7: Relationship between annual rainfall and runoff by NRCS model during calibration and validation period.

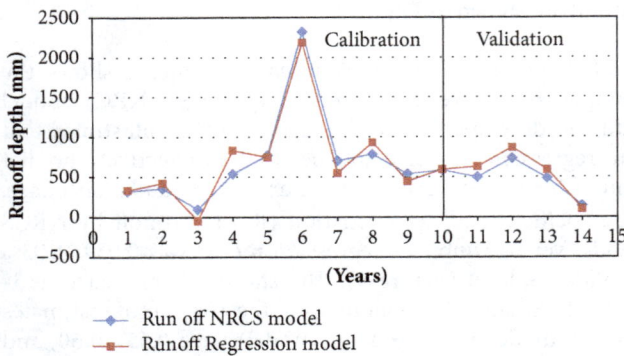

FIGURE 8: Comparison of annual runoff from NRCS model and regression model.

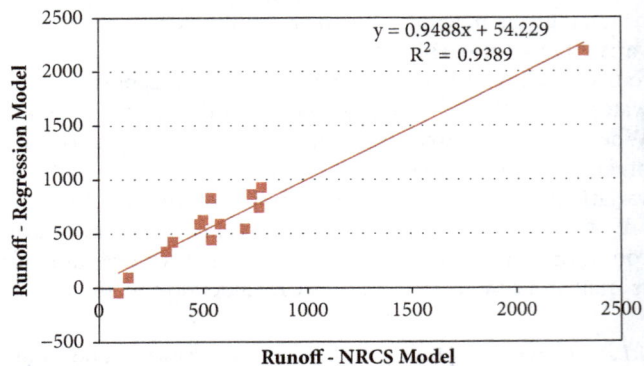

FIGURE 9: Scatter plot of annual runoff by NRCS model and regression model for Urapakkam watershed for 2000-01 to 2013-14.

4.2. *Water Balance Study.* Population density of Kanchipuram district is 892/km^2 [44]. Assuming the per capita water demand to be 100 liters/day for domestic purposes, the total domestic water requirement for the population of the watershed with 4.576 km^2 is estimated at 407 m^3/day, which is 0.148 MCM per annum.

TABLE 5: Computation of irrigation demand for paddy for the year 2008-09.

Item	Month						Season
	September	October	November	December	January	February	
Percentage of annual daytime hours, p		0.26	0.27	0.26	0.27	0.27	
Average Temperature T_{mean} (°C)		26.00	24.80	24.60	25.8	27.80	
Potential evapotranspiration Et_o (mm/day)		5.19	5.14	5.02	5.27	5.61	
Crop coefficient K_c		1.10	1.10	1.20	1.30	1.00	
Crop water requirement							
mm/day		5.71	5.66	6.03	6.84	5.61	
mm/month		171.26	175.38	186.82	191.65	174.00	
SAT, mm	200.00						
PSL, mm/month @4 mm/day		120.00	124.00	124.00	112.00	124.00	
WL, mm							
P_e, mm	48.60	499.80	167.80	0.00	0.00	0.00	
Irrigation demand, mm	151.40	0.00	131.58	310.82	303.65	298.00	**1195.45**

Irrigation is practiced in a small area of 10 ha in the watershed. The average irrigation demand for the watershed area is computed to be around 1195 mm. By considering an efficiency of 70%, the actual irrigation water requirement is 1708 mm. The total irrigation demand for 10 ha of paddy crop in the watershed is estimated as 0.171 MCM. Details of the computation of irrigation demand for paddy are presented in Table 5.

The current total domestic and irrigation water requirement of the watershed is 0.320 MCM. After meeting all the water demands of the watershed there is sufficient water available even if only 50% of the total available rainwater can be effectively harvested.

5. Conclusions

Accurate estimation of runoff is essential for the effective management of water resources especially in semiarid regions like Tamil Nadu. NRCS model is a widely accepted model for the estimation of runoff for small watersheds. It is extensively used by engineers and hydrologists but the lack of sufficient data may give poor results. With the understanding of available water in a watershed, irrigation scheduling, crop rotation, and cropping pattern can be planned.

Total volume of runoff available for Urapakkam watershed from NRCS model for a representative year is 2.46 MCM. A simple regression model was developed to estimate depth of annual runoff for the watershed which is given as runoff (mm) = 0.736 × rainfall (mm) − 467.8. Statistical tests confirm that there is no significant difference between the runoff estimates from the two models. Hence the study confirms that it is advantageous to have calibrated regression

models for easy computation of runoff and for the effective management of water resources for ungauged basins.

Water balance study shows that 1.23 MCM of water can be effectively stored in the watershed after losses by improving the drainage and capacity of the storages. The total domestic and irrigation demand is estimated at 0.320 MCM. Hence apart from meeting the present domestic and irrigation demand in the watershed, additional demand due to future development and demand of adjoining areas can also be met from the water available in the watershed if properly harvested.

References

[1] I. A. Shiklomanov, "Appraisal and Assessment of world water resources," *Water International*, vol. 25, no. 1, pp. 11–32, 2000.

[2] FAO, "FAO Activities on Promoting Fertigation," in *Proceedings of the Regional Workshop on Guidelines for Efficient Fertilizers Use through Modern Irrigation*, FAO Regional Office for the Near East, Cairo, Egypt, 1998.

[3] W. Cosgrove and F. Rijsberman, *World Water Vision: Making Water Everybodys Business*, World Water Council, Earthscan Publications, 2000.

[4] Ministry of Water Resources. 2012. Per Capita Water Availability. Press Information Bureau, Government of India, 26-April 2012, http://pib.nic.in/newsite/PrintRelease.aspx?relid=82676.

[5] S. K. Gupta and R. D. Deshpande, "Water for India in 2050: First-order assessment of available options," *Current Science*, vol. 86, no. 9, pp. 1216–1224, 2004.

[6] Soil Conservation Service (SCS). Hydrology, National Engineering Handbook, Supplement A, Section 4, Chapter 10, Soil Conservation Service, U.S.D.A., Washington, DC, USA, 1972.

[7] L. K. Sherman, "The unit hydrograph method," in *Physics of the Earth*, O. E. Meinzer, Ed., pp. 514–525, Dover Publications Inc., New York, NY, USA, 1949.

[8] V. Mockus, "Estimation of total (peak rates of) surface runoff for individual storms," in *Exhibit A of Appendix B, Interim Survey Report Grand (Neosho) River Watershed, U.S.D.A*, 1949.

[9] S. K. Mishra and V. P. Singh, "Another look at SCS-CN method," *Journal of Hydrologic Engineering*, vol. 4, no. 3, pp. 257–264, 1999.

[10] S. K. Mishra, M. K. Jain, P. K. Bhunya, and V. P. Singh, "Field applicability of the SCS-CN-based Mishra-Singh general model and its variants," *Water Resources Management*, vol. 19, no. 1, pp. 37–62, 2005.

[11] K. X. Soulis, J. D. Valiantzas, N. Dercas, and P. A. Londra, "Investigation of the direct runoff generation mechanism for the analysis of the SCS-CN method applicability to a partial area experimental watershed," *Hydrology and Earth System Sciences*, vol. 13, no. 5, pp. 605–615, 2009.

[12] P. Koteshwaram and S. M. A. Alvi, "Secular trends and periodicities in rainfall at west coast stations in India," *Current Science*, vol. 38, no. 10, pp. 229–231, 1969.

[13] K. V. Ramamurthy, M. K. Sonam, and S. S. Muley, "Long term variation in the rainfall over upper Narmada catchment," *Mausam*, vol. 38, no. 3, pp. 313–318, 1987.

[14] R. Jha and R. K. Jaiswal, "Analysis of rainfall pattern in a catchment," *IAH Journal of Hydrology (India), Roorkee*, 1992.

[15] F. H. S. Chiew, M. J. Stewardson, and T. A. McMahon, "Comparison of six rainfall-runoff modelling approaches," *Journal of Hydrology*, vol. 147, no. 1-4, pp. 1–36, 1993.

[16] E. M. Lungu, "Detection of changes in rainfall and runoff patterns," *Journal of Hydrology*, vol. 147, no. 1-4, pp. 153–167, 1993.

[17] J. Meher and R. Jha, "Time series analysis of rainfall data over Mahanadi River basin," *Emerging Science*, vol. 3, no. 6, pp. 22–32, 2011.

[18] K. X. Soulis and J. D. Valiantzas, "SCS-CN parameter determination using rainfall-runoff data in heterogeneous watersheds-the two-CN system approach," *Hydrology and Earth System Sciences*, vol. 16, no. 3, pp. 1001–1015, 2012.

[19] R. Kabiri, A. Chan, and R. Bai, "Comparison of SCS and Green-Ampt Methods in Surface Runoff-Flooding Simulation for Klang Watershed in Malaysia," *Open Journal of Modern Hydrology*, vol. 03, no. 03, pp. 102–114, 2013.

[20] S. Mohan and M. Abraham, "Derivations of simple site-specific recharge-precipitation relationships: A case study from the Cuddalore Basin, India," *Environmental Geosciences*, vol. 17, no. 1, pp. 37–44, 2010.

[21] R. Viji, P. R. Prasanna, and R. Ilangovan, "GIS based SCS-CN method for estimating runoff in Kundahpalam watershed, Nilgries district, Tamilnadu," *Earth Sciences Research Journal*, vol. 19, no. 1, pp. 59–64, 2015.

[22] P. K. Gupta, S. Punalekar, S. Panigrahy, A. Sonakia, and J. S. Parihar, "Runoff modeling in an agro-forested watershed using remote sensing and gis," 17, no. 11, pp. 1255–1267, 2012.

[23] R. Chatterjee, "Groundwater resources estimation — case studies from India," *Journal of the Geological Society of India*, vol. 77, no. 2, pp. 201–204, 2011.

[24] Central Ground Water Board (CGWB). Water Balance Studies in Upper Yamuna Basin – Terminal Report - Project Findings and Recommendations. Chandigarh: Central Ground Water Board 2000.

[25] G. Thakuriah and R. Saikia, "Estimation of Surface Runoff using NRCS Curve number procedure in Buriganga Watershed, Assam, India - A Geospatial Approach," *International Research Journal of Earth Sciences*, vol. 2, no. 5, pp. 1–7, 2014.

[26] V. K. Dadhwal, "Remote sensing and GIS for agricultural crop acreage and yield estimation," in *Proceedings of the International Archives of Photogrammetry and Remote Sensing, XXXII*, pp. 58–67, 1999.

[27] G. B. Geena and P. N. Ballukraya, "Estimation of runoff for Red hills watershed using SCS method and GIS," *Indian Journal of Science and Technology*, vol. 4, no. 8, pp. 899–902, 2011.

[28] K. H. Syed, D. C. Goodrich, D. E. Myers, and S. Sorooshian, "Spatial characteristics of thunderstorm rainfall fields and their relation to runoff," *Journal of Hydrology*, vol. 271, no. 1-4, pp. 1–21, 2003.

[29] X. Zhan and M.-L. Huang, "ArcCN-Runoff: An ArcGIS tool for generating curve number and runoff maps," *Environmental Modeling and Software*, vol. 19, no. 10, pp. 875–879, 2004.

[30] K. Geetha, S. K. Mishra, T. I. Eldho, A. K. Rastogi, and R. P. Pandey, "Modifications to SCS-CN method for long-term hydrologic simulation," *Journal of Irrigation and Drainage Engineering*, vol. 133, no. 5, pp. 475–486, 2007.

[31] V. Kumar, M. Babu, T. V. Praveen et al., "Analysis of the Runoff for Watershed Using SCS-CN Method and Geographic Information Systems," *International Journal of Engineering Science and Technology*, 2010.

[32] R. Pradhan, P. P. Mohan, M. K. Ghose et al., "Estimation of Rainfall Runoff using Remote Sensing and GIS in and around Singtam, East Sikkim," *International Journal of Geomatics and Geosciences*, vol. 1, article 466, 2010.

[33] A. Panahi, "Evaluating SCS-CN method in estimating the amount of runoff in Soofi Chay basin using GIS," *Life Science Journal*, vol. 10, no. 5, pp. 271–277, 2013.

[34] S. Satheeshkumar, S. Venkateswaran, and R. Kannan, "Rainfall–runoff estimation using SCS–CN and GIS approach in the Pappiredipatti watershed of the Vaniyar sub basin, South India," *Modeling Earth Systems and Environment*, vol. 3, no. 1, 2017.

[35] IMD, *Temperature of Kanchipuram District*, Regional Meteorological Centre, Chennai, India, 2013, Temperature of Kanchipuram District. Regional Meteorological Centre.

[36] ENVIS 2015. Kanchipuram District. ENVIS Centre Tamilnadu, http://www.tnenvis.nic.in/files/KANCHIPURAM%20.pdf.

[37] AIS and LUS, *Watershed Atlas of India, Department of Agriculture and Co-operation. All India Soil and Landuse Survey*, IARI Campus, New Delhi, India, 1990.

[38] V. T. Chow, D. R. Maidment, and L. W. Mays, *Applied Hydrology*, McGraw-Hill, New York, New York, USA, 1988.

[39] Natural Resources Conservation Service (NRCS), "Chapter 7-Hydrologic Soil Groups," in *Hydrology National Engineering Handbook, title 630*, Washington, DC, USA, 2007.

[40] United States Department of Agriculture, Soil Conservation Service (USDA-SCS). Technical Release 55, Urban hydrology for small watersheds, USA, 1986.

[41] A. G. Bluman, *Elemtary Statistics*, McGraw Hill, New York, NY, USA, 6th edition, 2008.

[42] C. Brouwer and M. Heibloem, "Irrigation Water Needs- Irrigation Water Management," in *Training Manual No. 3*, Food and Agriculture Organisation, Rome, Italy, 1986.

15

3D Mapping of the Submerged *Crowie* Barge using Electrical Resistivity Tomography

Kleanthis Simyrdanis [1,2] **Ian Moffat,**[1] **Nikos Papadopoulos,**[2] **Jarrad Kowlessar,**[1] **and Marian Bailey**[1]

[1]*Archaeology, College of Humanities, Arts and Social Sciences, Flinders University, Adelaide, SA, Australia*
[2]*Laboratory of Geophysical-Satellite Remote Sensing, Institute for Mediterranean Studies, Rethymno, Greece*

Correspondence should be addressed to Kleanthis Simyrdanis; ksimirda@ims.forth.gr

Academic Editor: Yun-tai Chen

This study explores the applicability and effectiveness of electrical resistivity tomography (ERT) as a tool for the high-resolution mapping of submerged and buried shipwrecks in 3D. This approach was trialled through modelling and field studies of *Crowie*, a paddle steamer barge which sunk at anchor in the Murray River at Morgan, South Australia, in the late 1950s. The mainly metallic structure of the ship is easily recognisable in the ERT data and was mapped in 3D both subaqueously and beneath the sediment-water interface. The innovative and successful use of ERT in this case study demonstrates that 3D ERT can be used for the detailed mapping of submerged cultural material. It will be particularly useful where other geophysical and diver based mapping techniques may be inappropriate due to shallow water depths, poor visibility, or other constraints.

1. Introduction

During the last few years there has been an increasing trend of employing the electrical resistivity method in subaqueous areas for geological and environmental studies [1, 2] and underwater geological mapping [3]. A number of methodological studies on the use of ERT in the aquatic environment have been undertaken, including investigating the most appropriate electrode position and array type. Orlando [4] used numerical simulation modelling to investigate the resolution of underwater resistivity surveys, employing both floating and submerged electrodes. Her results showed that floating cables give low resolution data when the contrast between the resistivity of water and sediment layer is too small. Submarine electrical resistivity data can be collected utilising different electrode arrays such as Dipole-Dipole or Pole-Dipole which are the most widely used ones due to their simple geometry and high data density. In addition, Orlando discusses the Schlumberger array which gives comparable results [4] and Rucker et al. recommend a Pole-Pole array

for the collection of marine resistivity data [3]. Baumgartner and Christensen [5] describe a particular array where probes are vertically aligned to increase the investigation depth for mapping resistivity anomalies located at greater water depth. Chiang et al. [6, 7] studied the efficiency of a marine towed electrical resistivity method and its sensitivity using 2D numerical simulation modelling as a well-suited method for methane hydrate exploration at a shallow depth beneath the seafloor.

However the use of electrical resistivity method in submarine archaeology prospection studies is not common. Ranieri et al. ([8]: 11) studied the development and application of an integrated methodology for the 3D rendering of geoelectrical data of buried and submerged archaeological features, in complex environments such as coastal areas. A comprehensive feasibility study was also undertaken, investigating the efficiency of ERT in reconstructing submerged archaeological relics (e.g., building foundations) in shallow seawater environment through numerical modelling and field experimentation [9]. The study showed that floating and

FIGURE 1: Photos of the *Crowie* barge (courtesy photograph of the State Library of South Australia, previously published in [13]: 138).

submerged survey modes can be used with equal success in cases of relatively shallow marine environments when the water depth does not exceed 1 m. In deeper marine environments, the submerged mode survey is recommended for outlining isolated targets. Valid a priori information, in terms of the seawater resistivity and thickness, is important and can greatly improve the inversion results for the data captured with the submerged ERT mode. On the other hand, erroneous information can cause severe distortions in the inversion ERT models and misleading interpretations. Passaro et al. used multibeam bathymetry, seismic, geoelectric, and magnetic methods to investigate archaeological targets protruding from very shallow water [10]. Passaro ([11]: 3) also used electrical resistivity survey to detect a shipwreck in Salerno, southern Italy.

Despite this interest, ERT so far appears not to have been used to map a shipwreck in detail. ERT is particularly useful in cases when the water is too turbid for visual surveys and the water depth is too shallow for acoustic techniques. Furthermore, since many ships are metal-based structured, they become easily detectable with ERT when the target is located in an electrically high resistivity environment such as sand.

2. *Crowie* Shipwreck

The present study undertook a high-resolution survey of the barge *Crowie*, which is submerged in the Murray River (South Australia). *Crowie* was a commercial cargo barge, built in 1911 by Captain J. G. Arnold. It is claimed to be one of the largest barges ever built for the Murray River ([12]: 63; in [13]: 137), and its enormous size is recorded in historical accounts as causing several accidents when traveling up the river. *Crowie* was eventually superseded by other transport methods and was abandoned near the Morgan wharf where it sank at

its mooring, sometime between 1946 and 1950 ([13]: 137) (Figure 1).

The dimensions of *Crowie* and its structural composition are well recorded in historical records as well as from previous multibeam and side-scan sonar surveys (summarised by [13]) (Figure 2). The barge is 45 m long, with a 9 m beam and depth from the base of the hold to the deck of 2.5 m ([13]: 140). It was built using a bottom-based construction technique. This led to a flat bottom and straight sides amenable to carry the maximum possible amount of cargo ([13]: 141). The vessel had one deck, a carvel built sharp stern, straight stem, and a composite construction consisting of iron frames and topsides overlain by timber. It had nine internal bulkheads ([13]: 139).

The vessel sits upright in a known location immediately upstream of the Morgan wharf and downstream of the wreck of the PS *Corowa*. The vessel slightly protrudes from the water during low water periods and extends to a depth of several meters. The extremely turbid water of the Murray makes a visual assessment of the wreck impossible; however geophysical data and inspection of the wreck by touch suggest it is largely intact below the deck-level for its entire length ([13]: 139). The frames of the vessel are intact and easily identifiable. Any iron topsides that were on the vessel at the time of wrecking have corroded away; however metal below a depth of 15 cm below the waterline is believed to be largely intact ([13]: 139).

The large amount of metal used in the construction of the *Crowie* vessel is likely to give it a highly conductive electrical signature, which should contrast significantly the fresh water and sand which host the wreck. For that reason, electrical resistivity tomography can be considered an appropriate method to apply to this survey, in order not only to detect the specific "target" and define its dimensions (as the previous methods successfully accomplished) but also to estimate its

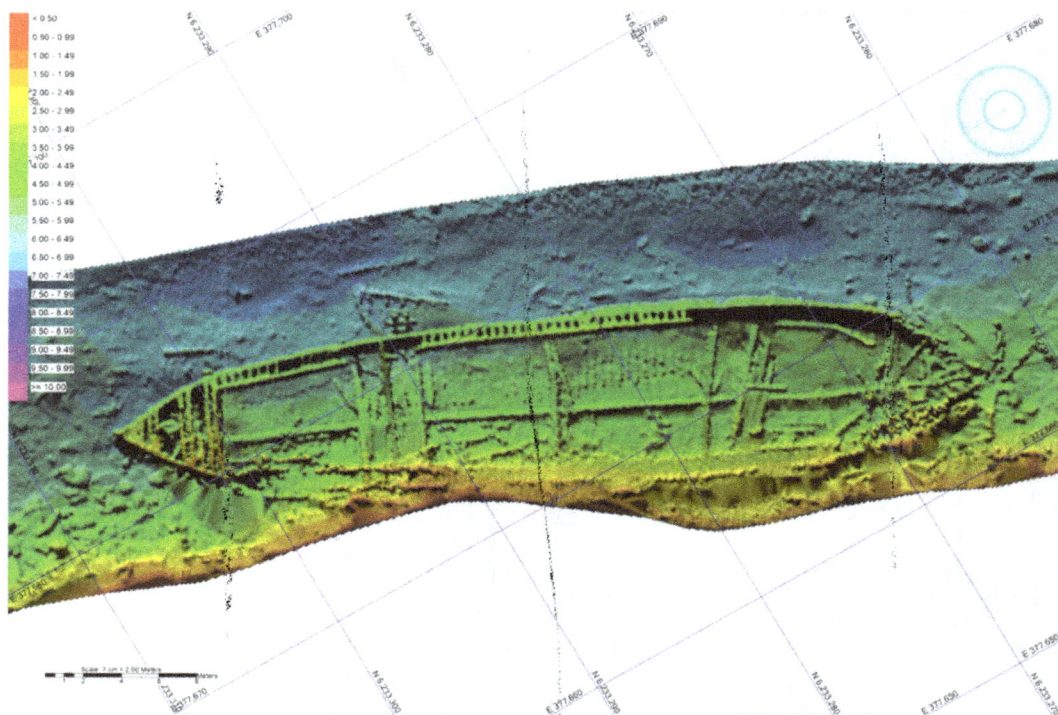

FIGURE 2: Multibeam image of the *Crowie* barge (courtesy image of Gareth Carpenter, previously published in [13]: 141).

burial depth and confirm the nature of the material that the barge is composed of.

3. Methodology

Initially the viability of the proposed field survey was validated using data produced from a three-dimensional numerical simulation model. The ERT numerical experiments were performed using 3D forward and inversion algorithms within the "Res3DMod" and "Res3DInvx64" software [14]. A finite-element modelling subroutine was used to calculate the apparent resistivity values, and a nonlinear smoothness-constrained least-squares optimisation technique was used to calculate the resistivity of the model blocks [15]. The model was constructed with parameters approximating those expected for the field site.

The indicative model used for the numerical simulations included a water layer $D = 6$ m thick (under the air half-space) and three homogenous medium layers $D_1 = 3.5$ m, $D_2 = 1$ m, and $D_3 = 1.5$ m as shown in Figure 3. The resistivity value of the water layer was set to $\rho_{water} = 28$ ohm·m and for the homogeneous layers was set to $\rho_1 = 300$ ohm·m, $\rho_2 = 1500$ ohm·m, and $\rho_3 = 150$ ohm·m, respectively. The above resistivity values were estimated based on the local geological map and descriptions of the lithology of units within the Murray Basin [16–18]. A conductive target, which simulated a metallic ship structure (shape was defined according to the multibeam image results in [13]: 141), was embedded in the first layer with dimensions 5.5 m width and 12.5 m length. Internal divisions were also included denoting

the different compartments of the upper part of the barge. The resistivity value for the target's structure was set to $\rho_{target} = 0.05$ ohm·m.

The grid (10×15 m) for the numerical simulation was constructed using 672 electrodes (21 lines of 32 electrodes) with spacing $a = 0.5$ m in both directions (X, Y). The electrodes were submerged in the water bottom 6 m below the water surface. A Pole-Dipole (including the combination of forward and reverse measurements where the electrode B was placed at a theoretically infinite distance) and maximum separation distance $N = 11a$ ("a" unit electrode spacing) was selected and 29618 apparent resistivity measurements were simulated. Data were corrupted with 10% Gaussian noise so that the simulation data were approaching the real noise levels that would be expected in the field. A priori information for constraining the inversion procedure was included, taking into consideration the water depth and the resistivity of the water. The inversion algorithm was terminated after seven iterations unless some other criteria were met (e.g., slow convergence rate of less than 3%, RMS error smaller than noise level).

The outer structure of the ship was reproduced quite accurately within the dimensions of the constitutive model's dimensions (Figure 4). The reconstructed resistivity depth sections clearly depict the outer structure of the metal target, with resistivity values close to 0.05–0.2 ohm·m from 0 m to the depth of 1 m below water bottom (depth layers D1–D4). The inner divisions of the barge are not well resolved except for the main axis of the barge which is more obvious at depth slices D3 to D5. Although the target extends no more than

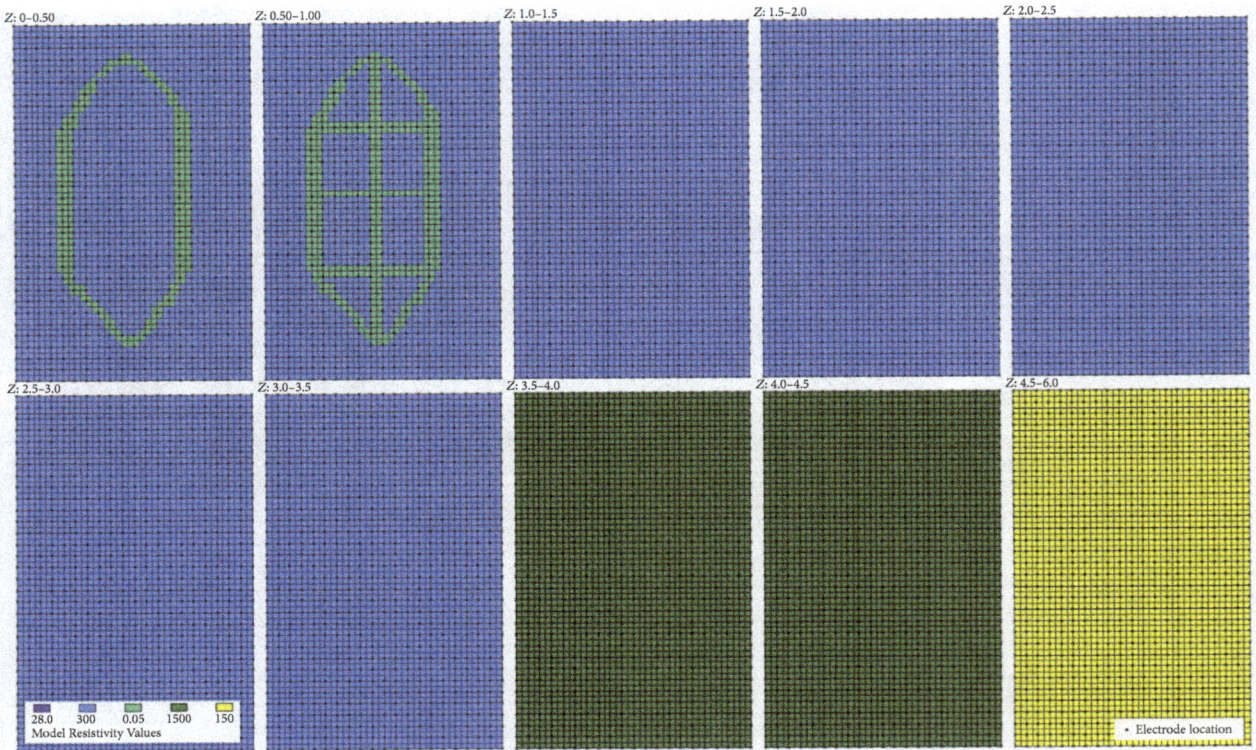

FIGURE 3: Constitutive parameters of 3D model.

1 m in depth (D4) below the water bottom (according to the model), the inversion result depicts the target (with distorted shape) continuing to 5 m below the water bottom (D5 to D11). The upper background highly resistive layer (0 m to 3.5 m, D1–D10) is represented with a range of gradually increased resistivity values from 150 to 300 ohm·m. The deeper layers (D11-D12) are represented with values close to 1600 ohm·m. The numerical limitations of the modelling and inversion procedures cannot efficiently cope with the magnitude of the resistivity contrast between the conductive target and the resistive background. Hence, the RMS error was high with values reaching almost 30% after 6 iterations.

4. Field Case Study

Following the theoretical modelling of this feature, field survey was undertaken on *Crowie*. The barge *Crowie* was submerged near the bank of the Murray River in Morgan (South Australia) (Figure 5). The exact position of the target is described in Roberts et al. [13].

The study area is located in the Murray Basin, epicratonic sedimentary basin of Paleocene to Quaternary age [18]. This basin contains a sedimentary sequence up to 600 m thick including freshwater, marine, coastal, and continental sediments [17, 18]. At the survey location the stratigraphy includes the Oligocene aged Ettrick Formation, the Early-Middle Miocene aged Mannum Limestone, and a number of

Quaternary aged aeolian units ([16]; see Figure 6). The Quaternary aeolian sediments are estimated to have a resistivity value of ρ_{sand} = 300 ohm·m, the Mannum Limestone has a value of $\rho_{limestone}$ = 1500 ohm·m, and the Marly Ettrick Formation has a value of ρ_{marl} = 150 ohm·m.

4.1. Data Acquisition (Grid, Equipment). The survey grid was set as a 60 by 15.5 m rectangle approximately parallel to the riverbank as shown in Figure 7. The four corners of the grid were established using heavy rocks as anchors, making sure that they would be unmovable and stable throughout the fieldwork, with taut ropes running vertically from the anchor to a float on the surface. At the extremities of the grid, measuring tapes were floated as guidelines between these anchors for determining the position of each survey line.

The data were acquired using 58 parallel lines (length of each line 15.5 m) oriented perpendicular to the bank and equally spaced (L = 1 m apart). The sensors (32 electrodes), equally spaced (a = 0.5 m apart) on each survey line, were submerged on the water bottom on top of the sunken ship perpendicular to its length.

A ZZ FlashRES-Universal resistivity meter was used for injecting and measuring the potential throughout the survey. The water conductivity value and temperature were measured daily with a YSI Pro-1030 hand-held conductivity meter. The average value of conductivity (ρ = 28 ohm·m) was used as a priori information for the inversion procedure. A Pole-Dipole array was used for data collection with a total of #1498

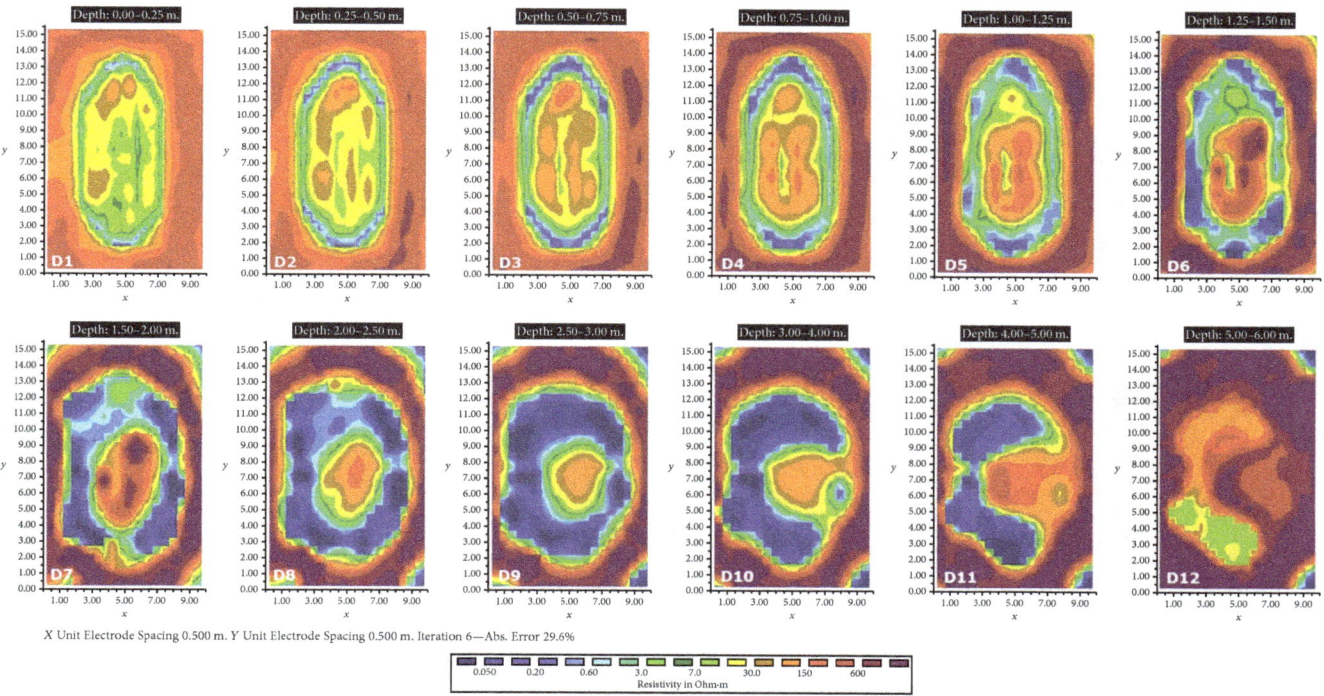

FIGURE 4: 3D inversion depth slices. Resistivity values mapped on a logarithmic scale. Both axes are given in meters.

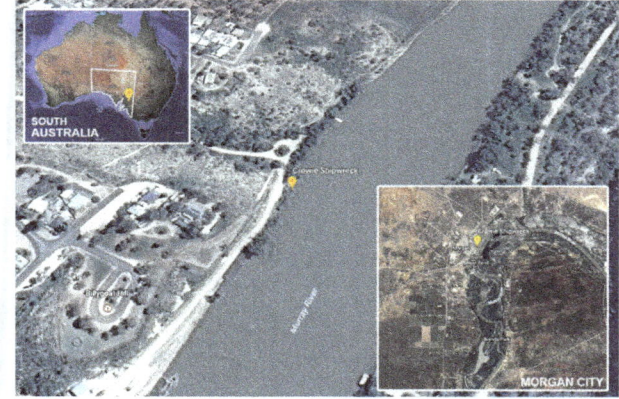

FIGURE 5: Map of the survey area in Morgan (South Australia).

measurements (consisting of 749 forward and 749 reverse measurements). The distance "a" between the dipole "MN" varied between 0.5 and 4 m. The distance factor "n" between electrode "A" and dipole "MN" varied from 1 to 8a.

4.2. Data Acquisition (Bathymetry). A total station survey was conducted over the wreck site using a Leica TS16 to measure the river depth in the location of the ERT survey. The position of the total station was georectified based on two static GPS points collected with a CHC 90+ GPS postprocessed using the Auspos service. The position of all bathymetry points were recorded using a prism mounted on a 7.37-meter long pole which was placed on the river bottom by an operator in a small boat. The bathymetric survey recorded 116 points on the bottom of the river around the study area. The points were distributed around the survey area spaced out to ensure reasonable site coverage. The water level was recorded using a prism held at the water surface at 8 different locations. To calculate the bathymetry of the study area, a surface was interpolated using the Inverse Distance Weighting (IDW) surface interpolation function in ESRI ArcGIS software package. The IDW interpolation is a local neighborhood interpolation method, which estimates unknown values based on the distance and values of nearby known points. This interpolation method was chosen due to the limited number of points selected and the heterogeneity of the measured values (which is thought to be based on debris on the river bottom). Unlike kriging or other geostatistical surface interpolations, IDW should be less affected by the potential outliers in the data set [19]. The surface was calculated using a variable search radius with 12 points and no maximum distance selected.

The surface produced can be seen in Figure 8(a). No bounding box was used for the surface interpolation, so the values produced are only accurate within the 60 × 15.5 m survey grid. The resulting surface provided elevations based on the Australian Height Datum for a continuous surface covering the study area. To calculate the water depth, the raster calculator tool was used to create a new raster surface, by subtracting the elevation of the river bottom from the average water surface elevation. This produced a continuous surface describing the depth of the water for the site.

The surface describing water depth for the site can be seen in Figure 8(b). Individual ERT lines were surveyed with a total

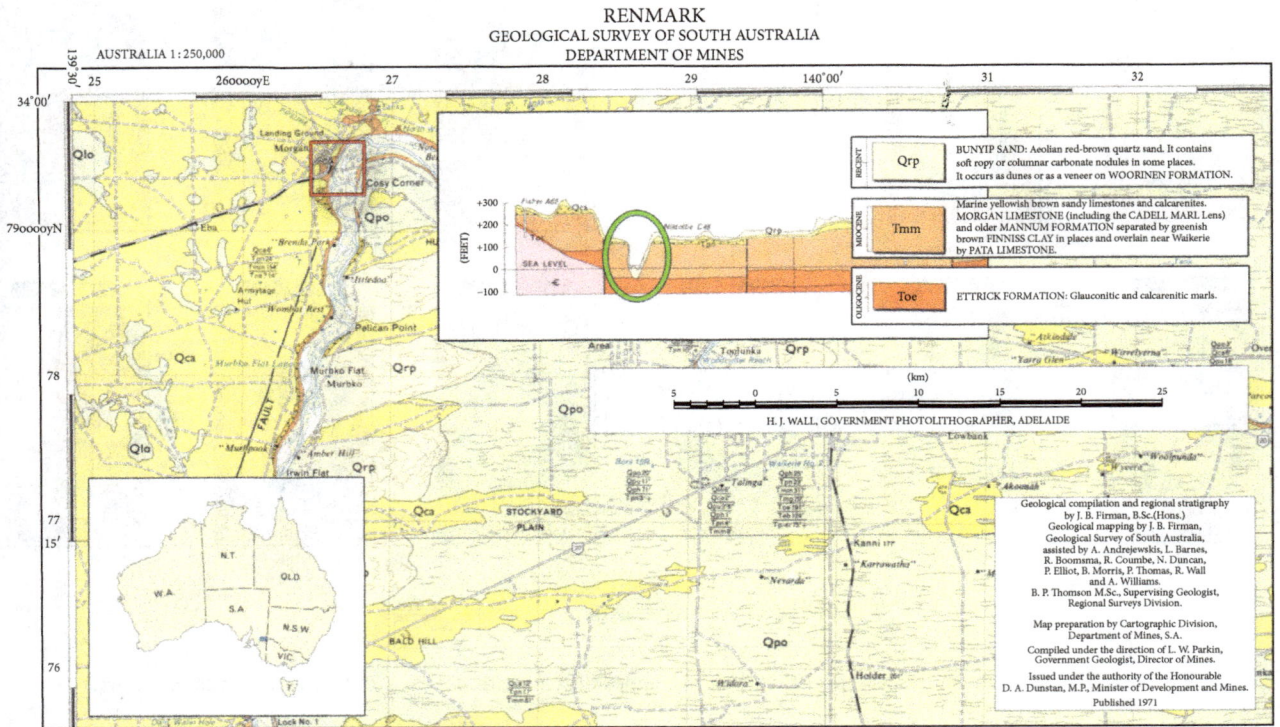

FIGURE 6: Geological map of the broader survey area (modified from [16]).

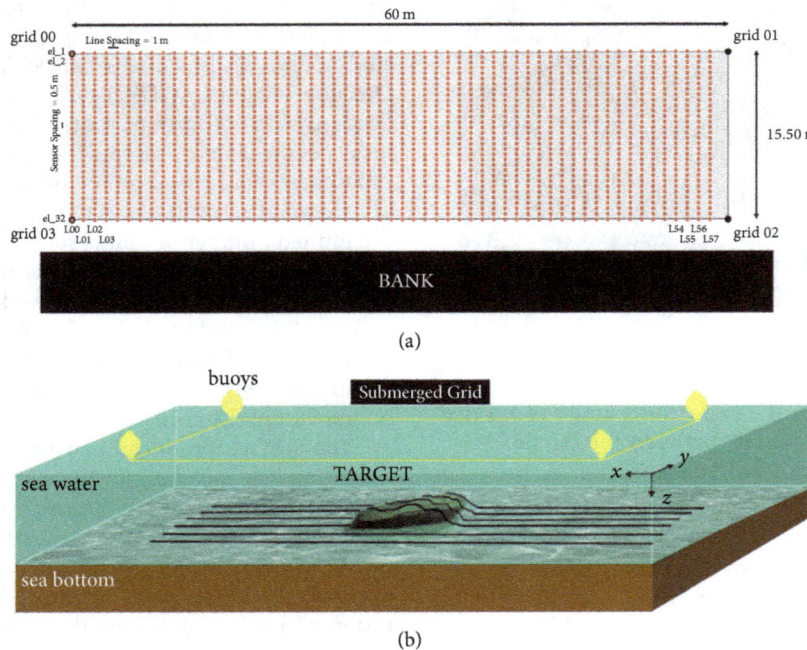

(a)

(b)

FIGURE 7: Selected survey grid. Red dots indicating electrode position (a). Cables submerged at the river (b).

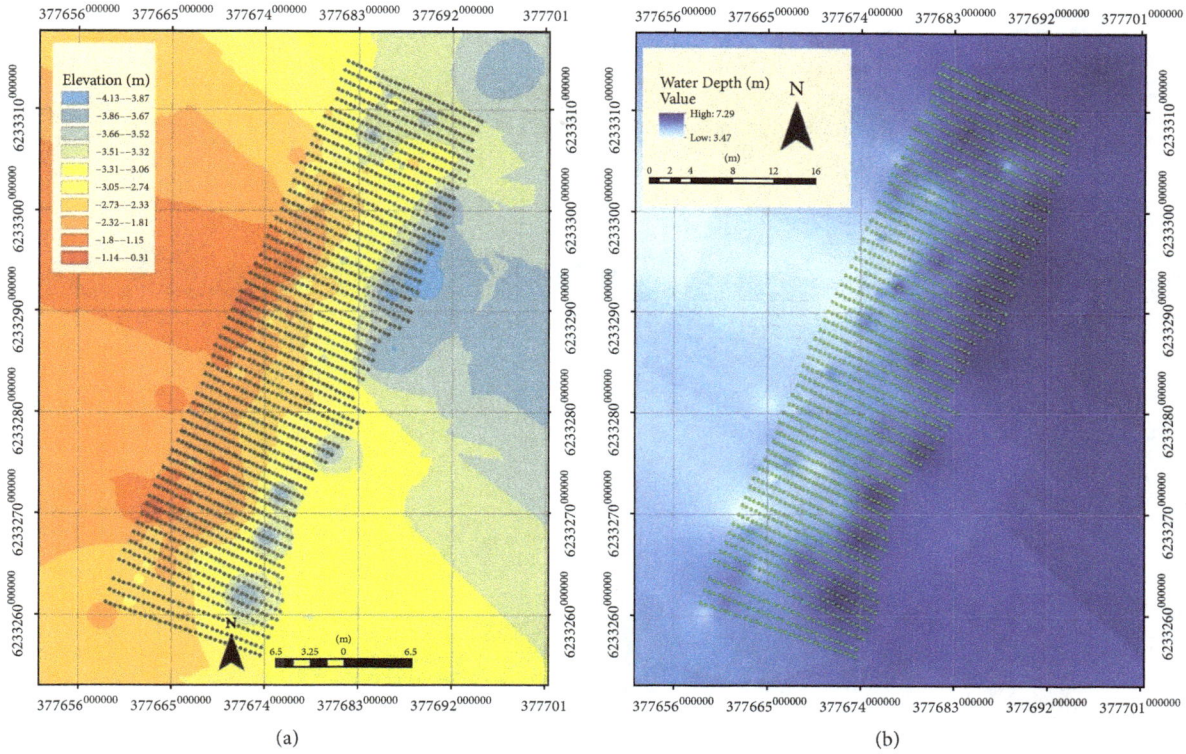

FIGURE 8: Elevation (a) and bathymetry (b) model. The dots represent the electrode positions.

station marking the start and end position of each line. The individual sensor locations were estimated by evenly spacing a number of points along the length of the measured line, using the "Points Along Line" tool in ESRI ArcGIS software package. An attribute was calculated for all sensors based on the water depth for use in later ERT processing. The elevation (above sea level) for the survey area was between −0.31 m and −4.13 m which corresponds to water depths between 3.47 m and 7.29 m.

4.3. Data Processing. In total, 112,984 raw data measurements were acquired during the field survey from survey lines L00 until L57. Data were exported after despiking outlier values (due to high geometric factor, small potential values, or values collected with insufficient current) and removing data with high % RMS error using a spreadsheet. The resistance values varied from 0.01 to 30 V/I, as depicted in Figure 9(a). Specifically, the first 12 lines (L00–L12) had much higher group of resistance values (from 0.5 to 30 V/I) and the lines that were over the wreck (L13–L57) mainly have values less than 0.5 V/I. Regarding the apparent resistivity values (Figure 9(b)), the first lines (L00–L12) have values ranging from 50 to 400 ohm·m, and the rest (L13–L57) have lower values from 1 to 20 ohm·m, reflecting the presence of the wreck in these lines.

After the preprocessing of the raw data, the a priori information was incorporated in each line. Before adding the water column depth in each survey line, the slope was smoothed in both directions (X, Y): On X direction the vertical distance should be less than the electrode spacing $(a = 0.5 \text{ m})$ and on Y direction the vertical distance should be less than the line spacing $(L = 1 \text{ m})$. The a priori information was completed after the resistivity value of the river water $(\rho_{\text{water}} = 28 \text{ ohm·m})$ was added at the end of each survey line file.

High initial value of the damping factor (factor = 10) was selected (close to the surface) in order to cope with the high electrical contrast between the high conductive metallic target and the highly resistive background. The inversion for each line was completed using L2-norm (smooth) method for the data and L1-norm (robust) for the model since the target consists of sharp boundaries. Each 2D survey line was filtered individually (according to the %RMS error) and final 3D model images were produced after inverting all the 58 merged lines.

4.4. Inversion Results. In Figure 10, 12 different depth slices (D1–D12) represent the resistivity distribution at different depths below the river bottom where the first slice D1 is at ~0.10 m and the last one D12 at ~5.5 m, below the river bottom. The overall resistivity values (displayed on a logarithmic scale) varied from 0.05 to 1500 ohm·m.

At the NNE side of the grid from $x = 1 \text{ m}$ to $x = 12 \text{ m}$, high resistivity values of 1000 ohm·m are present in all depth slices (D1–D12) to a maximum depth of 6 m. From $x = 12 \text{ m}$ to $x = 57 \text{ m}$ the background resistivity value varies from 30

Resistance (Volt/Amp)

(a)

Ap. Resist. (ohm·m)

(b)

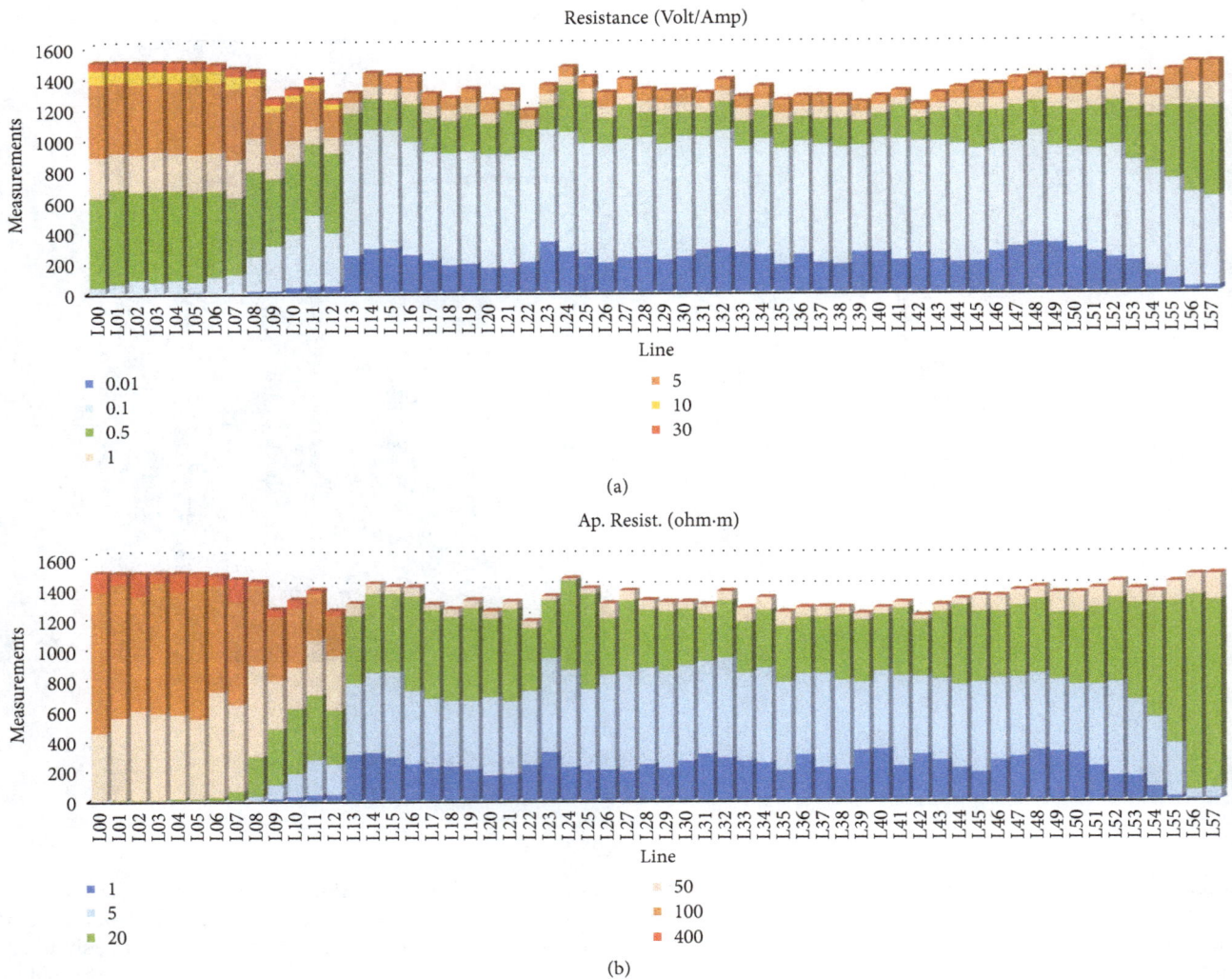

FIGURE 9: Histograms of resistance (a) and apparent resistivity (b) value per survey line.

to 50 ohm·m, from 0 to 1 m depth ($D1$–$D4$), and gradually increases to 1500 ohm·m into deeper layers till 6 m depth ($D5$–$D12$).

Until the first 0.5 m of depth ($D1$, $D2$), the exterior of the hull is difficult to discriminate. After this depth the boundary becomes more resolved to a depth of 1.5 m ($D3$–$D6$) with very low resistivity values that vary from 0.05 to 0.2 ohm·m. These boundaries are confined between $x = 11$ m and $x = 55$ m and between $y = 3$ m and $y = 13$ m presenting a structure of 44 m length and 10 m width, respectively. From the depth of 1.5 to 4 m ($D7$–$D10$) the conductive area becomes wider, presumably showing the ship's deck. After the depth of 4 m ($D11$) the conductive area gradually disappears until it completely fades away after 6 m depth ($D12$). This is an indication either that the metallic part of the barge is no longer present or that the maximum depth of the barge is 6 m.

Apart from the external boundaries of the ship, internal sections are also resolved in the inversion depth images. At the depth of ~0.6 m ($D3$), a bulkhead (long metallic bar) can be seen running parallel to the centre line of the ship from

bow to stern from $x = 20$ m to $x = 50$ m and at $y = 7$ m. Furthermore, at $x = 22$ m, 32 m, and 41 m at a depth of ~1.7 m ($D7$) internal sections, perpendicular to the main axis of the ship, are visible as well. A highly resistive anomaly is observed at a depth of 2.5 to 5 m ($D9$–$D11$) located at $x = 34$ m to $x = 42$ m and at $y = 5$ m which is interpreted as a damaged part of the vessel, as validated by the multibeam data (Figure 2).

At the ESE part of the survey grid from $x = 11$ m to $x = 47$ m and $y = 14$ m to $y = 16$ m from a depth of 3 to 6 m ($D10$–$D12$) a low resistivity area parallel to the main ship axis can be seen with resistivity values close to 0.1 ohm·m. This has been interpreted as a detached part of the ship lying nearby although it may also be an artefact of the inversion process given its location near the boundary of the survey area. The RMS error is estimated as less than 30%, which may be attributed to the high contrast between the resistivity values of the anomaly related to *Crowie* and the surrounding substrate.

The inversion model values were exported into commercial 3D graphics software (Voxler). A three-dimensional

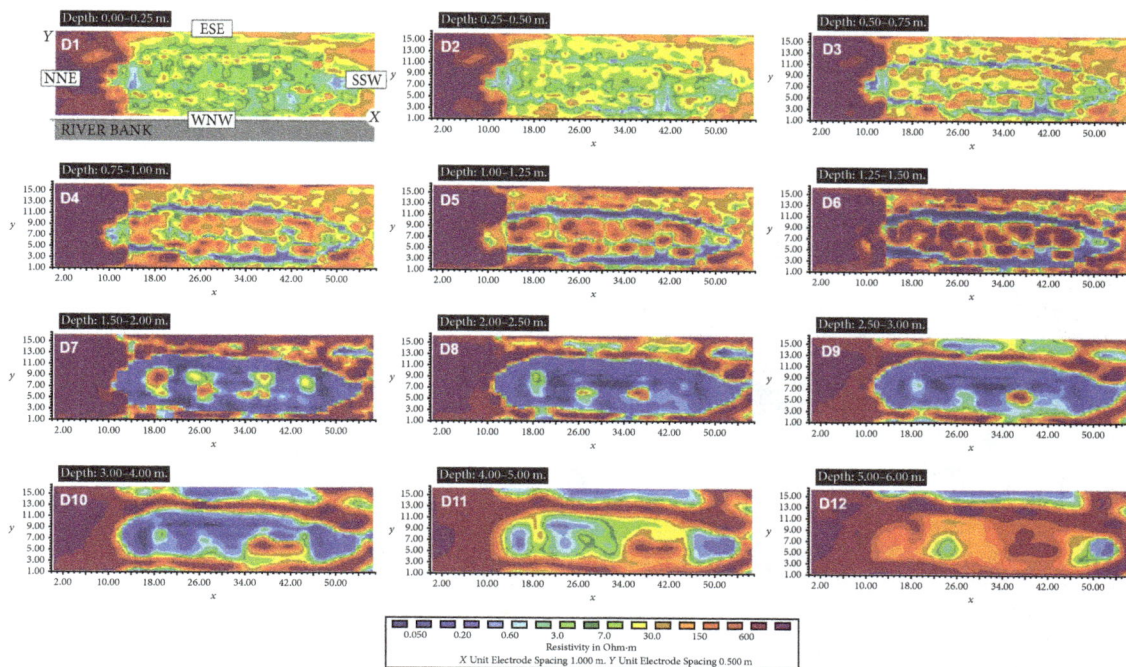

FIGURE 10: 3D inversion depth slices. Resistivity values mapped on a logarithmic scale. Both axes are given in meters.

representation of the barge can be seen in Figure 11, which also includes the bathymetry of the area. The resistivity values are mapped on a logarithmic scale. In Figure 11(a), the water layer (light blue) and the topography (light brown) are shown, using the volume rendered plot mode. In Figure 11(b) selected orthoimages (slices through the model) in all directions (X, Y, and Z) are depicted, indicating with dark blue colours the ships' lower resistive areas. The overall volume with the target in it is shown in Figure 11(c) with highly resistive areas denoted by red colour and the target itself is shown in Figure 11(d) depicted with green colour using the isosurface mode of plotting. The isovalue used for ships' metal structure representation is ρ = 0.06 ohm·m. Apart from the outer boundaries of the ship, the internal bulkhead can be seen as well, across the long axis of the ship.

5. Discussion

Overall, this study demonstrates that ERT is a suitable technique for investigating shipwrecks in aquatic environments. Despite this potential, some aspects of underwater ERT are more challenging than terrestrial surveys. For example, establishing the survey grid is a time consuming procedure as it is more difficult to set the four corners of the grid at the right position and keep them fixed throughout the survey period. Furthermore, placing the cable on the right position at each different line (if it is floating) is not an easy operation because it depends mainly on the current and wave conditions and any obstacles that might be protruding from the bottom. Any movement of the cable during the data acquisition can increase the noise levels, so extra stabilisation technique is required for some surveys. On the other hand, no

metal probes are required since the cable itself is sufficient to propagate the current into the water and the river sediments. After the data acquisition, extra time for the data processing routine is needed since the a priori information including bathymetry and water resistivity should be included for each different survey line. The same arrays and equipment can be used for terrestrial and underwater ERT surveys; however aquatic surveys are more demanding on batteries due to the very conductive environment.

From an archaeological point of view, underwater ERT provides a new means of investigating material culture items that would otherwise be impossible due to turbid water (visual methods) or shallow water depths (acoustic methods).

6. Conclusion

The shipwreck *Crowie* was successfully mapped using ERT deployed underwater in the Murray River. Data from a model simulating the barge structure were obtained and analysed before undertaking the field survey, so that the optimum survey parameters (such as the appropriate array, probe spacing) could be selected according to the particular characteristics of this target and survey area. A priori information, including water column depth and water conductivity, were incorporated throughout the data processing for more accurate model representation.

The field survey involved 58 parallel lines being placed underwater on top of the wreck. Following processing and inversion, the resistivity distribution in the 3D inversion images clearly shows a conductive target shaped like the structure of a ship. The external boundaries of the ship, due to the low resistivity values (metallic parts), can be

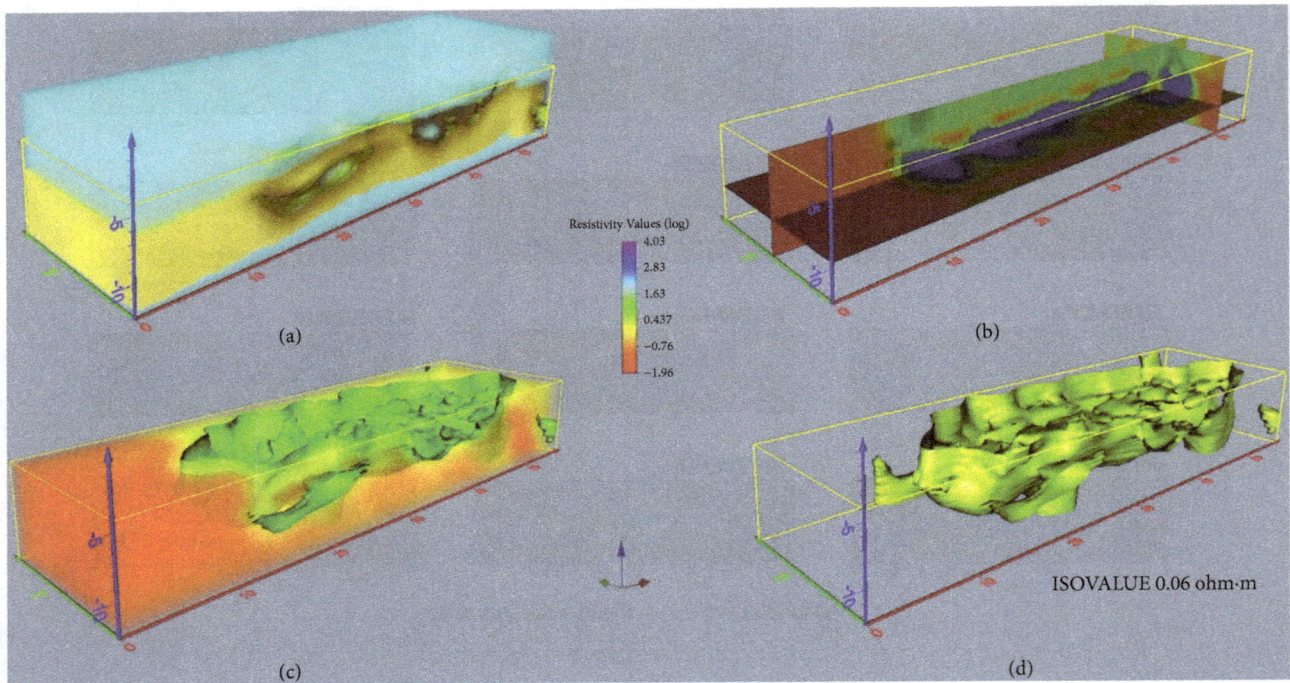

FIGURE 11: 3D model representation of the metallic structure of the *Crowie* barge. Resistivity values mapped on a logarithmic scale.

distinguished from the highly resistive background (sand and limestone sediments) and they are in good correlation with the historical record. The outer formation of the barge is accurately defined (in accordance with the modelling results) indicating the efficacy of the survey methodology. Apart from the successful imaging of the exterior of the ship, internal compartments of the barge are also visible where they present a suitable resistivity contrast.

Using the ERT method in this novel way to map a submerged barge provides important new information about this geophysical technique and about the wreck of *Crowie*. This result demonstrates that several important characteristics of a submerged ship can be defined, such as the burial depth of the wreck, the outer limits, overall dimensions, and crucially the shape of the ship. The excellent results obtained from *Crowie* suggest many future marine archaeological applications of the ERT method.

Acknowledgments

This project was supported by an Endeavour Fellowship to Dr. Kleanthis Simyrdanis funded by the Australian Government. Lee Rippon, John Naumann, Celeste Jordan, Belinda Duke, Shannon Sullivan, and Anika Johnstone contributed to the field work for the project. Thanks are due to Lisa and Barry from the Morgan Waterfront Marina and ZZ Resistivity Imaging for their support of this research. Dr. Ian Moffat is the recipient of Australian Research Council Discovery Early Career Award funded by the Australian Government (DE160100703).

References

[1] C. Colombero, C. Comina, F. Gianotti, and L. Sambuelli, "Waterborne and on-land electrical surveys to suggest the geological evolution of a glacial lake in NW Italy," *Journal of Applied Geophysics*, vol. 105, pp. 191–202, 2014.

[2] G. Tassis, P. Tsourlos, J. Rønning, and T. Dahlin, "Detection and Characterization of Fracture Zones in Bedrock: Possibilities and Limitations," in *Proceedings of the Near Surface Geoscience 2014 - First Applied Shallow Marine Geophysics Conference*, Athens, Greece, September 2014.

[3] D. F. Rucker, G. E. Noonan, and W. J. Greenwood, "Electrical resistivity in support of geological mapping along the Panama Canal," *Engineering Geology*, vol. 117, no. 1-2, pp. 121–133, 2011.

[4] L. Orlando, "Some considerations on electrical resistivity imaging for characterization of waterbed sediments," *Journal of Applied Geophysics*, vol. 95, pp. 77–89, 2013.

[5] F. Baumgartner and N. B. Christensen, "Analysis and application of a non-conventional underwater geoelectrical method in Lake Geneva, Switzerland," *Geophysical Prospecting*, vol. 46, no. 5, pp. 527–541, 1998.

[6] C.-W. Chiang, T.-N. Goto, C.-C. Chen, and S.-K. Hsu, "Efficiency of a marine towed electrical resistivity method," *Terrestrial, Atmospheric and Oceanic Sciences*, vol. 22, no. 4, pp. 443–446, 2011.

[7] C.-W. Chiang, T.-N. Goto, H. Mikada, C.-C. Chen, and S.-K. Hsu, "Sensitivity of deep-towed marine electrical resistivity imaging using two-dimensional inversion: A case study on methane hydrate," *Terrestrial, Atmospheric and Oceanic Sciences*, vol. 23, no. 6, pp. 725–732, 2012.

[8] G. Ranieri, F. Loddo, A. Godio et al., "Reconstruction of

archaeological features in a mediterranean coastal environment using non-invasive techniques," 2009.

[9] K. Simyrdanis, N. Papadopoulos, J.-H. Kim, P. Tsourlos, and I. Moffat, "Archaeological investigations in the shallow seawater environment with electrical resistivity tomography," *Near Surface Geophysics*, vol. 13, no. 6, pp. 601–611, 2015.

[10] S. Passaro, F. Budillon, S. Ruggieri et al., "Integrated geophysical investigation applied to the definition of buried and outcropping targets of archaeological relevance in very shallow water," *Italian Journal of Quaternary Sciences*, vol. 22, no. 1, pp. 33–38, 2009.

[11] S. Passaro, "Marine electrical resistivity tomography for shipwreck detection in very shallow water: A case study from Agropoli (Salerno, southern Italy)," *Journal of Archaeological Science*, vol. 37, no. 8, pp. 1989–1998, 2010.

[12] R. Parsons, *Ships of the inland rivers: An Outline History And Details of All Known Paddle Ships, Barges And Other Vessels Trading on The Murray-Darling System*, Gould Genealogy, Modbury, Australia, 3rd edition, 2005.

[13] A. Roberts, W. van Duivenvoorde, M. Morrison et al., "'They call 'im Crowie': an investigation of the Aboriginal significance attributed to a wrecked River Murray barge in South Australia," *International Journal of Nautical Archaeology*, vol. 46, no. 1, pp. 132–148, 2017.

[14] M. H. Loke and J. W. Lane, "Inversion of data from electrical resistivity imaging surveys in water-covered areas," *Exploration Geophysics*, vol. 35, no. 4, pp. 266–271, 2004.

[15] C. deGroot-Hedlin and S. Constable, "Occam's inversion to generate smooth, two-dimensional models from magnetotelluric data," *Geophysics*, vol. 55, no. 12, pp. 1613–1624, 1990.

[16] J. B. Firman et al., *Renmark 1:250000 Geological Map, Geological Survey of South Australia*, 1971.

[17] P. A. Rogers, "Continental Sediments of the Murray Basin," in *The geology of South Australia*, J. F. Drexel and W. V. Preiss, Eds., vol. 54, pp. 252–254, The Phanerozoic, South Australia, Australia, 1995.

[18] P. A. Rogers, J. M. Lindsay, N. F. Alley, S. R. Barnett, K. Lablack, and G. LKwitko, "Murray Basin," in *The geology of South Australia*, J. F. Drexel and W. V. Preiss, Eds., vol. 2, pp. 157–163, The Phanerozoic, South Australia, Australia, 1995.

[19] V. Chaplot, F. Darboux, H. Bourennane, S. Leguédois, N. Silvera, and K. Phachomphon, "Accuracy of interpolation techniques for the derivation of digital elevation models in relation to landform types and data density," *Geomorphology*, vol. 77, no. 1-2, pp. 126–141, 2006.

Permissions

List of Contributors

Eliezer Manguelle-Dicoum
Faculty of Science, University of Yaounde 1, Yaounde, Cameroon

Marcelin Mouzong Pemi
Faculty of Science, University of Yaounde 1, Yaounde, Cameroon
Department of Renewable Energy, Higher Technical Teachers' Training College (HTTTC), University of Buea, Cameroon

Joseph Kamguia
National Institute of Cartography (NIC), Yaounde, Cameroon

Severin Nguiya
Faculty of Industrial Engineering, University of Douala, Cameroon

Abderrahmane Nekkache Ghenim and Abdesselam Megnounif
"Eau et Ouvrage dans Leur Environnement" Laboratory, Tlemcen University, BP 230, 13000 Tlemcen, Algeria

Kishan Singh Rawat
Centre for Remote Sensing and Geoinformatics, Sathyabama Institute of Science and Technology, Chennai 600119, India

Shashi Vind Mishra
Division of Environmental Sciences, Indian Agricultural Research Institute, New Delhi 110012, India

Sudhir Kumar Singh
K. Banerjee Centre of Atmospheric and Ocean Studies, IIDS, Nehru Science Centre, University of Allahabad, Allahabad, India

Upasna Chandarana Kothari and Moe Momayez
Mining & Geological Engineering, University of Arizona, Tucson, AZ, USA

Sukir Maryanto
Department of Physics, Faculty of Mathematics and Science, University of Brawijaya, Malang 65145, Indonesia

Ika Karlina Laila Nur Suciningtyas and Cinantya Nirmala Dewi
Postgraduate Program of Physics, Faculty of Mathematics and Science, University of Brawijaya, Malang 65145, Indonesia

Arief Rachmansyah
Department of Civil Engineering, Faculty of Engineering, University of Brawijaya, Malang 65145, Indonesia

Jason E. French
Department of Earth and Atmospheric Sciences, University of Alberta, 1-26 Earth Science Building, Edmonton, Alberta, Canada T6G 2E3

David F. Blake
NASA Ames Research Center, Exobiology Branch, MS 239-4, Moffett Field, CA 94035-1000, USA

Kazuya Ishitsuka
Division of Sustainable Resources Engineering, Hokkaido University, Sapporo 065-0068, Japan

Shinichiro Iso
School of Creative Science and Engineering, Waseda University, Tokyo 169-8555, Japan
Fukada Geological Institute, Tokyo 113-0021, Japan

Toshifumi Matsuoka
Fukada Geological Institute, Tokyo 113-0021, Japan

Kyosuke Onishi
Geology and Geotechnical Engineering Research Group, Public Works Research Institute, Tsukuba 305-8516, Japan

J. Ahuja
Energy Research Centre, Panjab University, Chandigarh 160014, India

U. Gupta and R. K. Wanchoo
Dr. S. S. Bhatnagar University Institute of Chemical Engineering & Technology, Panjab University, Chandigarh 160014, India

Isabel Bernal, Hernando Tavera, Wilfredo Sulla, Luz Arredondo and Javier Oyola
Instituto Geofísico del Perú (IGP), Lima, Peru

S. Y. Moussavi Alashloo, D. P. Ghosh, Y. Bashir and W. Y. Wan Ismail
Center of Seismic Imaging, Universiti Teknologi PETRONAS, 32610 Seri Iskandar, Malaysia

Carlos Sotomayor-Beltran
Image Processing Research Laboratory (INTI-Lab), Universidad de Ciencias y Humanidades, Lima 39, Peru

A. K. Tiwari, S. P. Maurya and N. P. Singh
Department of Geophysics, Faculty of Science, Banaras Hindu University, Varanasi 221005, India

Jaime Urrutia-Fucugauchi and Ligia Pérez-Cruz
Programa Universitario de Perforaciones en Océanos y Continentes, Departamento de Geomagnetismo y Exploración Geofísica, Instituto de Geofísica, Universidad Nacional Autónoma de México, Delegación Coyoacán, 04510 Mexico City, DF, Mexico

Marykutty Abraham
Sathyabama University, Chennai, India

Riya Ann Mathew
Anna University, Chennai, India

Ian Moffat, Jarrad Kowlessar and Marian Bailey
Archaeology, College of Humanities, Arts and Social Sciences, Flinders University, Adelaide, SA, Australia

Kleanthis Simyrdanis
Archaeology, College of Humanities, Arts and Social Sciences, Flinders University, Adelaide, SA, Australia
Laboratory of Geophysical-Satellite Remote Sensing, Institute for Mediterranean Studies, Rethymno, Greece

Nikos Papadopoulos
Laboratory of Geophysical-Satellite Remote Sensing, Institute for Mediterranean Studies, Rethymno, Greece

Index

www.ingramcontent.com/pod-product-compliance
Lightning Source LLC
Chambersburg PA
CBHW082012190326
41458CB00010B/3161